Texts in Computer Science

Series Editor

Orit Hazzan ⓘ, Faculty of Education in Technology and Science, Technion—Israel Institute of Technology, Haifa, Israel

Titles in this series now included in the Thomson Reuters Book Citation Index!

'Texts in Computer Science' (TCS) delivers high-quality instructional content for undergraduates and graduates in all areas of computing and information science, with a strong emphasis on core foundational and theoretical material but inclusive of some prominent applications-related content. TCS books should be reasonably self-contained and aim to provide students with modern and clear accounts of topics ranging across the computing curriculum. As a result, the books are ideal for semester courses or for individual self-study in cases where people need to expand their knowledge. All texts are authored by established experts in their fields, reviewed internally and by the series editors, and provide numerous examples, problems, and other pedagogical tools; many contain fully worked solutions.

The TCS series is comprised of high-quality, self-contained books that have broad and comprehensive coverage and are generally in hardback format and sometimes contain color. For undergraduate textbooks that are likely to be more brief and modular in their approach, require only black and white, and are under 275 pages, Springer offers the flexibly designed Undergraduate Topics in Computer Science series, to which we refer potential authors.

Gerard O'Regan

Mathematical Foundations of Software Engineering

A Practical Guide to Essentials

 Springer

Gerard O'Regan
Mallow, Cork, Ireland

ISSN 1868-0941   ISSN 1868-095X (electronic)
Texts in Computer Science
ISBN 978-3-031-26214-2   ISBN 978-3-031-26212-8   (eBook)
https://doi.org/10.1007/978-3-031-26212-8

This Springer imprint is published by the registered company Springer Nature Switzerland AG
The registered company address is: Gewerbestrasse 11, 6330 Cham, Switzerland

To
Present and past staff of Coláiste Chríost Rí,
Cork.

Preface

Overview

The objective of this book is to give the reader a flavour of the mathematical foundations of software engineering. The rich applications of mathematics to software engineering includes its applications to error detection and correcting codes with finite field theory; the field of cryptography which uses the results of number theory; the modelling of telecommunication networks with graph theory; the application of discrete mathematics and proof techniques to the software correctness field (especially safety critical systems using formal methods and model checking); the application of financial mathematics to the banking and insurance fields; and the application of calculus and vectors to traditional engineering applications.

Organization and Features

Chapter 1 introduces software engineering and discusses both traditional and Agile software engineering. Chapter 2 examines which mathematics is needed in software engineering, including the core mathematics that all software engineers should be familiar with, as well as specific mathematics for the particular software engineering domains such as the safety critical field; to traditional engineering applications; and to the financial sector.

Chapter 3 discusses the mathematical prerequisites, and we discuss fundamental building blocks in mathematics including sets, relations, and functions. A set is a collection of well-defined objects, and it may be finite or infinite. A relation between two sets A and B indicates a relationship between members of the two sets and is a subset of the Cartesian product of the two sets. A function is a special type of relation such that for each element in A there is at most one element in the codomain B. We discuss the fundamentals of number theory including prime number theory and the greatest common divisor and least common multiple of two numbers, and we provide a short introduction to trigonometry.

Chapter 4 presents a short introduction to algorithms, where an algorithm is a well-defined procedure for solving a problem. It consists of a sequence of steps that takes a set of values as input and produces a set of values as output. An algorithm is an exact specification of how to solve the problem, and it explicitly defines the procedure so that a computer program may implement the solution in some programming language.

Chapter 5 discusses algebra, and we discuss simple and simultaneous equations, including the method of elimination and the method of substitution to solve simultaneous equations. We show how quadratic equations may be solved by factorization, completing the square or using the quadratic formula. We present the laws of logarithms and indices. We discuss various structures in abstract algebra, including monoids, groups, rings, integral domains, fields, and vector spaces.

Chapter 6 discusses mathematical induction and recursion. Induction is a common proof technique in mathematics, and there are two parts to a proof by induction (the base case and the inductive step). We discuss strong and weak inductions, and we discuss how recursion is used to define sets, sequences, and functions. This leads us to structural induction, which is used to prove properties of recursively defined structures.

Chapter 7 discusses graph theory where a graph $G = (V, E)$ consists of vertices and edges. It is a practical branch of mathematics that deals with the arrangements of vertices and edges between them, and it has been applied to practical problems such as the modelling of computer networks, determining the shortest driving route between two cities, and the travelling salesman problem.

Chapter 8 discusses sequences and series and permutations and combinations. Arithmetic and geometric sequences and series are discussed.

Chapter 9 presents a short history of logic, and we discuss Greek contributions to syllogistic logic, stoic logic, fallacies, and paradoxes. Boole's symbolic logic and its application to digital computing are discussed, and we consider Frege's work on predicate logic.

Chapter 10 provides an introduction to propositional and predicate logic. Propositional logic may be used to encode simple arguments that are expressed in natural language and to determine their validity. The nature of mathematical proof is discussed, and we present proof by truth tables, semantic tableaux, and natural deduction. Predicate logic allows complex facts about the world to be represented, and new facts may be determined via deductive reasoning. Predicate calculus includes predicates, variables, and quantifiers, and a predicate is a characteristic or property that the subject of a statement can have.

Chapter 11 presents some advanced topics in logic including fuzzy logic, temporal logic, intuitionistic logic, undefined values, and the applications of logic to AI. Fuzzy logic is an extension of classical logic that acts as a mathematical model for vagueness. Temporal logic is concerned with the expression of properties that have time dependencies, and it allows temporal properties about the past, present, and future to be expressed. Intuitionism was a controversial theory on the foundations of mathematics based on a rejection of the law of the excluded middle and an insistence on constructive existence. We discuss approaches to deal with undefined values.

Chapter 12 discusses language theory and includes a discussion on grammars, parse trees, and derivations from a grammar. The important area of programming language semantics is discussed, including axiomatic, denotational, and operational semantics.

Chapter 13 discusses automata theory, including finite-state machines, pushdown automata, and Turing machines. Finite-state machines are abstract machines that are in only one state at a time, and the input symbol causes a transition from the current state to the next state. Pushdown automata have greater computational power, and they contain extra memory in the form of a stack from which symbols may be pushed or popped. The Turing machine is the most powerful model for computation, and this theoretical machine is equivalent to an actual computer in the sense that it can compute exactly the same set of functions.

Chapter 14 discusses computability and decidability. The Church–Turing thesis states that anything that is computable is computable by a Turing machine. Church and Turing showed that mathematics is not decidable. In other words, there is no mechanical procedure (i.e., algorithm) to determine whether an arbitrary mathematical proposition is true or false, and so the only way to determine the truth or falsity of a statement is try to solve the problem.

Chapter 15 discusses software reliability and dependability and covers topics such as software reliability and software reliability models, the cleanroom methodology, system availability, safety and security critical systems, and dependability engineering.

Chapter 16 discusses formal methods, which consist of a set of mathematical techniques to rigorously specify and derive a program from its specification. Formal methods may be employed to rigorously state the requirements of the proposed system; they may be employed to derive a program from its mathematical specification; and they may provide a rigorous proof that the implemented program satisfies its specification. They have been mainly applied to the safety critical field.

Chapter 17 presents the Z specification language, which is one of the most widely used formal methods. It was developed at Oxford University in the UK.

Chapter 18 discusses model checking which is an automated technique such that given a finite-state model of a system and a formal property, then it systematically checks whether the property is true of false in a given state in the model. It is an effective technique to identify potential design errors, and it increases the confidence in the correctness of the system design.

Chapter 19 discusses the nature of proof and theorem proving, and we discuss automated and interactive theorem provers. We discuss the nature of mathematical proof and formal mathematical proof.

Chapter 20 discusses cryptography, which is an important application of number theory. The codebreaking work done at Bletchley Park in England during the Second World War is discussed, and the fundamentals of cryptography, including private and public key cryptosystems, are discussed.

Chapter 21 presents coding theory and is concerned with error detection and error correction codes. The underlying mathematics includes abstract mathematics such as group theory, rings, fields, and vector spaces.

Chapter 22 discusses statistics which is an empirical science that is concerned with the collection, organization, analysis, interpretation, and presentation of data. We discuss sampling; the average and spread of a sample; the abuse of statistics; frequency distributions; variance and standard deviation; correlation and regression; statistical inference and hypothesis testing.

Chapter 23 discusses probability which is a branch of mathematics that is concerned with measuring uncertainty and random events. We discuss discrete and continuous random variables; probability distributions such as the binomial and normal distributions; variance and standard deviation; confidence intervals; tests of significance; the central limit theorem; and Bayesian statistics.

Chapter 24 discusses data science, which is a multidisciplinary field that extracts knowledge from data sets that consist of structured and unstructured data, and large data sets may be analysed to extract useful information. Data science may be regarded as a branch of statistics as it uses many concepts from the field, and in order to prevent errors occurring during data analysis it is essential that both the data and models are valid.

Chapter 25 provides a short introduction to calculus and provides a high-level overview of limits, continuity, differentiation, and integration. Chapter 26 presents applications of the calculus in determining velocity, acceleration, area, and volume, as well as a short discussion on Fourier series, Laplace transforms, and differential equations.

Chapter 27 discusses matrices including 2×2 and general $n \times m$ matrices. Various operations such as the addition and multiplication of matrices are considered, and the determinant and inverse of a square matrix are discussed. The application of matrices to solving a set of linear equations using Gaussian elimination is considered.

Chapter 28 discusses complex numbers and quaternions. Complex numbers are of the form $a + bi$ where a and b are real numbers and $i^2 = -1$. Quaternions are a generalization of complex numbers to quadruples that satisfy the quaternion formula $i^2 = j^2 = k^2 = -1$. Chapter 29 discusses vectors, where a vector is represented as a directed line segment such that the length represents the magnitude of the vector and the arrow indicates the direction of the vector.

Chapter 30 discusses basic financial mathematics, and we discuss simple and compound interest, annuities, and mortgages. Chapter 31 discusses operations research which is a multidisciplinary field that is concerned with the application of mathematical and analytic techniques to assist in decision making. It employs techniques such as mathematical modelling, statistical analysis, and mathematical optimization as part of its goal to achieve optimal (or near optimal) solutions to complex decision-making problems.

Finally, Chap. 32 discusses a selection of software tools to support mathematics for software engineering, and we discuss Microsoft Excel, Minitab, Python, the R statistical software environment, and Mathematica.

Audience

The audience of this book includes software engineering students who wish to become familiar with foundation mathematics for software engineering and mathematicians who are curious as to how mathematics is applied in the software engineering field. The book will also be of interest to the motivated general reader.

Mallow, Cork, Ireland

Gerard O'Regan

Acknowledgments

I am deeply indebted to friends and family who supported my efforts in this endeavour. My thanks, as always, goes to the team at Springer for their professional work. I would like to pay a special thanks to past and present staff of Coláiste Chríost Rí, Cork, who provided a solid education to the author, and stimulated his interest in mathematics, science, and the wider world. In particular, I would like to thank Máirtín O'Fathaigh and Pádraig O'Scanlán (Irish), Cathal O'Corcaire (French), Tony Power and Kevin Cummins (English), Mr. O'Callaghan and Mr. O'Brien (Mathematics), Seamus Lankford (History and Céilí dancing on a Sunday night), Mr. O'Leary (Geography), Mr. Brennan (Music), Jim Cremin (Science and Chemistry), Noel Brett (Physics), and Mr. Desmond (Economics). Also, my thanks goes to Br. Pius, Richard Tobin, and Br. Columcille. Ta cuid de na daoine seo imithe at shlí na fírinne anois, ach ba mhaith liom buíochas a ghabháil leo.

Mallow, Cork, Ireland

Gerard O'Regan

Contents

Abbreviations

ACL	A Computational Logic
ACM	Association for Computing Machinery
AES	Advanced Encryption Standard
AI	Artificial Intelligence
AMN	Abstract Machine Notation
ATP	Automated Theorem Proving
BCH	Bose, Chauduri and Hocquenghem
BCS	British Computer Society
BI	Business Intelligence
BNF	Backus Naur Form
CCS	Calculus Communicating Systems
CIA	Central Intelligence Agency
CICS	Customer Information Control System
CMG	Computer Management Group
CMM	Capability Maturity Model
CMMI®	Capability Maturity Model Integration
COBOL	Common Business Oriented Language
COPQ	Cost Of Poor Quality
COTS	Customised Off The Shelf
CPO	Complete Partial Order
CSP	Communicating Sequential Processes
CTL	Computational Tree Logic
CVPA	Cost Volume Profit Analysis
DES	Data Encryption Standard
DPDA	Deterministic PDA
DSA	Digital Signature Algorithm
DSDM	Dynamic Systems Development Method
DSS	Digital Signature Standard
ESA	European Space Agency
FC	Fixed Cost
FSM	Finite State Machine
FV	Future Value

GCD	Greatest Common Divisor
GCHQ	General Communications Headquarters
GDPR	General Data Protection Regulation
GSM	Global System for Mobile communications
GUI	Graphical User Interface
HOL	Higher Order Logic
IBM	International Business Machines
IEC	International Electrotechnical Commission
IEEE	Institute of Electrical and Electronics Engineers
ISO	International Standards Organization
ITP	Interactive Theorem Proving
JAD	Joint Application Development
KLOC	Thousand Lines Of Code
LCM	Least Common Multiple
LEM	Law of Excluded Middle
LISP	List Processing
LP	Linear Programming
LPF	Logic of Partial Functions
LT	Logic Theorist
LTL	Linear Time Logic
MIT	Massachusetts Institute of Technology
MOD	Ministry of Defence
MSL	Mars Science Laboratory
MTBF	Mean time between failure
MTTF	Mean time to failure
NATO	North Atlantic Treaty Organization
NBS	National Bureau of Standards
NFA	Non deterministic Finite Automaton
NIST	National Institute of Standards and Technology
NP	Non Polynomial
NQTHM	New Quantified Theorem Prover
NSA	National Security Agency
OR	Operation Research
OTTER	Organized Techniques for Theorem proving and Effective Research
PDA	Pushdown Automata
PL/1	Programming Language 1
PVS	Prototype Verification System
RAD	Rapid Application Development
RRA	Royals Royce and Associates
RSA	Rivest, Shamir and Adleman
RUP	Rational Unified Process
SAM	Semi-automated mathematics
SCAMPI	Standard CMM Appraisal Method for Process Improvement
SECD	Stack, Environment, Code, Dump
SEI	Software Engineering Institute

SRI	Stanford Research Institute
TC	Total Cost
TDD	Test Driven Development
TM	Turing Machine
TPS	Theorem Proving System
TR	Total Revenue
TVC	Total Variable Cost
UAT	User Acceptance Testing
UML	Unified Modelling Language
UMTS	Universal Mobile Telecommunications System
VDM	Vienna Development Method
VDM♣	Irish School of VDM
XOR	Exclusive OR
Y2K	Year 2000
ZB	Zetta Byte

List of Figures

List of Tables

Fundamentals of Software Engineering

1

1.1 Introduction

The approach to software development in the 1950s and 1960s has been described as the *"Mongolian Hordes Approach"* by Brooks [1].[1] The "method" or lack of method was applied to projects that were running late, and it involved adding many inexperienced programmers to the project, with the expectation that this would allow the project schedule to be recovered. However, this approach was

---

[1] The "Mongolian Hordes" management myth is the belief that adding more programmers to a software project that is running late will allow catch-up. In fact, as Brooks says adding people to a late software project makes it later.

© The Author(s), under exclusive license to Springer Nature Switzerland AG 2023
G. O'Regan, *Mathematical Foundations of Software Engineering*,
Texts in Computer Science, https://doi.org/10.1007/978-3-031-26212-8_1

deeply flawed as it led to programmers with inadequate knowledge of the project attempting to solve problems, and they inevitably required significant time from the other project team members.

This resulted in the project being delivered even later, as well as subsequent problems with quality (i.e., the approach of throwing people at a problem does not work). The philosophy of software development back in the 1950/60s was characterized by:

> The completed code will always be full of defects.
> The coding should be finished quickly to correct these defects.
> Design as you code approach.

This philosophy accepted defeat in software development, and suggested that irrespective of a solid engineering approach, that the completed software would always contain lots of defects, and that it therefore made sense to code as quickly as possible, and to then identify the defects that were present, and to correct them as quickly as possible to solve a problem.

In the late 1960s it was clear that the existing approaches to software development were deeply flawed, and that there was an urgent need for change. The NATO Science Committee organized two famous conferences to discuss critical issues in software development [2]. The first conference was held at Garmisch, Germany, in 1968, and it was followed by a second conference in Rome in 1969. Over 50 people from 11 countries attended the Garmisch conference, including Edsger Djkstra, who did important theoretical work on formal specification and verification. The NATO conferences highlighted problems that existed in the software sector in the late 1960s, and the term *"software crisis"* was coined to refer to these. There were problems with budget and schedule overruns, as well as the quality and reliability of the delivered software.

The conference led to the birth of *software engineering* as a discipline in its own right, and the realization that programming is quite distinct from science and mathematics. Programmers are like engineers in that they build software products, and they therefore need education in traditional engineering as well as the latest technologies. The education of a classical engineer includes product design and mathematics. However, often computer science education places an emphasis on the latest technologies, rather than on the important engineering foundations of designing and building high-quality products that are safe for the public to use.

Programmers therefore need to learn the key engineering skills to enable them to build products that are safe for the public to use. This includes a solid foundation on design and on the mathematics required for building safe software products. Mathematics plays a key role in classical engineering, and in some situations, it may also assist software engineers in the delivery of high-quality software products. Several mathematical approaches to assist software engineers are described in [3].

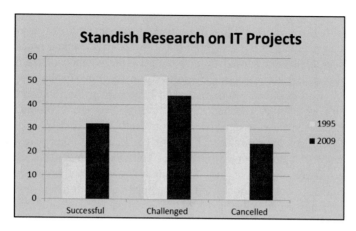

Fig. 1.1 Standish report—results of 1995 and 2009 survey

There are parallels between the software crisis in the late 1960s, and serious problems with bridge construction in the nineteenth century. Several bridges collapsed, or were delivered late or over-budget, since people involved in their design and construction did not have the required engineering knowledge. This led to bridges that were poorly designed and constructed, leading to their collapse and loss of life, as well as endangering the lives of the public.

This led to legislation requiring engineers to be licensed by the Professional Engineering Association prior to practicing as engineers. This organization specified a core body of knowledge that the engineer is required to possess, and the licensing body verifies that the engineer has the required qualifications and experience. This helps to ensure that only personnel competent to design and build products do so. Engineers have a professional responsibility to ensure that the products are properly built and are safe for the public to use.

The Standish group has conducted research (Fig. 1.1) on the extent of problems with IT projects since the mid-1990s. These studies were conducted in the USA, but there is no reason to believe that European or Asian companies perform any better. The results indicate serious problems with on-time delivery of projects, and projects being cancelled prior to completion.[2] However, the comparison between 1995 and 2009 suggests that there have been some improvements with a greater percentage of projects being delivered successfully, and a reduction in the percentage of projects being cancelled.

Fred Brooks argues that software is inherently complex, and that there is no *silver bullet* that will resolve all the problems associated with software development such as schedule or budget overruns [1, 4]. Poor software quality can lead to

---

[2] These are IT projects covering diverse sectors including banking, telecommunications, etc., rather than pure software companies. Software companies following maturity frameworks such as the CMMI generally achieve more consistent results.

defects in the software that may adversely impact the customer, and even lead to loss of life. It is therefore essential that software development organizations place sufficient emphasis on quality throughout the software development CMM process.

The Y2K problem was caused by a two-digit representation of dates, and it required major rework to enable legacy software to function for the new millennium. Clearly, well-designed programs would have hidden the representation of the date, which would have required minimal changes for year 2000 compliance. Instead, companies spent vast sums of money to rectify the problem.

The quality of software produced by some companies is impressive.[3] These companies employ mature software processes and are committed to continuous improvement. There is a lot of industrial interest in software process maturity models for software organizations, and various approaches to assess and mature software companies are described in [5, 6].[4] These models focus on improving the effectiveness of the management, engineering and organization practices related to software engineering, and in introducing best practice in software engineering. The disciplined use of the mature software processes by the software engineers enables high-quality software to be consistently produced.

1.2   What Is Software Engineering?

Software engineering involves the multiperson construction of multiversion programs. The IEEE 610.12 definition of Software Engineering is:

> Software engineering is the application of a systematic, disciplined, quantifiable approach to the development, operation, and maintenance of software; that is, the application of engineering to software, and the study of such approaches.

Software engineering includes:

1. Methodologies to design, develop, and test software to meet customers' needs.
2. Software is engineered. That is, the software products are properly designed, developed, and tested in accordance with engineering principles.
3. Quality and safety are properly addressed.
4. Mathematics may be employed to assist with the design and verification of software products. The level of mathematics employed will depend on the *safety*

---

[3] I recall projects at Motorola that regularly achieved 5.6 σ-quality in a L4 CMM environment (i.e., approx. 20 defects per million lines of code. This represents very high quality).

[4] Approaches such as the CMM or SPICE (ISO 15504) focus mainly on the management and organizational practices required in software engineering. The emphasis is on defining software processes that are fit for purpose and consistently following them. The process maturity models focus on what needs to be done rather how it should be done. This gives the organization the freedom to choose the appropriate implementation to meet its needs. The models provide useful information on practices to consider in the implementation.

critical nature of the product. Systematic peer reviews and rigorous testing will often be sufficient to build quality into the software, with heavy *mathematical techniques reserved for safety and security critical software.*

5. Sound project management and quality management practices are employed.
6. Support and maintenance of the software is properly addressed.

Software engineering is not just programming. It requires the engineer to state precisely the requirements that the software product is to satisfy, and then to produce designs that will meet these requirements. The project needs to be planned and delivered on time and budget. The requirements must provide a precise description of the problem to be solved, i.e., *it should be evident from the requirements what is and what is not required.*

The requirements need to be rigorously reviewed to ensure that they are stated clearly and unambiguously and reflect the customer's needs. The next step is then to create the design that will solve the problem, and it is essential to validate the correctness of the design. Next, the software code to implement the design is written, and peer reviews and software testing are employed to verify and validate the correctness of the software.

The verification and validation of the design is rigorously performed for safety critical systems, and it is sometimes appropriate to employ mathematical techniques for these systems. However, it will usually be sufficient to employ peer reviews or software inspections as these methodologies provide a high degree of rigour. This may include approaches such as Fagan inspections [7], Gilb inspections [8], or Prince 2's approach to quality reviews [9].

The term *"engineer"* is a title that is awarded on merit in classical engineering. It is generally applied only to people who have attained the necessary education and competence to be called engineers, and who base their practice on classical engineering principles. The title places responsibilities on its holder to behave professionally and ethically. Often in computer science the term *"software engineer"* is employed loosely to refer to anyone who builds things, rather than to an individual with a core set of knowledge, experience, and competence.

Several computer scientists (such as Parnas[5]) have argued that computer scientists should be educated as engineers to enable them to apply appropriate scientific principles to their work. They argue that computer scientists should receive a solid foundation in mathematics and design, to enable them to have the professional competence to perform as engineers in building high-quality products that are safe for the public to use. The use of mathematics is an integral part of the engineer's work in other engineering disciplines, and so the *software engineer* should be able to use mathematics to assist in the modelling or understanding of the behaviour or properties of the proposed software system.

---

[5] Parnas has made important contributions to computer science. He advocates a solid engineering approach with the extensive use of classical mathematical techniques in software development. He also introduced information hiding in the 1970s, which is now a part of object-oriented design.

Software engineers need education[6] on specification, design, turning designs into programs, software inspections, and testing. The education should enable the software engineer to produce well-structured programs that are fit for purpose.

Parnas has argued that software engineers have responsibilities as professional engineers.[7] They are responsible for designing and implementing high-quality and reliable software that is safe to use. They are also accountable for their decisions and actions[8] and have a responsibility to object to decisions that violate professional standards. Engineers are required to behave professionally and ethically with their clients. The membership of the professional engineering body requires the member to adhere to the code of ethics[9] of the profession. Engineers in other professions are licensed, and therefore Parnas argues that a similar licensing approach be adopted for professional software engineers[10] to provide confidence that they are competent for the assignment. Professional software engineers are required to follow best practice in software engineering and the defined software processes.[11]

Many software companies invest heavily in training, as the education and knowledge of its staff are essential to delivering high-quality products and services.

---

[6] Software companies that are following approaches such as the CMM or ISO 9001 consider the education and qualification of staff prior to assigning staff to performing specific tasks. The appropriate qualifications and experience for the specific role are considered prior to appointing a person to carry out the role. Many companies are committed to the education and continuous development of their staff, and on introducing best practice in software engineering into their organization.

[7] The ancient Babylonians used the concept of accountability, and they employed a code of laws (known as the Hammurabi Code) c. 1750 B.C. It included a law that stated that if a house collapsed and killed the owner then the builder of the house would be executed.

[8] However, it is unlikely that an individual programmer would be subject to litigation in the case of a flaw in a program causing damage or loss of life. A comprehensive disclaimer of responsibility for problems rather than a guarantee of quality accompanies most software products. Software engineering is a team-based activity involving many engineers in various parts of the project, and it would be potentially difficult for an outside party to prove that the cause of a particular problem is due to the professional negligence of a particular software engineer, as there are many others involved in the process such as reviewers of documentation and code and the various test groups. Companies are more likely to be subject to litigation, as a company is legally responsible for the actions of their employees in the workplace, and a company is a wealthier entity than one of its employees. The legal aspects of licensing software may protect software companies from litigation. However, greater legal protection for the customer can be built into the contract between the supplier and the customer for bespoke-software development.

[9] Many software companies have a defined code of ethics that employees are expected to adhere. Larger companies will wish to project a good corporate image and to be respected worldwide.

[10] The British Computer Society (BCS) has introduced a qualification system for computer science professionals that it used to show that professionals are properly qualified. The most important of these is the BCS Information Systems Examination Board (ISEB) which allows IT professionals to be qualified in service management, project management, software testing, and so on.

[11] Software companies that are following the CMMI or ISO 9001 standards will employ audits to verify that the processes and procedures have been followed. Auditors report their findings to management and the findings are addressed appropriately by the project team and affected individuals.

Employees receive professional training related to the roles that they are performing, such as project management, software design and development, software testing, and service management. The fact that the employees are professionally qualified increases confidence in the ability of the company to deliver high-quality products and services. A company that pays little attention to the competence and continuous development of its staff will obtain poor results and suffer a loss of reputation and market share.

1.3   Challenges in Software Engineering

The challenge in software engineering is to deliver high-quality software on time and on budget to customers. The research done by the Standish Group was discussed earlier in this chapter, and the results of their 1998 research (Fig. 1.2) on project cost overruns in the USA indicated that 33% of projects are between 21 and 50% overestimate, 18% are between 51 and 100% over estimate, and 11% of projects are between 101 and 200% overestimate.

The accurate estimation of project cost, effort and schedule is a challenge in software engineering. Therefore, project managers need to determine how good their estimation process actually is and to make appropriate improvements. The use of software metrics is an objective way to do this, and improvements in estimation will be evident from a reduced variance between estimated and actual effort (see Chap. 10). The project manager will determine and report the actual versus estimated effort and schedule for the project.

Risk management is an important part of project management, and the objective is to identify potential risks early and throughout the project, and to manage them

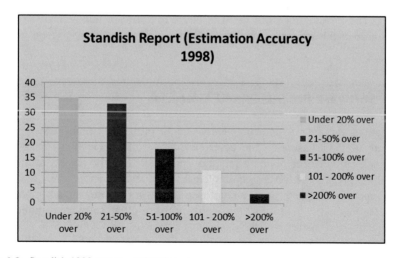

Fig. 1.2 Standish 1998 report—estimation accuracy

appropriately. The probability of each risk occurring and its impact is determined, and the risks are managed during project execution.

Software quality needs to be properly planned to enable the project to deliver a quality product. Flaws with poor quality software may lead to a negative perception of the company and may potentially lead to damage to the customer relationship with a subsequent loss of market share.

There is a strong economic case to building quality into the software, as less time is spent in reworking defective software. The cost of poor quality (COPQ) should be measured, and targets set for its reductions. It is important that lessons are learned during the project and acted upon appropriately. This helps to promote a culture of continuous improvement.

Several high-profile software failures are discussed in [6]. These include the millennium bug (Y2K) problem; the floating-point bug in the Intel microprocessor; the European Space Agency Ariane-5 disaster, and so on. These failures led to embarrassment for the organizations, as well as the associated cost of replacement and correction.

The millennium bug was due to the use of two digits to represent dates rather than four digits. The solution involved finding and analysing all code that that had a Y2K impact; planning and making the necessary changes; and verifying the correctness of the changes. The worldwide cost of correcting the millennium bug is estimated to have been in billions of dollars.

The Intel Corporation was slow to acknowledge the floating-point problem in its Pentium microprocessor, and in providing adequate information on its impact to its customers. It incurred a large financial cost in replacing microprocessors for its customers. The Ariane-5 failure caused major embarrassment and damage to the credibility of the European Space Agency (ESA). Its maiden flight ended in failure on 4 June 1996, after a flight time of just 40s.

These failures indicate that quality needs to be carefully considered when designing and developing software. The effect of software failure may be large costs to correct the software, loss of credibility of the company, or even loss of life.

1.4   Software Processes and Lifecycles

Organizations vary by size and complexity, and the processes employed will reflect the nature of their business. The development of software involves many processes such as those for defining requirements; processes for project estimation and planning; processes for design, implementation, testing, and so on.

It is important that the processes employed are fit for purpose, and a key premise in the software quality field is that the quality of the resulting software is influenced by the quality and maturity of the underlying processes, and compliance to them. Therefore, it is necessary to focus on the quality of the processes as well as the quality of the resulting software.

There is, of course, little point in having high-quality processes unless their use is institutionalized in the organization. That is, all employees need to follow the processes consistently. This requires that the employees are trained on the processes, and that process discipline is instilled with an appropriate audit strategy that ensures compliance to them. Data will be collected to improve the process. The software process assets in an organization generally consist of:

- A software development policy for the organization
- Process maps that describe the flow of activities
- Procedures and guidelines that describe the processes in more detail
- Checklists to assist with the performance of the process
- Templates for the performance of specific activities (e.g., design, testing)
- Training materials.

The processes employed to develop high-quality software generally include:

- Project Management Process
- Requirements Process
- Design Process
- Coding Process
- Peer Review Process
- Testing Process
- Supplier Selection and Management processes
- Configuration Management Process
- Audit Process
- Measurement Process
- Improvement Process
- Customer Support and Maintenance processes.

The software development process has an associated lifecycle that consists of various phases. There are several well-known lifecycles employed such as the waterfall model [10]; the spiral model [11], the Rational Unified Process [12] and the Agile methodology [13] which has become popular in recent years. The choice of a particular software development lifecycle is determined from the needs of the specific project. The various lifecycles are described in more detail in the following sections.

1.4.1 Waterfall Lifecycle

The waterfall model (Fig. 1.3) starts with requirements gathering and definition. It is followed by the system specification (with the functional and non-functional requirements), the design and implementation of the software, and comprehensive testing. The testing generally includes unit, system, and user acceptance testing.

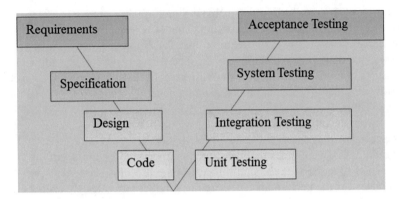

Fig. 1.3 Waterfall V lifecycle model

The waterfall model is employed for projects where the requirements can be identified early in the project lifecycle or are known in advance. We are treating the waterfall model as the "V" lifecycle model, with the left-hand side of the "V" detailing requirements, specification, design, and coding and the right-hand side detailing unit tests, integration tests, system tests, and acceptance testing. Each phase has entry and exit criteria that must be satisfied before the next phase commences. There are several variations to the waterfall model.

Many companies employ a set of templates to enable the activities in the various phases to be consistently performed. Templates may be employed for project planning and reporting; requirements definition; design; testing; and so on. These templates may be based on the IEEE standards or industrial best practice.

1.4.2 Spiral Lifecycles

The spiral model (Fig. 1.4) was developed by Barry Boehm in the 1980s [11], and it is useful for projects where the requirements are not fully known at project initiation, or where the requirements evolve as a part of the development lifecycle. The development proceeds in several spirals, where each spiral typically involves objectives and an analysis of the risks, updates to the requirements, design, code, testing, and a user review of the iteration or spiral.

The spiral is, in effect, a reusable prototype with the business analysts and the customer reviewing the current iteration and providing feedback to the development team. The feedback is analysed and used to plan the next iteration. This approach is often used in joint application development, where the usability and look and feel of the application is a key concern. This is important in web-based development and in the development of a graphical user interface (GUI). The implementation of part of the system helps in gaining a better understanding of the requirements of the system, and this feeds into subsequent development

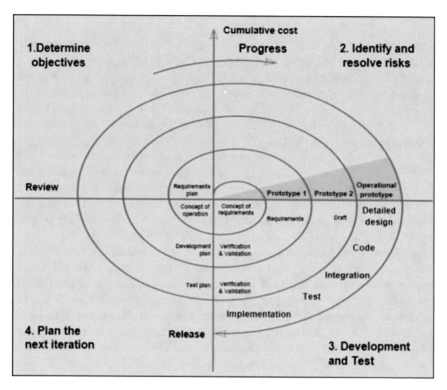

Fig. 1.4 SPIRAL lifecycle model ... Public Domain

cycles. The process repeats until the requirements and the software product are fully complete.

There are several variations of the spiral model including Rapid Application Development (RAD); Joint Application Development (JAD) models; and the Dynamic Systems Development Method (DSDM) model. The Agile methodology (discussed in Chap. 14) has become popular in recent years, and it employs sprints (or iterations) of 2–4 weeks duration to implement a number of user stories. A sample spiral model is shown in Fig. 1.4.

There are other lifecycle models such as the iterative development process that combines the waterfall and spiral lifecycle model. An overview of Cleanroom is presented in Chap. 11, and the methodology was developed by Harlan Mills at IBM. It includes a phase for formal specification, and its approach to software testing is based on the predicted usage of the software product, which allows a software reliability measure to be calculated. The Rational Unified Process (RUP) was developed by Rational, and it is discussed in the next section.

1.4.3 Rational Unified Process

The *Rational Unified Process* [12] was developed at the Rational Corporation (now part of IBM) in the late 1990s. It uses the Unified Modelling Language (UML) as a tool for specification and design, where UML is a visual modelling language for software systems that provides a means of specifying, constructing, and documenting the object-oriented system. It was developed by James Rumbaugh, Grady Booch, and Ivar Jacobson, and it facilitates the understanding of the architecture and complexity of the system.

RUP is *use case driven, architecture centric, iterative,* and *incremental,* and includes cycles, phases, workflows, risk mitigation, quality control, project management, and configuration control (Fig. 1.5). Software projects may be very complex, and there are risks that requirements may be incomplete, or that the interpretation of a requirement may differ between the customer and the project team. RUP is a way to reduce risk in software engineering.

Requirements are gathered as use cases, where the *use cases describe the functional requirements from the point of view of the user of the system.* They describe what the system will do at a high level and ensure that there is an appropriate focus on the user when defining the scope of the project. *Use cases also drive the development process,* as the developers create a series of design and implementation models that realize the use cases. The developers review each successive model for conformance to the use-case model, and the test team verifies that the implementation correctly implements the use cases.

The software architecture concept embodies the most significant static and dynamic aspects of the system. The architecture grows out of the use cases and factors such as the platform that the software is to run on, deployment considerations, legacy systems, and the non-functional requirements.

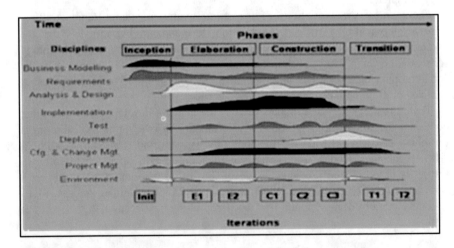

Fig. 1.5 Rational unified process

RUP decomposes the work of a large project into smaller slices or mini-projects, and *each mini-project is an iteration that results in an increment to the product*. The iteration consists of one or more steps in the workflow, and generally leads to the growth of the product. If there is a need to repeat an iteration, then all that is lost is the misdirected effort of one iteration, rather that the entire product. Another words, RUP is a way to mitigate risk in software engineering.

1.4.4 Agile Development

There has been a massive growth of popularity among software developers in lightweight methodologies such as *Agile*. This is a software development methodology that is more responsive to customer needs than traditional methods such as the waterfall model. *The waterfall development model is similar to a wide and slow-moving value stream*, and halfway through the project 100% of the requirements are typically 50% done. *However, for Agile development 50% of requirements are typically 100% done halfway through the project.*

This methodology has a strong collaborative style of working and its approach includes:

- Aims to achieve a narrow fast flowing value stream
- Feedback and adaptation employed in decision making
- User stories and sprints are employed
- Stories are either done are not done (no such thing as 50% done)
- Iterative and incremental development is employed
- A project is divided into iterations
- An iteration has a fixed length (i.e., time boxing is employed)
- Entire software development lifecycle is employed for the implementation of each story
- Change is accepted as a normal part of life in the Agile world
- Delivery is made as early as possible
- Maintenance is seen as part of the development process
- Refactoring and evolutionary design employed
- Continuous integration is employed
- Short cycle times
- Emphasis on quality
- Stand-up meetings
- Plan regularly
- Direct interaction preferred over documentation
- Rapid conversion of requirements into working functionality
- Demonstrate value early
- Early decision making.

Ongoing changes to requirements are considered normal in the Agile world, and it is believed to be more realistic to change requirements regularly throughout the project rather than attempting to define all the requirements at the start of the project. The methodology includes controls to manage changes to the requirements, and good communication and early regular feedback is an essential part of the process.

A story may be a new feature or a modification to an existing feature. It is reduced to the minimum scope that can deliver business value, and a feature may give rise to several stories. Stories often build upon other stories and the entire software development lifecycle is employed for the implementation of each story. *Stories are either done or not done, i.e., there is such thing as a story being 80% done.* The story is complete only when it passes its acceptance tests. Stories are prioritized based on a number of factors including:

- Business value of story
- Mitigation of risk
- Dependencies on other stories.

The scrum approach is an Agile method for managing iterative development, and it consists of an outline planning phase for the project followed by a set of sprint cycles (where each cycle develops an increment). *Sprint planning* is performed before the start of the iteration, and stories are assigned to the iteration to fill the available time. Each scrum sprint is of a fixed length (usually 2–4 weeks), and it develops an increment of the system. The estimates for each story and their priority are determined, and the prioritized stories are assigned to the iteration. *A short morning stand-up meeting is held daily* during the iteration, and attended by the scrum master, the project manager,[12] and the project team. It discusses the progress made the previous day, problem reporting and tracking, and the work planned for the day ahead. A separate meeting is held for issues that require more detailed discussion.

Once the iteration is complete the latest product increment is demonstrated to an audience including the product owner. This is to receive feedback and to identify new requirements. The team also conducts a retrospective meeting to identify what went well and what went poorly during the iteration. This is for continuous improvement of future iterations. Planning for the next sprint then commences. The scrum master is a facilitator who arranges the daily meetings and ensures that the scrum process is followed. The role involves removing roadblocks so that the team can achieve their goals and communicating with other stakeholders.

Agile employs pair programming and a collaborative style of working with the philosophy that two heads are better than one. This allows multiple perspectives in decision making and a broader understanding of the issues.

Software testing is very important and Agile generally employs automated testing for unit, acceptance, performance, and integration testing. Tests are run frequently with the goal of catching programming errors early. They are generally

---

[12] Agile teams are self-organizing, and the project manager role is generally not employed for small projects (< 20 staff).

run on a separate build server to ensure that all dependencies are checked. Tests are rerun before making a release. *Agile employs test-driven development with tests written before the code.* The developers write code to make a test pass with ideally developers only coding against failing tests. This approach forces the developer to write testable code.

Refactoring is employed in Agile as a design and coding practice. The objective is to change how the software is written without changing what it does. Refactoring is a tool for evolutionary design where the design is regularly evaluated, and improvements are implemented as they are identified. It helps in improving the maintainability and readability of the code and in reducing complexity. The automated test suite is essential in showing that the integrity of the software is maintained following refactoring.

Continuous integration allows the system to be built with every change. Early and regular integration allows early feedback to be provided. It also allows all of the automated tests to be run thereby identifying problems earlier. Agile is discussed in more detail in Chap. 14 of [14].

1.4.5 Continuous Software Development

Continuous software development is in a sense the successor to Agile and involves activities such as continuous integration, continuous delivery, continuous testing, and continuous deployment of the software. Its objective is to enable technology companies to accelerate the delivery of their products to their customers, thereby delivering faster business benefits as well as reshaping relationships with their customers.

Continuous integration is a coding philosophy with an associated set of practices where each developer submits their work as soon as it is finished, and several builds may take place during the day in response to the addition of significant change. The build has an associated set of unit and integration tests that are automated and are used to verify the integrity of the build, and this ensures that the addition of the new code is of a high quality. Continuous integration ensures that the developers receive immediate feedback on the software that they are working on.

Continuous delivery builds on the activities in continuous integration, where each code that is added to the build has automated unit and system tests conducted. Automated functional tests, regression tests, and possibly acceptance tests will be conducted, and once the automated tests pass the software is sent to a staging environment for deployment.

Continuous testing allows the test group to continuously test the most up to date version of the software, and it includes manual testing as well as user acceptance testing. It differs from conventional testing as the software is expected to change over time.

Continuous deployment allows changes to be delivered to end users quickly without human intervention, and it requires the completion of the automated delivery tests prior to deployment to production.

1.5   Activities in Software Development

There are various activities involved in software development including:

- Requirements Definition
- Design
- Implementation
- Software Testing
- Support and Maintenance

These activities are discussed in the following sections and cover both traditional software engineering and Agile.

1.5.1   Requirements Definition

The user (business) requirements specify what the customer wants and define what the software system is required to do (*as distinct from how this is to be done*). The requirements are the foundation for the system, and if they are incorrect, then the implemented system will be incorrect. *Prototyping may be employed* to assist in the definition and validation of the requirements. The process of determining the requirements, analysing, and validating them and managing them throughout the project lifecycle is termed *requirements engineering*.

The *user requirements* are determined from discussions with the customer to determine their actual needs, and they are then refined into the *system requirements*, which state the *functional* and *non-functional* requirements of the system. The specification of the user requirements needs to be unambiguous to ensure that all parties involved in the development of the system share a common understanding of what is to be developed and tested.

There is no requirements document as such in Agile, and the product backlog (i.e., the prioritized list of functionality of the product to be developed) is the closest to the idea of a requirements document in a traditional project. However, the written part of a user story in Agile is incomplete until the discussion of that story takes place. It is often useful to think of the written part of a story as a pointer to the real requirement, such as a diagram showing a workflow or the formula for a calculation. The Agile software development methodology argues that as requirements change so quickly that a requirements document is unnecessary, since such a document would be out of date as soon as it was written.

Requirements gathering in traditional software engineering involve meetings with the stakeholders to gather all relevant information for the proposed product. The stakeholders are interviewed, and requirements workshops conducted to elicit the requirements from them. An early working system (prototype) is often used to identify gaps and misunderstandings between developers and users. The prototype may serve as a basis for writing the specification.

The requirements workshops are used to discuss and prioritize the requirements, as well as identifying and resolving any conflicting requirements. The collected information is consolidated into a coherent set of requirements. Changes to the requirements may occur during the project, and these need to be controlled. It is essential to understand the impacts (e.g., schedule, budget, and technical) of a proposed change to the requirements prior to its approval.

Requirements verification is concerned with ensuring that the requirements are properly implemented (i.e., building it right) in the design and implementation. *Requirements validation* is concerned with ensuring that the right requirements are defined (building the right system), and that they are precise, complete, and reflect the actual needs of the customer.

The requirements are validated by the stakeholders to ensure that they are those desired, and to establish their feasibility. This may involve several reviews of the requirements until all stakeholders are ready to approve the requirements document. Other validation activities include reviews of the prototype and the design, and user acceptance testing.

The requirements for a system are generally documented in a natural language such as "English". Other notations that are employed include the visual modelling language UML [15], and formal specification languages such as VDM or Z for the safety critical field.

The specification of the system requirements of the product is essentially a statement of what the software development organization will provide to meet the business (user) requirements. That is, the detailed business requirements are a statement of what the customer wants, whereas the specification of the system requirements is a statement of what will be delivered by the software development organization.

It is essential that the system requirements are valid with respect to the user requirements, and they are reviewed by the stakeholders to ensure their validity. Traceability may be employed to show that the business requirements are addressed by the system requirements.

There are two categories of system requirements: namely, functional and non-functional requirements. The *functional requirements* define the functionality that is required of the system, and it may include screen shots, report layouts or desired functionality specified as use cases. The *non-functional requirements* will generally include security, reliability, availability, performance, and portability requirements, as well as usability and maintainability requirements.

1.5.2 Design

The design of the system consists of engineering activities to describe the architecture or structure of the system, as well as activities to describe the algorithms and functions required to implement the system requirements. It is a creative process concerned with how the system will be implemented, and its activities include architecture design, interface design, and data structure design. There are often several possible design solutions for a particular system, and the designer will need to decide on the most appropriate solution.

Refactoring is employed in Agile as a design and coding practice. The objective is to change how the software is written without changing what it does. Refactoring is a tool for evolutionary design where the design is regularly evaluated, and improvements are implemented as they are identified. It helps in improving the maintainability and readability of the code and in reducing complexity. The automated test suite is essential in demonstrating that the integrity of the software is maintained following refactoring.

The design may be specified in various ways such as graphical notations that display the relationships between the components making up the design. The notation may include flow charts, or various UML diagrams such as sequence diagrams, state charts, and so on. Program description languages or pseudocode may be employed to define the algorithms and data structures that are the basis for implementation.

Function-oriented design is historical, and it involves starting with a high-level view of the system and refining it into a more detailed design. The system state is centralized and shared between the functions operating on that state.

Object-oriented design is based on the concept of *information hiding* developed by Parnas [16]. The system is viewed as a collection of objects rather than functions, with each object managing its own state information. The system state is decentralized, and an object is a member of a class. The definition of a class includes attributes and operations on class members, and these may be inherited from super classes. Objects communicate by exchanging messages

It is essential to verify and validate the design with respect to the system requirements, and this may be done by traceability of the design to the system requirements and design reviews.

1.5.3   Implementation

This phase is concerned with implementing the design in the target language and environment (e.g., C++ or Java), and it involves writing or generating the actual code. The development team divides up the work to be done, with each programmer responsible for one or more modules. The coding activities often include code reviews or walkthroughs to ensure that quality code is produced, and to verify its correctness. The code reviews will verify that the source code conforms to the coding standards and that maintainability issues are addressed. They will also verify that the code produced is a valid implementation of the software design.

The development of a new feature in Agile begins with writing a suite of test cases based on the requirements for the feature. The tests fail initially, and so the first step is to write some code that enables the new test cases to pass. This new code may be imperfect (it will be improved later). The next step is to ensure that the new feature works with the existing features, and this involves executing all new and existing test cases.

This may involve modification of the source code to enable all of the tests to pass and to ensure that all features work correctly together. The final step is refactoring the code, and this involves cleaning up and restructuring the code, and improving its structure and readability. The test cases are rerun during the refactoring to ensure that the functionality is not altered in any way. The process repeats with the addition of each new feature.

Software reuse provides a way to speed up the development process. Components or objects that may be reused need to be identified and handled accordingly. The implemented code may use software components that have either being developed internally or purchased off the shelf. Open-source software has become popular in recent years, and it allows software developed by others to be used (*under an open-source license*) in the development of applications.

The benefits of software reuse include increased productivity and a faster time to market. There are inherent risks with customized-off-the shelf (COTS) software, as the supplier may decide to no longer support the software, or there is no guarantee that software that has worked successfully in one domain will work correctly in a different domain. It is therefore important to consider the risks as well as the benefits of software reuse and open-source software.

1.5.4 Software Testing

Software testing is employed to verify that the requirements have been correctly implemented, and that the software is fit for purpose, as well as identifying defects present in the software. There are various types of testing that may be conducted including *unit testing, integration testing, system testing, performance testing, and user acceptance testing*. These are described below:

Unit and Integration Testing
Unit testing is performed by the programmer on the completed unit (or module) and prior to its integration with other modules. The programmer writes these tests, and the objective is to show that the code satisfies the design. The unit test case is generally documented, and it should include the test objective and the expected results.

Code coverage and branch coverage metrics are often generated to give an indication of how comprehensive the unit testing has been. These metrics provide visibility into the number of lines of code executed, as well as the branches covered during unit testing. The developer executes the unit tests; records the results; corrects any identified defects, and retests the software.

Test driven development (TDD) is employed in the Agile world, and this involves writing the unit test cases (and possibly other test cases) before the code, and the code is then written to pass the defined test cases. These tests are automated in the Agile world and are run with every build.

Integration testing is performed on the integrated system once all of the individual units work correctly in isolation. The objective is to verify that all of the modules and their interfaces work correctly together, and to identify and resolve any issues. Modules that work correctly in isolation may fail when integrated with other modules. The developers generally perform this type of testing. These tests are automated in the Agile world.

System and Performance Testing

The purpose of system testing is to verify that the implementation is valid with respect to the system requirements. It involves the specification of system test cases, and the execution of the test cases will verify that the system requirements have been correctly implemented. An independent test group generally conducts this type of testing, and the system tests are traceable to the system requirements.

The purpose of performance testing is to ensure that the performance of the system satisfies the non-functional requirements. It may include *load performance testing*, where the system is subjected to heavy loads over a long period of time, and *stress testing*, where the system is subjected to heavy loads during a short time interval. *Performance testing often involves the simulation of many users* using the system and involves measuring the response times for various activities.

Any system requirements that have been incorrectly implemented will be identified, and defects logged and reported to the developers. System testing may also include security and usability testing. The preparation of the test environment may involve ordering special hardware and tools, and needs to be set up early in the project.

User Acceptance Testing

UAT testing is usually performed under controlled conditions at the customer site, and its operation will closely resemble the real-life behaviour of the system. The customer will see the product in operation and will judge whether the system is fit for purpose. The objective is to demonstrate that the product satisfies the business requirements and meets the customer expectations. Upon its successful completion the customer is happy to accept the product.

1.5.5 Support and Maintenance

Software systems often have a long lifetime, and the software needs to be continuously enhanced over its lifetime to meet the evolving needs of the customers. This may involve regular new releases with new functionality and corrections to known defects.

Any problems that the customer identifies with the software are reported as per the customer support and maintenance agreement. The support issues will require investigation, and the issue may be *a defect in the software, an enhancement to the software,* or *due to a misunderstanding.* An appropriate solution is implemented to resolve, and testing is conducted to verify that the solution is correct, and that the changes made have not adversely affected other parts of the system. A postmortem may be conducted to learn lessons from the defect,[13] and to take corrective action to prevent a reoccurrence.

The goal of building a correct and reliable software product the first time is difficult to achieve, and the customer is always likely to find some issues with the released software product. It is accepted today that quality needs to be built into each step in the development process, with the role of software inspections and testing to identify as many defects as possible prior to release and minimize the risk that serious defects will be found postrelease.

The effective in-phase inspections of the deliverables will influence the quality of the resulting software and lead to a corresponding reduction in the number of defects. The testing group plays a key role in verifying that the system is correct, and in providing confidence that the software is fit for purpose and ready to be released. The approach to software correctness involves testing and retesting, until the testing group believe that all defects have been eliminated. Dijkstra [17] comments on testing are well-known:

> Testing a program demonstrates that it contains errors, never that it is correct.

That is, irrespective of the amount of time spent testing, it can never be said with absolute confidence that all defects have been found in the software. Testing provides increased confidence that the program is correct, and statistical techniques may be employed to give a measure of the software reliability.

Some mature organizations have a quality objective of three defects per million lines of code, which was introduced by Motorola as part of its six-sigma (6σ) program. It was originally applied it to its manufacturing businesses and subsequently applied to its software organizations. The goal is to reduce variability in manufacturing processes and to ensure that the processes performed within strict process control limits.

1.6   Software Inspections

Software inspections are used to build quality into software products. There are a number of well-known approaches such as the Fagan Methodology [7]; Gilb's approach [8]; and Prince 2's approach.

Fagan inspections were developed by Michael Fagan of IBM. It is a seven-step process that identifies and removes errors in work products. The process mandates that requirement documents, design documents, source code, and test plans are

---

[13] This is essential for serious defects that have caused significant inconvenience to customers (e.g., a major telecom outage). The software development organization will wish to learn lessons to determine what went wrong in its processes that prevented the defect from been identified during peer reviews and testing. Actions to prevent a reoccurrence will be identified and implemented.

all formally inspected by experts independent of the author of the deliverable to ensure quality.

There are various *roles* defined in the process including the *moderator* who chairs the inspection. The *reader's* responsibility is to read or paraphrase the deliverable, and *the author* is the creator of the deliverable and has a special interest in ensuring that it is correct. The *tester* role is concerned with the test viewpoint.

The inspection process will consider whether the design is correct with respect to the requirements, and whether the source code is correct with respect to the design. Software inspections play an important role in building quality into software and in reducing the cost of poor quality in the organization.

1.7   Software Project Management

The timely delivery of quality software requires good management and engineering processes. Software projects have a history of being delivered late or over budget, and good project management practices include the following activities:

- Estimation of cost, effort, and schedule for the project
- Identifying and managing risks
- Preparing the project plan
- Preparing the initial project schedule and key milestones
- Obtaining approval for the project plan and schedule
- Staffing the project
- Monitoring progress, budget, schedule, effort, risks, issues, change requests, and quality
- Taking corrective action
- Replanning and rescheduling
- Communicating progress to affected stakeholders
- Preparing status reports and presentations.

The project plan will contain or reference several other plans such as the project quality plan; the communication plan; the configuration management plan; and the test plan.

Project estimation and scheduling are difficult as often software projects are breaking new ground and may differ from previous projects. That is, previous estimates may often not be a good basis for estimation for the current project. Often, unanticipated problems can arise for technically advanced projects, and the estimates may often be optimistic. Gantt charts are often employed for project scheduling, and these show the work breakdown for the project, as well as task dependencies and allocation of staff to the various tasks.

The effective management of risk during a project is essential to project success. Risks arise due to uncertainty and the risk management cycle involves[14] risk identification; risk analysis and evaluation; identifying responses to risks; selecting and planning a response to the risk; and risk monitoring. The risks are logged, and the likelihood of each risk arising, and its impact is then determined. The risk is assigned an owner and an appropriate response to the risk determined. Project management is discussed in more detail in Chap. 4 of [14].

1.8 CMMI Maturity Model

The CMMI is a framework to assist an organization in the implementation of best practice in software and systems engineering. It is an internationally recognized model for software process improvement and assessment and is used worldwide by thousands of organizations. It provides a solid engineering approach to the development of software, and it supports the definition of high-quality processes for the various software engineering and management activities.

It was developed by the Software Engineering Institute (SEI) who adapted the process improvement principles used in the manufacturing field to the software field. They developed the original CMM model and its successor the CMMI. The CMMI states *what the organization needs to do* to mature its processes rather than *how this should be done.*

The CMMI consists of five maturity levels with each maturity level consisting of several process areas. Each process area consists of a set of goals, and these goals are implemented by practices related to that process area. Level two is focused on management practices; level three is focused on engineering and organization practices; level four is concerned with ensuring that key processes are performing within strict quantitative limits; and level five is concerned with continuous process improvement. Maturity levels may not be skipped in the staged representation of the CMMI, as each maturity level is the foundation for the next level. The CMMI and Agile are compatible, and CMMI v1.3 supports Agile software development.

The CMMI allows organizations to benchmark themselves against other organizations. This is done by a formal SCAMPI appraisal conducted by an authorized lead appraiser. The results of the appraisal are generally reported back to the SEI, and there is a strict qualification process to become an *authorized lead appraiser.* An appraisal is useful in verifying that an organization has improved, and it enables the organization to prioritize improvements for the next improvement cycle. The CMMI is discussed in more detail in Chap. 20 of [14].

---

[14] These are the risk management activities in the Prince2 methodology.

1.9   Formal Methods

Dijkstra and Hoare have argued that the way to develop correct software is to derive the program from its specifications using mathematics, and to employ *mathematical proof* to demonstrate its correctness with respect to the specification. This offers a rigorous framework to develop programs adhering to the highest quality constraints. However, in practice mathematical techniques have proved to be cumbersome to use, and their widespread use in industry is unlikely at this time.

The *safety–critical area* is one domain to which mathematical techniques have been successfully applied. There is a need for extra rigour in the safety and security critical fields, and mathematical techniques can demonstrate the presence or absence of certain desirable or undesirable properties (e.g., *"when a train is in a level crossing, then the gate is closed"*).

Spivey [18] defines a *"formal specification"* as the use of mathematical notation to describe in a precise way the properties which an information system must have, without unduly constraining the way in which these properties are achieved. It describes *what* the system must do, as distinct from *how* it is to be done. This abstraction away from implementation enables questions about what the system does to be answered, independently of the detailed code. Further, the unambiguous nature of mathematical notation avoids the problem of ambiguity in an imprecisely worded natural language description of a system.

The formal specification thus becomes the key reference point for the different parties concerned with the construction of the system and is a useful way of promoting a common understanding for all those concerned with the system. The term *"formal methods"* is used to describe a formal specification language, and a method for the design and implementation of computer systems.

The specification is written precisely in a mathematical language. The derivation of an implementation from the specification may be achieved via *stepwise refinement*. Each refinement step makes the specification more concrete and closer to the actual implementation. There is an associated *proof obligation* that the refinement be valid, and that the concrete state preserves the properties of the more abstract state. Thus, assuming the original specification is correct and the proofs of correctness of each refinement step are valid, then there is a very high degree of confidence in the correctness of the implemented software.

Formal methods have been applied to a diverse range of applications, including circuit design, artificial intelligence, specification of standards, specification and verification of programs, etc. They are described in more detail in Chap. 16.

1.10   Review Questions

1. Discuss the research results of the Standish Group the current state of IT project delivery?

2. What are the main challenges in software engineering?
3. Describe various software lifecycles such as the waterfall model and the spiral model.
4. Discuss the benefits of Agile over conventional approaches. List any risks and disadvantages?
5. Describe the purpose of the CMMI? What are the benefits?
6. Describe the main activities in software inspections.
7. Describe the main activities in software testing.
8. Describe the main activities in project management?
9. What are the advantages and disadvantages of formal methods?

1.11 Summary

The birth of software engineering was at the NATO conference held in 1968 in Germany. This conference highlighted the problems that existed in the software sector in the late 1960s, and the term "*software crisis*" was coined to refer to these. The conference led to the realization that programming is quite distinct from science and mathematics, and that software engineers need to be properly trained to enable them to build high-quality products that are safe to use.

The Standish group conducts research on the extent of problems with the delivery of projects on time and budget. Their research indicates that it remains a challenge to deliver projects on time, on budget and with the right quality.

Programmers are like engineers in the sense that they build products. Therefore, programmers need to receive an appropriate education in engineering as part of their training. The education of traditional engineers includes training on product design and an appropriate level of mathematics.

Software engineering involves multiperson construction of multiversion programs. It is a systematic approach to the development and maintenance of the software, and it requires a precise statement of the requirements of the software product, and then the design and development of a solution to meet these requirements. It includes methodologies to design, develop, implement, and test software as well as sound project management, quality management, and configuration management practices. Support and maintenance of the software need to be properly addressed.

Software process maturity models such as the CMMI have become popular in recent years. They place an emphasis on understanding and improving the software process to enable software engineers to be more effective in their work.

References

1. Brooks F (1975) The mythical man month. Addison Wesley
2. Naur P, Randell B (1975) Software engineering. Petrocelli. IN. Buxton. Report on two NATO conferences held in Garmisch, Germany (October 1968) and Rome, Italy (October 1969)

3. O'Regan G (2006) Mathematical approaches to software quality. Springer
4. Brooks F (1986) No silver bullet. Essence and accidents of software engineering. Information processing. Elsevier, Amsterdam
5. O'Regan G (2010) Introduction to software process improvement. Springer
6. O'Regan G (2014) Introduction to software quality. Springer Verlag
7. Fagan M (1976) Design and code inspections to reduce errors in software development. IBM Syst J 15(3)
8. Gilb T, Graham D (1994) Software inspections. Addison Wesley
9. (2004) Managing successful projects with PRINCE2. Office of Government Commerce, UK
10. Royce W (1970) The software lifecycle model (waterfall model). In: Proceedings of WEST-CON
11. Boehm B (1988) A spiral model for software development and enhancement. Computer
12. Rumbaugh J et al (1999) The unified software development process. Addison Wesley
13. Alliance A, Manifesto for Agile software development. http://agilemanifesto.org
14. O'Regan G (2022) Concise guide to software engineerin (2nd edn). Springer
15. Jacobson I, Booch G, Rumbaugh J (1999) The unified software modelling language user guide. Addison-Wesley
16. Parnas D (1972) On the criteria to be used in decomposing systems into modules. Commun ACM 15(12).
17. Dijkstra EW (1972) Structured programming. Academic
18. Spivey JM (1992) The Z notation. A reference manual. Prentice Hall International Series in Computer Science

Software Engineering Mathematics

2

2.1 Introduction

The computer sector in the 1960s was dominated by several large mainframe computer manufacturers. Computers were large, expensive and difficult to use for a non-specialist. The software used on the mainframes of the 1960s was proprietary, and the hardware of manufacturers was generally incompatible with one another. It was usually necessary to rewrite all existing software application programs for a new computer if a business decided to change to a new manufacturer or upgrade to a more powerful machine from its existing manufacturer.

Software projects tended to be written once off for specific customers, and large projects were often characterized by under estimation and over expectations. There was a very small independent software sector in the 1960s, with software and training included as part of the computer hardware delivered to the customers. IBM's dominant position in the market led to antitrust inquiries by the US Justice Department, and this led IBM to "unbundle" its software and services from its hardware sales. It then began charging separately for software, training and hardware, and this led to the creation of a multi-billion-dollar software industry, and to a major growth of software suppliers.

© The Author(s), under exclusive license to Springer Nature Switzerland AG 2023   27
G. O'Regan, *Mathematical Foundations of Software Engineering*,
Texts in Computer Science, https://doi.org/10.1007/978-3-031-26212-8_2

We discussed the two NATO conferences in the late 1960s that led to the birth of software engineering as a discipline in its own right (see Chap. 1), and the realization that programming is quite different from science and mathematics. Mathematics may be employed to assist with the design and verification of software products. However, the level of mathematics employed will depend on the safety critical nature of the product, as systematic peer reviews and testing are often sufficient.

Software engineers today work in many different domains such as telecommunications field; the banking and insurance fields; the general software sector; utilities; the medical device field; and the pharmaceutical sector. There is specialized knowledge required for each field and the consequence of a software failure varies between these fields (e.g., the defective software of the Therac-25 radiation machine led to several fatalities [1]).

It is essential that the software engineer has the required education and knowledge to perform his/her role effectively, and this includes knowledge of best practice in software engineering as well as the specialized knowledge required for the specific field that the software engineer is working in. The software engineer's education provides the necessary foundation in software engineering, but this will generally be supplemented with specific training for that sector on commencing employment.

Software engineering requires the engineer to state precisely the requirements that the software product is to satisfy, and then to produce designs that will meet these requirements. Engineers provide a precise description of the problem to be solved; they then proceed to producing a design and validating its correctness; finally, the design is implemented and testing is performed to verify the correctness of the implementation with respect to the requirements. The software requirements needs to be unambiguous and should clearly state what is and what is not required.

Classical engineers produce the product design and then analyse their design for correctness. They use mathematics in their analysis, as this is the basis of confirming that the specifications are met. The level of mathematics employed will depend on the particular application and calculations involved. The term *"engineer"* is generally applied only to people who have attained the necessary education and competence to be called engineers, and who base their practice on mathematical and scientific principles. Often in computer science the term engineer is employed rather loosely to refer to anyone who builds things, rather than to an individual with a core set of knowledge, experience, and competence.

Parnas argues that computer scientists should have the right education to apply scientific and mathematical principles to their work. This includes mathematics and design, to enable them to be able to build high-quality and safe products. He advocates a solid engineering approach to the teaching of mathematics with an emphasis on its application to developing and analysing product designs. He argues that software engineers need education on engineering mathematics; specification and design; converting designs into programs; software inspections, and testing. The education should enable the software engineer to produce well-designed programs that will correctly implement the requirements.

Software engineers may work in domains where just basic mathematics is required to do their work, or they may be employed in a sector where substantial mathematics is required. It is important that software engineers receive the right education in software engineering mathematics so that they have the right tools in their toolbox to apply themselves successfully to their work.

2.2 Early Software Engineering Mathematics

Robert Floyd was born in New York in 1936, and he did pioneering work on software engineering from the 1960s (Fig. 2.1). He made important contributions to the theory of parsing; the semantics of programming languages; program verification; and methodologies for the creation of efficient and reliable software.

Mathematics and Computer Science were regarded as two completely separate disciplines in the 1960s, and software development was based on the assumption that the completed code would always contain defects. It was therefore better and more productive to write the code as quickly as possible, and to then perform debugging to find the defects. Programmers then corrected the defects, made patches, and re-tested and found more defects. This continued until they could no longer find defects. Of course, there was always the danger that defects remained in the code that could give rise to software failures.

Floyd believed that there was a way to construct a rigorous proof of the correctness of the programs using mathematics. He showed that mathematics could be used for program verification, and he introduced the concept of *assertions* that provided a way to verify the correctness of programs.

Flowcharts were employed in the 1960s to explain the sequence of basic steps for computer programs. Floyd's insight was to build upon flowcharts and to apply *an invariant assertion to each branch* in the flowchart. These assertions state the essential relations that exist between the variables at that point in the flow chart. An example relation is "$R = Z > 0, X = 1, Y = 0$". He devised a general flowchart language to apply his method to programming languages. The language essentially contains boxes linked by flow of control arrows [2].

Fig. 2.1 Robert Floyd

Fig. 2.2 Branch assertions in flowcharts

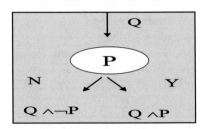

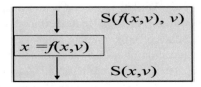

Fig. 2.3 Assignment assertions in flowcharts

Consider the assertion Q that is true on entry to a branch where the condition at the branch is P. Then, the assertion on exit from the branch is $Q \wedge \neg P$ if P is false and $Q \wedge P$ otherwise (Fig. 2.2).

The use of assertions may be employed in an assignment statement. Suppose x represents a variable and v represents a vector consisting of all the variables in the program. Suppose $f(x, v)$ represents a function or expression of x and the other program variables represented by the vector v. Suppose the assertion $S(f(x, v), v)$ is true before the assignment $x = f(x, v)$. Then the assertion $S(x, v)$ is true after the assignment (Fig. 2.3). This is given by:

Floyd used flowchart symbols to represent entry and exit to the flowchart. This included entry and exit assertions to describe the program's entry and exit conditions.

Floyd's technique showed how a computer program is a sequence of logical assertions. Each assertion is true whenever control passes to it, and statements appear between the assertions. The initial assertion states the conditions that must be true for execution of the program to take place, and the exit assertion essentially describes what must be true when the program terminates.

Floyd's insight was his recognition that if it can be shown that the assertion immediately following each step is a consequence of the assertion immediately preceding it, then the assertion at the end of the program will be true, provided the appropriate assertion was true at the beginning of the program.

He published an influential paper, "*Assigning Meanings to Programs*", in 1967 [2], and this paper influenced Hoare's work on preconditions and post-conditions leading to Hoare logic [3]. Floyd's paper also presented a formal grammar for flowcharts, together with rigorous methods for verifying the effects of basic actions like assignments.

Fig. 2.4 C. A. R. Hoare

Hoare logic is a formal system of logic used for programming semantics and for program verification. It was developed by C. A. R. Hoare (Fig. 2.4) and was originally published in Hoare's 1969 paper *"An axiomatic basis for computer programming"* [3]. Hoare and others have subsequently refined it, and it provides a logical methodology for precise reasoning about the correctness of computer programs.

Hoare was influenced by Floyd's [2] paper that applied assertions to flowcharts, and he recognized that this provided an effective method for proving the correctness of programs. He built upon Floyd's approach to cover the familiar constructs of high-level programming languages.

This led to the axiomatic approach to defining the semantics of every statement in a programming language, and the approach consists of axioms and proof rules. He introduced what has become known as the Hoare triple, and this describes how the execution of a fragment of code changes the state. A Hoare triple is of the form:

$$P\{Q\}R$$

where P and R are assertions and Q is a program or command. The predicate P is called the *precondition*, and the predicate R is called the *postcondition*.

Definition 2.1 (*Partial Correctness*) The meaning of the Hoare triple above is that whenever the predicate P holds of the state before the execution of the command or program Q, then the predicate R will hold after the execution of Q. The brackets indicate partial correctness as if Q does not terminate then R can be any predicate. R may be chosen to be false to express that Q does not terminate.

Total correctness requires Q to terminate, and at termination R is true. Termination needs to be proved separately. Hoare logic includes axioms and rules of inference rules for the constructs of imperative programming language.

Hoare and Dijkstra were of the view that the starting point of a program should always be the specification, and that the proof of the correctness of the program should be developed along with the program itself.

That is, the starting point is the mathematical specification of what a program is to do, and mathematical transformations are applied to the specification until it is turned into a program that can be executed. The resulting program is then known to be correct by construction.

2.3   Debate on Mathematics in Software Engineering

The debate concerning the level of use of mathematics in software engineering is still ongoing. Many practitioners are against the use of mathematics and avoid its use. They tend to employ methodologies such as software inspections and testing to improve confidence in the correctness of the software. They argue that in the current competitive industrial environment where time to market is a key driver that the use of such formal mathematical techniques would seriously impact the market opportunity. Industrialists often need to balance conflicting needs such as quality, cost, and delivering on time. They argue that the commercial necessities require methodologies and techniques that allow them to achieve their business goals effectively.

The other camp argues that the use of mathematics is essential in the delivery of high-quality and reliable software, and that if a company does not place sufficient emphasis on quality, it will pay the price in terms of poor quality and loss of reputation.

It is generally accepted that mathematics and formal methods must play a role in the safety critical and security critical fields. Apart from that the extent of the use of mathematics is a hotly disputed topic. The pace of change in the world is extraordinary, and companies face immense competitive forces in a global market place.

It is unrealistic to expect companies to deploy mathematical techniques unless they have clear evidence that it will support them in delivering commercial products to the market place ahead of their competition, at the right price and with the right quality. Formal methods and other mathematical techniques need to prove that they can do this if they wish to be taken seriously in mainstream software engineering.

2.4   The Emergence of Formal Methods

Formal methods refer to various mathematical techniques used for the formal specification and development of software. They consist of a formal specification language and employ a collection of tools to support the syntax checking of the specification, as well as the proof of properties of the specification. They allow questions to be asked about what the system does independently of the implementation. The use of mathematical notation helps in ensuring precision in the description of a system.

The term "formal methods" is used to describe a formal specification language and a method for the design and implementation of computer systems. They may be employed at a number of levels starting with the formal specification only and developing the program informally, to formal specification and refinement with some program verification, and finally to full formal specification, refinement and verification.

The specification is written in a mathematical language, and the implementation may be derived from the specification via stepwise refinement. The refinement step makes the specification more concrete and closer to the actual implementation. There is an associated proof obligation to demonstrate that the refinement is valid, and that the concrete state preserves the properties of the abstract state. Thus, assuming that the original specification is correct and the proof of correctness of each refinement step is valid, then there is a very high degree of confidence in the correctness of the implemented software.

The mathematical analysis of the formal specification allows questions to be asked about what the system does, and these questions may be answered independently of the implementation. Mathematical notation is precise, and this helps to avoid the problem of ambiguity inherent in a natural language description of a system. The formal specification may be used to promote a common understanding for all stakeholders.

One of the earliest formal methods was VDM which was developed at the IBM research laboratory in Vienna in the 1970s. VDM emerged as part of their work into the specification of the semantics of the PL/1 programming language. Over time other formal specification languages such as Z and B were developed, as well as a plethora of specialized calculi such as CSP, CCS, and π-calculus, and various temporal logics and theorem provers have been developed, and the important area of model checking emerged.

However, despite the interest in formal methods in academia the industrial take up of formal methods has been quite limited, and they are mainly used in the safety critical and security critical fields. Formal methods have been criticized as being difficult to use, as being unintuitive and lacking industrial strength tool support, and so on.

However, formal methods should be regarded as a tool in the software engineer's toolbox to be used in important domains such as the safety critical field. They should therefore be included as part of the software engineer's education.

2.5   What Mathematics Do Software Engineers Need?

Mathematics plays a key role in classical engineering to assist with design and verification of products. It is therefore reasonable to apply appropriate mathematics in software engineering (especially for safety and security critical systems) to assure that the delivered systems conform to the requirements.

The extent to which mathematics should be used is controversial with strong views in both camps between those who advocate a solid engineering approach with mathematical rigorous and those who argue for a lighter approach with minimal mathematics (e.g., those in the Agile world). In many domains, rigorous peer reviews and testing will be sufficient to build quality into the software product, whereas in other more specialized areas (especially for safety and security critical applications), it is desirable to have the extra assurance that may be provided with mathematical techniques.

The domain in which the software engineer is working is also relevant, as specialized mathematical knowledge may be required to develop software for some domains. For example, a software engineer who is working on financial software engineering applications will require specialized knowledge of the calculation of simple and compound interest, annuities and so on in the banking domain, and knowledge of probability, statistics, calculus, and actuarial mathematics may be required in the insurance domain. That is, there is not a one size that fits all in the use of mathematics—the mathematics that the software engineer needs to employ depends on the particular domain that the software engineer is working in.

However, there is a core body of mathematical knowledge that the software engineer should possess, with more specialized mathematical knowledge required for specific domains. The core mathematics proposed for every software engineer includes arithmetic, algebra, logic, and trigonometry (Table 2.1).

Further, mathematics provides essential training in critical thinking and problem solving, allows the software engineer to perform a rigorous analysis of a particular situation, and avoids an over-reliance on intuition. Mathematical modelling provides a mathematical simplification of the real world and provides a way to explain a system as well as providing predictions. Engineers are taught how to apply mathematics in their work, and the emphasis is always on the application of mathematics to solve practical problems. Mathematics may be applied to solve practical problems and to develop products that are fit for purpose.

Table 2.1 Appropriate mathematics in software engineering

Area	Description
Core mathematics (reasoning/problem solving)	Arithmetic, algorithms, algebra, sets, relations and functions, sequences and series, trigonometry, coordinate systems, logic, graph theory, language theory, automata theory
Traditional engineering applications	Complex analysis, matrices, vectors, calculus, Fourier series, Laplace transforms
Financial software engineering (banking, insurance and business)	Simple and compound interest, probability, statistics, operations research, linear programming
Telecoms	Cryptography, coding theory
Safety/security critical	Software reliability and dependability, formal methods, Z specification language, logic, temporal logic, theorem provers, model checking
Robotics/computer graphics	Complex numbers, quaternions, vectors, matrices

Classical mathematics may be applied to software engineering and specialized mathematical methods and notations have also been developed. However, the successful delivery of a project requires a lot more than just the use of mathematics. It requires sound project management and quality management practices; the effective definition of the requirements; the management of changes to the requirements throughout the project; the management of risk; and so on (see the companion book [1]). A project that is not properly managed will suffer from schedule, budget, or cost overruns as well as problems with quality.

2.6 Review Questions

1. Why should mathematics be part of the education of software engineers?
2. What mathematics should software engineers know?
3. What is the role of mathematics in current software engineering?
4. Discuss the contributions of Floyd and Hoare.
5. Explain the difference between partial correctness and total correctness.
6. What are formal methods ? Explain their significance.
7. Explain the levels at which formal methods may be applied.

2.7 Summary

Classical engineering has a successful track record in building high-quality products that are safe for the public to use. It is therefore natural to consider using an engineering approach to developing software, and this involves identifying the customer requirements, carrying out a rigorous design to meet the requirements, developing and coding a solution to meet the design, and conducting appropriate inspections and testing to verify the correctness of the solution.

Mathematics plays a key role in classical engineering to assist with the design and verification of products. It makes sense to apply appropriate mathematics in software engineering (especially for safety critical systems) to assure that the delivered systems conform to the requirements. The extent to which mathematics should be used remains controversial.

There is a core body of mathematics that every software engineer should be familiar with, including arithmetic, algebra, and logic. The domain in which the software engineer is working is relevant, as specialized mathematical knowledge may be required by the software engineer for specific domains.

Mathematics is a tool of thought, and it provides essential training for critical thinking and problem solving for the modern software engineer.

References

1. O'Regan G (2022) Concise guide to software engineering, 2nd edn. Springer
2. Floyd R (1967) Assigning meanings to programs. Proc Symp Appl Math (19):19–32
3. Hoare CAR (1969) An axiomatic basis for computer programming. Commun ACM 12(10):576–585

Mathematical Prerequisites for Software Engineers

3

Key Topics

Sets

Relations

Functions

Natural numbers

Prime numbers

Fractions

Decimals

Percentages

Ratios

Proportions

Cartesian Coordinates

Pythagoras's Theorem

Periodic Functions

Degrees and Radians

Sine Rule

Cosine Rule

© The Author(s), under exclusive license to Springer Nature Switzerland AG 2023
G. O'Regan, *Mathematical Foundations of Software Engineering*,
Texts in Computer Science, https://doi.org/10.1007/978-3-031-26212-8_3

3.1   Introduction

This chapter sketches the mathematical prerequisites that software engineers should be familiar with, and we discuss fundamental concept such as sets, relations and functions, arithmetic, and trigonometry. Sets are collections of well-defined objects; relations indicate relationships between members of two sets A and B; and functions are a special type of relation where there is exactly (or at most)[1] one relationship for each element $a \in A$ with an element in B.

A set is a collection of well-defined objects that contains no duplicates. The term "*well defined*" means that for a given value it is possible to determine whether or not it is a member of the set. There are many examples of sets such as the set of natural numbers $\mathbb{N}$, the set of integer numbers $\mathbb{Z}$, and the set of rational numbers $\mathbb{Q}$. The natural numbers $\mathbb{N}$ is an infinite set consisting of the numbers $\{1, 2, ...\}$. Venn diagrams may be used to represent sets pictorially.

A binary relation $R(A, B)$ where A and B are sets is a subset of the Cartesian product $(A \times B)$ of A and B. The domain of the relation is A and the codomain of the relation is B. The notation aRb signifies that there is a relation between a and b and that $(a, b) \in R$. An n-ary relation $R(A_1, A_2, ... A_n)$ is a subset of $(A_1 \times A_2 \times ... \times A_n)$. However, an n-ary relation may also be regarded as a binary relation $R(A, B)$ with $A = A_1 \times A_2 \times ... \times A_{n-1}$ and $B = A_n$.

Functions may be total or partial. A total function $f: A \rightarrow B$ is a special relation such that for each element $a \in A$ there is exactly one element $b \in B$. This is written as $f(a) = b$. A partial function differs from a total function in that the function may be undefined for one or more values of A. The domain of a function (denoted by **dom** f) is the set of values in A for which the partial function is defined. The domain of the function is A if f is a total function. The codomain of the function is B.

Arithmetic (or number theory) is the branch of mathematics that is concerned with the study of numbers and their properties. It includes the study of the integer numbers, and operations on them, such as addition, subtraction, multiplication, and division.

Number theory studies various properties of integer numbers such as their parity and divisibility; their additive and multiplicative properties; whether a number is prime or composite; the prime factors of a number; the greatest common divisor and least common multiple of two numbers; and so on.

The natural numbers $\mathbb{N}$ consist of the numbers $\{1, 2, 3, ...\}$. The integer numbers are a superset of the set of natural numbers, and they consist of $\{... -2, -1, 0, 1, 2, ...\}$. The rational numbers $\mathbb{Q}$ are a superset of the set of integer numbers, and they consist of all numbers of the form $\{p/q$ where p and q are integers and $q \neq 0\}$. The real numbers $\mathbb{R}$ is a superset of the set of rational numbers, and they are defined to be the set of converging sequences of rational numbers. They

---

[1] We distinguish between total and partial functions. A total function $f: A \rightarrow B$ is defined for every element in A whereas a partial function may be undefined for one or more values in A.

contain the rational and irrational numbers. The complex numbers $\mathbb{C}$ consist of all numbers of the form $\{a + bi$ where $a, b \in \mathbb{R}$ and $i = \sqrt{-1}\}$, and they are a superset of the set of real numbers.

Number theory has many applications including cryptography and coding theory in computing. For example, the RSA public key cryptographic system relies on its security due to the infeasibility of the integer factorization problem for large numbers.

Trigonometry is concerned with the relationships between sides and angles of triangles, and the origin of the term is from the Greek words τρίγωνον (trigonon) meaning triangle and μετρον(metron) meaning measure. The origins of the field are from the Hellenistic world in the third century BC., but early work on angles had been done by the Sumerians and Babylonians.

Pythagoras's Theorem expresses the relationship between the hypotenuse of a right-angled triangle and the other two sides, and we define sine, cosine, and tangent for a right-angled triangle. The sine rule and cosine rule are invaluable in solving trigonometric problems, as well as various trigonometric identities. We discuss degrees and radians as well as sketching the curves of sine and cosine.

The Cartesian system was invented by Descartes in the seventeenth century, and it allows geometric shapes such as curves to be described by algebraic equations. We discuss both the two-dimensional plane and three-dimensional space.

3.2   Set Theory

A set is a fundamental building block in mathematics, and it is defined as a collection of well-defined objects. The elements in a set are of the same kind, and they are distinct with no repetition of the same element in the set.[2] Most sets encountered in computer science are finite, as computers can only deal with finite entities. Venn diagrams[3] are often employed to give a pictorial representation of a set, and to illustrate various set operations such as set union, intersection, and set difference.

There are many well-known examples of sets including the set of natural numbers denoted by $\mathbb{N}$; the set of integers denoted by $\mathbb{Z}$; the set of rational numbers is denoted by $\mathbb{Q}$; the set of real numbers denoted by $\mathbb{R}$; and the set of complex numbers denoted by $\mathbb{C}$.

A finite set may be defined by listing all its elements. For example, the set $A = \{2, 4, 6, 8, 10\}$ is the set of all even natural numbers less than or equal to 10. The order in which the elements are listed is not relevant, i.e., the set $\{2, 4, 6, 8, 10\}$ is the same as the set $\{8, 4, 2, 10, 6\}$.

---

[2] There are mathematical objects known as *multi-sets* or *bags* that allow duplication of elements. For example, a bag of marbles may contain three green marbles, two blue and one red marble.

[3] The British logician, John Venn, invented the Venn diagram. It provides a visual representation of a set and the various set theoretical operations. Their use is limited to the representation of two or three sets as they become cumbersome with a larger number of sets.

Sets may be defined by using a predicate to constrain set membership. For example, the set $S = \{n : \mathbb{N} : n \leq 10 \wedge n \bmod 2 = 0\}$ also represents the set $\{2, 4, 6, 8, 10\}$. That is, the use of a predicate allows a new set to be created from an existing set by using the predicate to restrict membership of the set. The set of even natural numbers may be defined by a predicate over the set of natural numbers that restricts membership to the even numbers. It is defined by:

$$\text{Evens} = \{x | x \in \mathbb{N} \wedge \text{even}(x)\}.$$

In this example, $even(x)$ is a predicate that is true if x is even and false otherwise. In general, $A = \{x \in E | P(x)\}$ denotes a set A formed from a set E using the predicate P to restrict membership of A to those elements of E for which the predicate is true.

The elements of a finite set S are denoted by $\{x_1, x_2, \ldots x_n\}$. The expression $x \in S$ denotes that the element x is a member of the set S, whereas the expression $x \notin S$ indicates that x is not a member of the set S.

A set S is a subset of a set T (denoted $S \subseteq T$) if whenever $s \in S$ then $s \in T$, and in this case the set T is said to be a superset of S (denoted $T \supseteq S$). Two sets S and T are said to be equal if they contain identical elements, i.e., $S = T$ if and only if $S \subseteq T$ and $T \subseteq S$. A set S is a proper subset of a set T (denoted $S \subset T$) if $S \subseteq T$ and $S \neq T$. That is, every element of S is an element of T and there is at least one element in T that is not an element of S. In this case, T is a proper superset of S (denoted $T \supset S$).

The empty set (denoted by $\emptyset$ or $\{\}$) represents the set that has no elements. Clearly $\emptyset$ is a subset of every set. The singleton set containing just one element x is denoted by $\{x\}$, and clearly $x \in \{x\}$ and $x \neq \{x\}$. Clearly, $y \in \{x\}$ if and only if $x = y$.

3.2.1   Set Theoretical Operations

Several set theoretical operations are considered in this section. These include the Cartesian product operation; the power set of a set; the set union operation; the set intersection operation; the set difference operation; and the symmetric difference operation.

Cartesian Product
The Cartesian product allows a new set to be created from existing sets. The Cartesian[4] product of two sets S and T (denoted $S \times T$) is the set of ordered pairs $\{(s, t) | s \in S, t \in T\}$. Clearly, $S \times T \neq T \times S$ and so the Cartesian product of two sets is not commutative. Two ordered pairs (s_1, t_1) and (s_2, t_2) are considered equal if and only if $s_1 = s_2$ and $t_1 = t_2$.

---

[4] Cartesian product is named after René Descartes who was a famous 17th French mathematician and philosopher. He invented the Cartesian coordinates system that links geometry and algebra, and allows geometric shapes to be defined by algebraic equations.

The Cartesian product may be extended to that of n sets S_1, S_2,..., S_n. The Cartesian product $S_1 \times S_2 \times \ldots \times S_n$ is the set of ordered n-tuples $\{(s_1, s_2, \ldots, s_n) | s_1 \in S_1, s_2 \in S_2, \ldots, s_n \in S_n\}$. Two ordered n-tuples $(s_1, s_2,\ldots, s_n)$ and $(s_1', s_2',\ldots, s_n')$ are considered equal if and only if $s_1 = s_1'$, $s_2, = s_2',\ldots, s_n = s_n'$.

The Cartesian product may also be applied to a single set S to create ordered n-tuples of S, i.e., $S^n = S \times S \times \ldots \times S (n$ times $)$.

Power Set

The power set of a set A (denoted $\mathbb{P}A$) denotes the set of subsets of A. For example, the power set of the set $A = \{1, 2, 3\}$ has eight elements and is given by:

$$\mathbb{P}A = \{\emptyset, \{1\}, \{2\}, \{3\}, \{1, 2\}, \{1, 3\}, \{2, 3\}, \{1, 2, 3\}\}.$$

There are $2^3 = 8$ elements in the power set of $A = \{1, 2, 3\}$ where the cardinality of A is 3. In general, there are $2^{|A|}$ elements in the power set of A.

Union and Intersection Operations

The union of two sets A and B is denoted by $A \cup B$. It results in a set that contains all of the members of A and of B and is defined by:

$$A \cup B = \{r | r \in A \text{ or } r \in B\}.$$

The intersection of two sets A and B is denoted by $A \cap B$. It results in a set containing the elements that A and B have in common and is defined by:

$$A \cap B = \{r | r \in A \text{ and } r \in B\}.$$

Union and intersection may be extended to more generalized union and intersection operations.

Set Difference Operations

The set difference operation $A \setminus B$ yields the elements in A that are not in B. It is defined by

$$A \setminus B = \{a | a \in A \text{ and } a \notin B\}$$

For A and B defined as $A = \{1, 2\}$ and $B = \{2, 3\}$ we have $A \setminus B = \{1\}$ and $B \setminus A = \{3\}$. Clearly, set difference is not commutative, i.e., $A \setminus B \neq B \setminus A$. Clearly, $A \setminus A = \emptyset$ and $A \setminus \emptyset = A$.

The symmetric difference of two sets A and B is denoted by $A \triangle B$ and is given by:

$$A \triangle B = A \setminus B \cup B \setminus A$$

The complement of a set A (with respect to the universal set U) is the elements in the universal set that are not in A. It is denoted by A^c (or A') and is defined as:

$$A^c = \{u | u \in U \text{ and } u \notin A\} = U \backslash A$$

3.2.2   Computer Representation of Sets

Sets are fundamental building blocks in mathematics, and so the question arises as to how is a set is stored and manipulated in a computer. The representation of a set M on a computer requires a change from the normal view that the order of the elements of the set is irrelevant, and we will need to assume a definite order in the underlying universal set $\mathcal{m}$ from which the set M is defined.

That is, a set is defined in a computer program with respect to an underlying universal set, and the elements in the universal set are listed in a definite order. Any set M arising in the program that is defined with respect to this universal set $\mathcal{m}$ is a subset of $\mathcal{m}$. Next, we show how the set M is stored internally on the computer.

The set M is represented in a computer as a string of binary digits $b_1 b_2 \ldots b_n$ where n is the cardinality of the universal set $\mathcal{m}$. The bits b_i (where i ranges over the values 1, 2, ... n) are determined according to the rule:

$b_i = 1$ if ith element of $\mathcal{m}$ is in M
$b_i = 0$ if ith element of $\mathcal{m}$ is not in M

For example, if $\mathcal{m} = \{1, 2, \ldots 10\}$ then the representation of $M = \{1, 2, 5, 8\}$ is given by the bit string 1100100100 where this is given by looking at each element of $\mathcal{m}$ in turn and writing down 1 if it is in M and 0 otherwise.

Similarly, the bit string 0100101100 represents the set $M = \{2, 5, 7, 8\}$, and this is determined by writing down the corresponding element in $\mathcal{m}$ that corresponds to a 1 in the bit string.

Clearly, there is a one-to-one correspondence between the subsets of $\mathcal{m}$ and all possible n-bit strings. Further, the set theoretical operations of set union, intersection, and complement can be carried out directly with the bit strings (provided that the sets involved are defined with respect to the same universal set). This involves a bitwise "or" operation for set union; a bitwise "and" operation for set intersection; and a bitwise "not" operation for the set complement operation.

3.3   Relations

A binary relation $R(A, B)$ where A and B are sets is a subset of $A \times B$, i.e., $R \subseteq A \times B$. The domain of the relation is A, and the codomain of the relation is B. The notation aRb signifies that $(a, b) \in R$.

A binary relation $R(A, A)$ is a relation between A and A (or a relation on A). This type of relation may always be composed with itself, and its inverse is also a binary relation on A. The identity relation on A is defined by $a\ i_A a$ for all $a \in A$.

A relation $R(A, B)$ may be represented pictorially. This is referred to as the graph of the relation, and it is illustrated in the diagram below. An arrow from x to y is drawn if (x, y) is in the relation. Thus for the height relation R given by $\{(a, p), (a, r), (b, q)\}$ an arrow is drawn from a to p, from a to r and from b to q to indicate that (a, p), (a, r) and (b, q) are in the relation R.

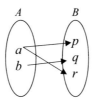

The pictorial representation of the relation makes it easy to see that the height of a is greater than the height of p and r; and that the height of b is greater than the height of q.

An n-ary relation $R(A_1, A_2, \ldots A_n)$ is a subset of $(A_1 \times A_2 \times \ldots \times A_n)$. However, an n-ary relation may also be regarded as a binary relation $R(A, B)$ with $A = A_1 \times A_2 \times \ldots \times A_{n-1}$ and $B = A_n$.

3.3.1 Reflexive, Symmetric and Transitive Relations

A binary relation on A may have additional properties such as being reflexive, symmetric or transitive. These properties are defined as

(i)   A relation on a set A is *reflexive* if $(a, a) \in R$ for all $a \in A$.
(ii) A relation R is *symmetric* if whenever $(a, b) \in R$ then $(b, a) \in R$.
(iii) A relation is *transitive* if whenever $(a, b) \in R$ and $(b, c) \in R$ then $(a, c) \in R$.

A relation that is reflexive, symmetric, and transitive is termed an *equivalence relation*.

Example 3.1 (*Equivalence relation*) The relation on the set of integers $\mathbb{Z}$ defined by $(a, b) \in R$ if $a - b = 2\,k$ for some $k \in \mathbb{Z}$ is an equivalence relation, and it partitions the set of integers into two equivalence classes, i.e., the even and odd integers.

Domain and Range of Relation
The domain of a relation $R\ (A, B)$ is given by $\{a \in A | \exists b \in B$ and $(a, b) \in R\}$. It is denoted by **dom** R. The domain of the relation $R = \{(a, p), (a, r), (b, q)\}$ is $\{a, b\}$.

The range of a relation R (A, B) is given by $\{b \in B | \exists a \in A$ and $(a, b) \in R\}$. It is denoted by **rng** R. The range of the relation $R = \{(a, p), (a, r), (b, q)\}$ is $\{p, q, r\}$.

Inverse of a Relation

Suppose $R \subseteq A \times B$ is a relation between A and B then the inverse relation $R^{-1} \subseteq B \times A$ is defined as the relation between B and A and is given by:

$$b R^{-1} a \text{ if and only if } a \ Rb$$

That is,

$$R^{-1} = \{(b, a) \in B \times A : (a, b) \in R\}$$

Example 3.2 Let R be the relation between $\mathbb{Z}$ and $\mathbb{Z}^+$ defined by mRn if and only if $m^2 = n$. Then $R = \{(m, n) \in \mathbb{Z} \times \mathbb{Z}^+ : m^2 = n\}$ and $R^{-1} = \{(n, m) \in \mathbb{Z}^+ \times \mathbb{Z} : m^2 = n\}$.

For example, $- 3 R 9, - 4 R 16, 0 R 0, 16 R^{-1} - 4, 9 R^{-1} - 3$, etc.

Partitions and Equivalence Relations

An equivalence relation on A leads to a partition of A, and vice versa for every partition of A there is a corresponding equivalence relation.

Let A be a finite set and let $A_1, A_2, ..., A_n$ be subsets of A such $A_i \neq \emptyset$ for all i, $A_i \cap A_j = \emptyset$ if $i \neq j$ and $A = \cup_i^n A_i = A_1 \cup A_2 \cup ... \cup A_n$.

The sets A_i partition the set A, and these sets are called the classes of the partition.

3.3.2   Composition of Relations

The composition of two relations $R_1(A, B)$ and $R_2(B, C)$ is given by $R_2 \circ R_1$ where $(a, c) \in R_2 \circ R_1$ if and only there exists $b \in B$ such that $(a, b) \in R_1$ and $(b, c) \in R_2$. The composition of relations is associative, i.e.,

$$(R_3 \circ R_2) \circ R_1 = R_3 \circ (R_2 \circ R_1)$$

The composition of $S \circ R$ is determined by choosing $x \in A$ and $y \in C$ and checking if there is a route from x to y in the graph. If so, we join x to y in $S \circ R$.

The union of two relations $R_1(A, B)$ and $R_2(A, B)$ is meaningful (as these are both subsets of $A \times B$). The union $R_1 \cup R_2$ is defined as $(a, b) \in R_1 \cup R_2$ if and only if $(a, b) \in R_1$ or $(a, b) \in R_2$.

Similarly, the intersection of R_1 and $R_2(R_1 \cap R_2)$ is meaningful and is defined as $(a, b) \in R_1 \cap R_2$ if and only if $(a, b) \in R_1$ and $(a, b) \in R_2$. The relation R_1 is a subset of $R_2(R_1 \subseteq R_2)$ if whenever $(a, b) \in R_1$ then $(a, b) \in R_2$.

The inverse of the relation R was discussed earlier and is given by the relation R^{-1} where $R^{-1} = \{(b, a) | (a, b) \in R\}$.

The composition of R and R^{-1} yields: $R^{-1} \circ R = \{(a,a)|a \in \text{dom } R\} = i_A$ and $R \circ R^{-1} = \{(b,b)|b \in \text{dom } R^{-1}\} = i_B$.

3.3.3 Binary Relations

A binary relation R on A is a relation between A and A, and a binary relation can always be composed with itself. Its inverse is a binary relation on the same set. The following are all relations on A:

$$R^2 = R \circ R$$
$$R^3 = (R \circ R) \circ R$$
$$R^0 = i_A \text{ (identity relation)}$$
$$R^{-2} = R^{-1} \circ R^{-1}$$

Example 3.3 Let R be the binary relation on the set of all people P such that $(a,b) \in R$ if a is a parent of b. Then the relation R^n is interpreted as:

R is the parent relationship
R^2 is the grandparent relationship
R^3 is the great grandparent relationship
R^{-1} is the child relationship
R^{-2} is the grandchild relationship
R^{-3} is the great grandchild relationship

This can be generalized to a relation R^n on A where $R^n = R \circ R \circ \ldots \circ R$ (n-times). The transitive closure of the relation R on A is given by:

$$R^* = \bigcup_{i=0}^{\infty} R^i = R^0 \cup R^1 \cup R^2 \cup \ldots R^n \cup \ldots$$

where R^0 is the reflexive relation containing only each element in the domain of R: i.e., $R^0 = i_A = \{(a,a)|a \in \text{dom } R\}$.

The positive transitive closure is similar to the transitive closure except that it does not contain R^0. It is given by:

$$R^+ = \bigcup_{i=1}^{\infty} R^i = R^1 \cup R^2 \cup \ldots \cup R^n \cup \ldots$$

$a \, R^+ \, b$ if and only if $a \, R^n \, b$ for some $n > 0$, i.e., there exists $c_1, c_2 \ldots c_n \in A$ such that

$$a R c_1, c_1 R c_2, \ldots, c_n R b.$$

3.4   Functions

A function $f: A \rightarrow B$ is a special relation such that for each element $a \in A$ there is exactly (or at most)[5] one element $b \in B$. This is written as $f(a) = b$.

A function is a relation but not every relation is a function. For example, the relation in the diagram below is not a function since there are two arrows from the element $a \in A$.

The domain of the function (denoted by **dom** f) is the set of values in A for which the function is defined. The domain of the function is A if f is a total function. The codomain of the function is B. The range of the function (denoted **rng** f) is a subset of the codomain and consists of:

$$\mathbf{rng}\ f = \{r | r \in B \text{ such that } f(a) = r \text{ for some } a \in A\}.$$

Functions may be partial or total. A *partial function* (or partial mapping) may be undefined for some values of A, and partial functions arise regularly in the computing field. *Total functions* are defined for every value in A and many functions encountered in mathematics are total.

Example 3.4 (*Functions*) Functions are an essential part of mathematics and computer science, and there are many well-known functions such as the trigonometric functions $\sin(x)$, $\cos(x)$, and $\tan(x)$; the logarithmic function $\ln(x)$; the exponential functions e^x; and polynomial functions.

(i) Consider the partial function $f: \mathbb{R} \rightarrow \mathbb{R} f(x) = 1/x$ (where $x \neq 0$).

Then, this partial function is defined everywhere except for $x = 0$.

(ii) Consider the function $f: \mathbb{R} \rightarrow \mathbb{R}$ where

$$f(x) = x^2$$

Then this function is defined for all $x \in \mathbb{R}$.

Partial functions often arise in computing as a program may be undefined or fail to terminate for several values of its arguments (e.g., infinite loops). Care is required to ensure that the partial function is defined for the argument to which it is to be applied.

---

[5] We distinguish between total and partial functions. A total function is defined for all elements in the domain, whereas a partial function may be undefined for one or more elements in the domain.

Example 3.5 Two partial functions f and g are equal if:

1. $\text{dom} f = \text{dom} g$
2. $f(a) = g(a)$ for all $a \in \text{dom} f$.

A function f is less defined than a function g ($f \subseteq g$) if the domain of f is a subset of the domain of g, and the functions agree for every value on the domain of f.

1. $\text{dom} f \subseteq \text{dom} g$
2. $f(a) = g(a)$ for all $a \in \text{dom} f$.

The composition of functions is similar to the composition of relations. Suppose $f: A \rightarrow B$ and $g: B \rightarrow C$ then $g \circ f: A \rightarrow C$ is a function, and it is written as $g \circ f(x)$ or $g(f(x))$ for $x \in A$.

The composition of functions is not commutative, and this can be seen by an example. Consider the function $f: \mathbb{R} \rightarrow \mathbb{R}$ such that $f(x) = x^2$ and the function $g: \mathbb{R} \rightarrow \mathbb{R}$ such that $g(x) = x + 2$. Then

$$g \circ f(x) = g(x^2) = x^2 + 2.$$
$$f \circ g(x) = f(x + 2) = (x + 2)^2 = x^2 + 4x + 4.$$

Clearly, $g \circ f(x) \neq f \circ g(x)$ and so composition of functions is not commutative. The composition of functions is associative, as the composition of relations is associative and every function is a relation. For $f: A \rightarrow B$, $g: B \rightarrow C$, and $h: C \rightarrow D$ we have:

$$h \circ (g \circ f) = (h \circ g) \circ f$$

A function $f: A \rightarrow B$ is *injective* (*one to one*) if

$$f(a_1) = f(a_2) \Rightarrow a_1 = a_2.$$

For example, consider the function $f: \mathbb{R} \rightarrow \mathbb{R}$ with $f(x) = x^2$. Then $f(3) = f(-3) = 9$ and so this function is not one to one.

A function $f: A \rightarrow B$ is *surjective* (*onto*) if given any $b \in B$ there exists an $a \in A$ such that $f(a) = b$. Consider the function $f: \mathbb{R} \rightarrow \mathbb{R}$ with $f(x) = x + 1$. Clearly, given any $r \in \mathbb{R}$ then $f(r - 1) = r$ and so f is onto.

A function is *bijective* if it is one to one and onto. That is, there is a one to one correspondence between the elements in A and B, and for each $b \in B$ there is a unique $a \in A$ such that $f(a) = b$.

The inverse of a relation was discussed earlier and the relational inverse of a function $f: A \rightarrow B$ clearly exists. The relational inverse of the function may or may not be a function.

However, if the relational inverse is a function it is denoted by $f^{-1}: B \to A$. A total function has an inverse if and only if it is bijective whereas a partial function has an inverse if and only if it is injective.

The identity function $1_A: A \to A$ is a function such that $1_A(a) = a$ for all $a \in A$. Clearly, when the inverse of the function exists then we have that $f^{-1} \circ f = 1_A$ and $f \circ f^{-1} = 1_B$.

Theorem 3.1 (Inverse of Function) *A total function has an inverse if and only if it is bijective.*

3.5   Arithmetic

The natural numbers $\mathbb{N}$ {1, 2, 3, ...} are used for counting things starting from the number one, and they form an ordered set $1 < 2 < 3 < ...$ and so on. The natural numbers are all positive (i.e., there are no negative natural numbers), and the number zero is not usually included as a natural number. However, the set of natural numbers including 0 is denoted by $\mathbb{N}_0$, and it is the set {0, 1, 2, 3, ...}.

The natural numbers are an ordered set and so given any pair of natural numbers (n, m) then either $n < m$, $n = m$ or $n > m$. There is no largest natural number as such, since given any natural number we can immediately determine a larger natural number (e.g., its successor). Each natural number has a unique successor and every natural number larger than one has a unique predecessor.

The addition of two numbers yields a new number, and the subtraction of a smaller number from a larger number yields the difference between them. Multiplication is the mathematical operation of scaling one number by another: for example: $3 * 4 = 4 + 4 + 4 = 12$.

Peano's axiomization of arithmetic is a formal axiomization of the natural numbers, and they include axioms for the successor of a natural number and the axiom of induction. The number zero is considered to be a natural number in the Peano system.

The natural numbers satisfy several nice algebraic properties such as closure under addition and multiplication (i.e., a natural number results from the addition or multiplication of two natural numbers); commutativity of addition and multiplication, i.e., $a + b = b + a$ and $a \times b = b \times a$; addition and multiplication are associative: $a + (b + c) = (a + b) + c$ and $a \times (b \times c) = (a \times b) \times c$. Further, multiplication is distributive over addition: $a \times (b + c) = a \times b + a \times c$.

A square number is an integer that is the square of another integer. For example, the number 4 is a square number since $4 = 2^2$. Similarly, the number 9 and the number 16 are square numbers. A number n is a square number if and only if one can arrange the n points in a square.

The square of an odd number is odd, whereas the square of an even number is even. This is clear since an even number is of the form $n = 2k$ for some k, and so $n^2 = 4k^2$ which is even. Similarly, an odd number is of the form $n = 2k + 1$ and so $n^2 = 4k^2 + 4k + 1$ which is odd.

A rectangular number n may be represented by a vertical and horizontal rectangle of n points. For example, the number 6 may be represented by a rectangle with length 3 and breadth 2, or a rectangle with length 2 and breadth 3. Similarly, the number 12 can be represented by a 4×3 or a 3×4 rectangle.

A triangular number n may be represented by an equilateral triangle of n points. It is the sum of k natural numbers from 1 to k. That is,

$$n = 1 + 2 + \cdots + k$$

Parity of Integers

The parity of an integer refers to whether the integer is odd or even. An integer n is odd if there is a remainder of one when it is divided by two (i.e., it is of the form $n = 2k + 1$). Otherwise, the number is even and of the form $n = 2k$.

The sum of two numbers is even if both are even or both are odd. The product of two numbers is even if at least one of the numbers is even.

Let a and b be integers with $a \neq 0$ then a is said to be a divisor of b (denoted by $a \mid b$) if there exists an integer k such that $b = ka$.

A divisor of n is called a *trivial divisor* if it is either 1 or n itself; otherwise it is called a *non-trivial divisor*. A *proper divisor* of n is a divisor of n other than n itself.

Properties of Divisors

(i) $a \mid b$ and $a \mid c$ then $a \mid b + c$
(ii) $a \mid b$ then $a \mid bc$
(iii) $a \mid b$ and $b \mid c$ then $a \mid c$.

A *prime number* is a natural number (> 1) whose only divisors are trivial. There are an infinite number of prime numbers.

The *fundamental theorem of arithmetic* states that every integer number can be factored as the product of prime numbers.

Pythagorean triples are combinations of three whole numbers that satisfy Pythagoras's equation $x^2 + y^2 = z^2$. There are an infinite number of such triples, and 3, 4, 5 is an example since $3^2 + 4^2 = 5^2$.

Theorem 3.2 (Division Algorithm) *For any integer a and any positive integer b there exists unique integers q and r such that:*

$$a = bq + r \quad 0 \leq r < b.$$

Theorem 3.3 (Irrationality of Square Root of Two) *The square root of two is an irrational number (i.e., it cannot be expressed as the quotient of two integer numbers).*

3.5.1 Fractions and Decimals

A simple fraction is of the form a/b where a and b are integers, with the number a above the bar termed the *numerator* and the number b below the bar termed the *denominator*. Each fraction may be converted to a decimal representation by dividing the numerator by the denominator, and the decimal representation may be to an agreed number of decimal places (as the decimal representation may not terminate).

The reverse operation of converting a decimal number to a fraction is straight forward, and involves determining the number of decimal places (n) of the number, and multiplying the number by the fraction $10^n/10^n$. The resulting fraction is then simplified.

For example, the conversion of the decimal number 0.25 to a fraction involves noting that we have 2 decimal places and so we multiply the decimal number 0.25 by $10^2/10^2$ (i.e., 100/100). This results in the fraction 25/100 which is then simplified to 1/4.

The addition of two fractions with the same denominator is straightforward as all that is involved is adding the numerators of both fractions together and simplifying. For example, $1/12 + 5/12 = (1+5)/12 = 6/12 = 1/2$.

The addition of fractions with different denominators is more difficult. One way to do this is to multiply both denominators of both fractions together to form a common denominator and then simplify. That is,

$$\frac{a}{m} + \frac{b}{n} = \frac{na + mb}{mn}$$

For example, $1/2 + 1/3 = (3.1 + 2.1)/3.2 = (3 + 2)/6 = 5/6$.

However, the usual approach when adding two fractions is to determine the least common multiple of both denominators, and then to convert each fraction into the equivalent fraction with the common LCM denominator, and then to add both numerators together and simplify. For example, consider

$$\frac{3}{4} + \frac{5}{6} =$$

First, the LCM of 4 and 6 is determined (see Sect. 3.5.4) and the LCM (4, 6) is the smallest multiple of both 4 and 6 and this is 12. We then convert both fractions into the equivalent fractions under the common LCM, i.e., we multiply the first fraction 3/4 by 3/3 and the second fraction 5/6 by 2/2 and this yields:

$$\frac{3}{4} + \frac{5}{6} = \frac{3.3}{12} + \frac{5.2}{12} = \frac{9 + 10}{12} = \frac{19}{12}$$

The multiplication of two numbers involves multiplying the numerators together and the denominators together and then simplifying:

$$\frac{a}{m} \times \frac{b}{n} = \frac{ab}{mn}$$

The division of one fraction by another involves inverting the divisor and multiplying and simplifying. That is,

$$\frac{a}{m} \div \frac{b}{n} = \frac{a}{m} \times \frac{n}{b} = \frac{an}{mb}.$$

3.5.2 Prime Number Theory

A positive integer $n > 1$ is called prime if its only divisors are n and 1. A number that is not a prime is called composite.

Theorem 3.4 (Fundamental Theorem of Arithmetic) *Every natural number $n > 1$ may be written uniquely as the product of primes:*

$$n = p_1^{\alpha_1} p_2^{\alpha_2} p_3^{\alpha_3} \dots p_k^{\alpha_k}$$

There are an infinite number of primes but, most integer numbers are composite and so a reasonable question to ask is how many primes are there less than a certain number. The prime distribution function (denoted by $\pi(x)$) is defined by:

$$\pi(x) = \sum_{p \leq x} 1 \quad (\text{where } p \text{ is prime})$$

The prime distribution function satisfies the following properties:

(i) $\lim\limits_{x \to \infty} \frac{\pi(x)}{x} = 0$

(ii) $\lim\limits_{x \to \infty} \pi(x) = \infty$

The first property expresses the fact that most integer numbers are composite, and the second property expresses the fact that there are an infinite number of prime numbers.

There is an approximation of the prime distribution function in terms of the logarithmic function $(x / \ln x)$ as follows:

$$\lim_{x \to \infty} \frac{\pi(x)}{x / \ln x} = 1 \quad (\text{Prime Number Theorem})$$

The approximation $x/\ln x$ to $\pi(x)$ gives an easy way to determine the approximate value of $\pi(x)$ for a given value of x. This result is known as the *Prime Number Theorem*, and it was originally conjectured by Gauss.

3.5.3 Greatest Common Divisors (GCD)

Let a and b be integers (not both zero) then the *greatest common divisor d* of a and b is a divisor of a and b (i.e., $d \mid a$ and $d \mid b$), and it is the largest such divisor (i.e., if $k \mid a$ and $k \mid b$ then $k \mid d$). It is denoted by gcd (a, b).

Properties of Greatest Common Divisors

(i) Let a and b be integers (not both zero) then exists integers x and y such that:

$$d = \gcd(a, b) = ax + by$$

(ii) Let a and b be integers (not both zero) then the set $S = \{ax + by$ where x, $y \in \mathbb{Z}\}$ is the set of all multiples of $d = $ gcd (a, b).

Relatively Prime
Two integers a, b are relatively prime if gcd $(a, b) = 1$.

Properties If p is a prime and $p \mid ab$ then $p \mid a$ or $p \mid b$.

Euclid's [6] algorithm is one of the oldest known algorithms, and it provides a procedure for finding the greatest common divisor of two numbers. It is described in Book VII of Euclid's Elements [1] and is discussed in more detail in Chap. 4.

Lemma 3.1 Let a, b, q, and r be integers with $b > 0$ and $0 \leq r < b$ such that $a = bq + r$. Then gcd $(a, b) = $ gcd (b, r).

Theorem 3.5 (Euclid's Algorithm) *Euclid's algorithm for finding the greatest common divisor of two positive integers a and b involves applying the division algorithm repeatedly as follows:*

$$
\begin{array}{ll}
a = bq_0 + r_1 & 0 < r_1 < b \\
b = r_1 q_1 + r_2 & 0 < r_2 < r_1 \\
r_1 = r_2 q_2 + r_3 & 0 < r_3 < r_2 \\
\cdots & \\
\cdots & \\
r_{n-2} = r_{n-1} q_{n-1} + r_n & 0 < r_n < r_{n-1} \\
r_{n-1} = r_n q_n &
\end{array}
$$

Then r_n is the greatest common divisor of a and b, i.e., gcd $(a, b) = r_n$.

---

[6] Euclid was a third century B.C. Hellenistic mathematician and is considered the father of geometry.

Lemma 3.2 *Let n be a positive integer greater than one then the positive divisors of n are precisely those integers of the form:*

$$d = p_1^{\beta_1} p_2^{\beta_2} p_3^{\beta_3} \ldots p_k^{\beta_k} \quad (\text{where } 0 \leq \beta_i \leq \alpha_i)$$

where the unique factorization of n is given by:

$$n = p_1^{\alpha_1} p_2^{\alpha_2} p_3^{\alpha_3} \ldots p_k^{\alpha_k}.$$

3.5.4 Least Common Multiple (LCM)

If m is a multiple of a and m is a multiple of b then it is said to be a *common multiple* of a and b. The least common multiple is the smallest of the common multiples of a and b, and it is denoted by LCM (a, b).

Properties If x is a common multiple of a and b then $m \mid x$. That is, every common multiple of a and b is a multiple of the least common multiple m.

Example 3.6 (*LCM of two numbers*) The LCM of two numbers is the smallest number that can be divided by both numbers. The LCM is calculated by first determining the prime factors of each number, and then multiplying each factor by the greatest number of times that it occurs in either number.

The procedure may be seen more clearly with the calculation of the LCM of 8 and 12, as $8 = 2^3$ and $12 = 2^2.3$. Therefore, the LCM $(8, 12) = 2^3.3 = 24$, since the greatest number of times that the factor 2 occurs is 3 and the greatest number of times that the factor 3 occurs is once.

3.5.5 Ratios and Proportions

Ratios and proportions are used to solve business problems such as computing inflation, currency exchange and taxation.

A *ratio* is a comparison of the relative values of numbers or quantities where the quantities are expressed in the same units. Business information is often based on a comparison of related quantities stated in the form of a ratio, and a ratio is usually written in the form of *num* 1 to *num* 2 or *num*1: *num*2 (e.g., 3 to 4 or 3:4).

The numbers appearing in a ratio are called the terms of the ratio, and the ratio is generally reduced to the lowest terms, e.g., the term 80:20 would generally be reduced to the ratio 4:1 with the common factor of 20 used to reduce the terms. If the terms contain decimals then the terms are each multiplied by the same number to eliminate the decimals and the term is then simplified.

One application of ratios is to allocate a quantity into parts by a given ratio (i.e., allocating a portion of a whole into parts).

Example 3.7 Consider a company that makes a profit of €180,000 which is to be divided between its three partners A, B, and C in the ratio 3:4:2. How much does each partner receive?

Solution

The total number of parts is $3 + 4 + 2 = 9$. That is, for every 9 parts A receives 3, B receives 4 and C receives 2. That is, A receives $3/9 = 1/3$ of the profits; B receives 4/9 of the profits and C receives 2/9 of the profits. That is,

$$A \text{ receives } 1/3 \times €180,000 = €60,000$$
$$B \text{ receives } 4/9 \times €180,000 = €80,000$$
$$C \text{ receives } 2/9 \times €180,000 = €40,000$$

A proportion is two ratios that are equal or equivalent (i.e., they have the same value and the same units). For example, the ratio 3:4 is the same as the ratio 6:8 and so they are the same proportion.

Often, an unknown term arises in a proportion and in such a case the proportions form a linear equation in one variable.

Example 3.8 Solve the proportion $2 : 5 = 8 : x$

Solution

$$\frac{2}{5} = \frac{8}{x}$$

Cross-multiplying we get

$$2 \times x = 8 \times 5$$
$$\Rightarrow 2x = 40$$
$$\Rightarrow x = 20$$

Example 3.9 A car travels 384 km on 32 L of petrol. How far does it travel on 25 L of petrol?

Solution

We let x represent the unknown distance that is travelled on 25 L of petrol. We then write the two ratios as a proportion, and solve the simple equation to find the unknown value x.

$$\frac{384}{32} = \frac{x}{25}$$

Cross-multiplying we get:

$$25 \times 384 = x \times 32$$
$$\Rightarrow 32x = 9600$$
$$\Rightarrow x = 300 \text{ km (for 25L of petrol)}$$

3.5.6 Percentages

Percent means "per hundred" and the symbol % indicates parts per hundred (i.e., a percentage is a fraction where the denominator is 100, which provides an easy way to compare two quantities). A percentage may be represented as a decimal or as a fraction, and Table 3.1 shows the representation of 25% as a percentage, decimal, and fraction:

Percentages are converted to decimals by moving the decimal point two places to the left (e.g., 25% = 0.25). Conversely, the conversion of a decimal to a percentage involves moving the decimal point two places to the right and adding the percentage symbol.

A percentage is converted to a fraction by dividing it by 100 and then simplifying (e.g., 25% = 25/100 = 1/4). Similarly, a fraction can be converted to a decimal by dividing the numerator by the denominator, and then moving the decimal point two places to the right and adding the percent symbol.

The value of the percentage of a number is calculated by multiplying the rate by the number to yield the new value. For example, 80% of 50 is given by $0.8 \times 50 = 40$. That is, the value of the new number is given by:

$$\text{New number} = \text{rate} \times \text{original number}$$

The rate (or percentage) that a new number is with respect to the original number is given by:

$$\text{Rate} = \frac{\text{New Number}}{\text{Original Number}} \times 100$$

For example, to determine what percentage of €120 that €15 is we apply the formula to get:

$$\text{Rate} = \frac{15}{120} \times 100 = 12.5\%$$

Suppose that 30% of the original number is 15 and we wish to find the original number. Then we let x represent the original number and we form a simple equation with one unknown:

$$0.3x = 15$$
$$\Rightarrow x = 15/0.3$$
$$\Rightarrow x = 50$$

Table 3.1 Percentage, decimal, and fraction

Percentage	Decimal	Fraction
25%	0.25	25/100 = 1/4

In general, when we are given the rate and the new number we may determine the original number from the formula:

$$\text{Original Number} = \frac{\text{New Number}}{\text{Rate}}$$

Example 3.10 Barbara is doing renovations on her apartment. She has budgeted 25% of the renovation costs for new furniture. The cost of the new furniture is €2200 and determine the total cost of the renovations to her apartment.

Solution

We let x represent the unknown cost of renovation and we form the simple equation with one unknown:

$$0.25x = 2200$$
$$\Rightarrow x = €2200/0.25$$
$$\Rightarrow x = €8800$$

Example 3.11 (i) What number is 25% greater than 40? (ii) What number is 20% less than 40?

Solution

We let x represent the new number.

For the first case $x = 40 + 0.25(40) = 40 + 10 = 50$.

For the second case $x = 40 - 0.2(40) = 40 - 8 = 32$.

Often, we will wish to determine the rate of increase or decrease as in the following example.

Example 3.12 Determine the percentage that 280 is greater than 200. (ii) Determine the percentage that 170 is less than 40?

Solution

(i) The amount of change is $280 - 200 = 80$. The rate of change is therefore $80/200 * 100 = 40\%$ (a 40% increase).

(ii) The amount of change is $170 - 200 = -30$. The rate of change is therefore $-30/200 * 100 = -15\%$ (a 15% decrease).

Example 3.13 Lilly increased her loan payments by 40% and now pays €700 back on her loan. What was her original payment?

Solution

Let the original amount be x and we form a simple equation of one unknown.

$$x + 0.4x = 700$$

$$1.4x = 700$$
$$x = 700/1.4$$
$$x = 500$$

For more detailed information on basic arithmetic see [2].

3.6 Trigonometry

Trigonometry is the branch of mathematics that deals with the measurement of sides and angles of triangles and their relationship with each other, and it has many practical applications in science and engineering.

One well-known theorem from geometry is *Pythagoras's Theorem* which states that for any right-angled triangle that the square of the *hypotenuse* (i.e., the side opposite the right angle) is equal to the sum of the squares of the other two sides (Fig. 3.1).

That is:

$$c^2 = a^2 + b^2$$

3.6.1 Definition of Sine, Cosine, and Tangent

The sine, cosine, and tangent of the angle θ in the right-angled triangle below are defined as:

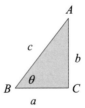

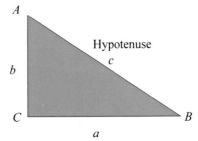

Fig. 3.1 Right-angled triangle

$$\text{Sin}\,\theta = \frac{\text{opposite}}{\text{hypotenuse}} = \frac{b}{c}$$

$$\text{Cos}\,\theta = \frac{\text{adjacent}}{\text{hypotenuse}} = \frac{a}{c}$$

$$\text{Tan}\,\theta = \frac{\text{opposite}}{\text{adjacent}} = \frac{b}{a}$$

The secant, cosecant, and cotangent are defined in terms of sin, cos, and tan as follows:

$$\text{Csc}\,\theta = 1/\sin\theta$$
$$\text{Sec}\,\theta = 1/\text{Cos}\theta$$
$$\text{Cot}\,\theta = 1/\text{Tan}\theta$$

The trigonometric ratios may be expressed using the unit circle centred on the origin of the plane.

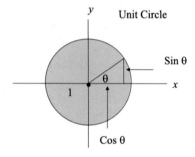

That is, every point on the unit circle is of the form $(\text{Cos}\,\theta, \text{Sin}\,\theta)$ for some angle θ, and so $\text{Cos}^2\,\theta + \text{Sin}^2\,\theta = 1$. Every point on a circle with radius r is of the form $(r\text{Cos}\,\theta, r\text{Sin}\,\theta)$.

3.6.2 Sine and Cosine Rules

The *sine rule* states that for any tringle ABC with sides a, b, c and opposite angles A, B, and C that:

$$\frac{\text{Sin}\,A}{a} = \frac{\text{Sin}\,B}{b} = \frac{\text{Sin}\,C}{c}$$

The sine rule may be used to solve problems where:

(i) One side and any two angles are given, or

(ii) Two sides and one angle (where angle is opposite one of the sides)

The *cosine rule* states that for any tringle *ABC* with sides *a*, *b*, *c* and opposite angles *A*, *B* and *C* that:

$$a^2 = b^2 + c^2 - 2bc \cos A$$

or,

$$b^2 = a^2 + c^2 - 2ac \cos B$$

or,

$$c^2 = a^2 + b^2 - 2ac \cos C$$

The cosine rule may be used to solve problems where:

(i) Two sides and the included angle are given, or
(ii) Three sides are given.

The *area* of any tringle *ABC* with sides *a*, *b*, *c* and angles *A*, *B*, and *C* is given by:

$$\text{Area} = 1/2\,ab\,\sin C = 1/2\,ac\,\sin B = 1/2\,bc\,\sin A.$$

Example 3.14 Solve the following triangle (using the sine rule and determine its area).

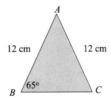

Solution

As this is an isosceles triangle we note that angle $< BCA$ is also 65°, and so angle $< BAC = 180 - 130 = 50°$. We thus have one unknown side $BC\ (= x)$ and so we apply the sine rule to get:

$$\frac{\sin 50}{x} = \frac{\sin 65}{12}$$
$$\Rightarrow \frac{0.766}{x} = \frac{0.9063}{12}$$

Thus $x = 10.14$ cm.

The area of the triangle is given by:

$$\text{Area} = 1/2\, ac\, \text{Sin} B$$
$$= 0.5 * 12 * 10.14 * \text{Sin} 65$$
$$= 55.14 \,\text{cm}^2.$$

Example 3.15 Solve the following triangle (using the cosine rule and determine its area).

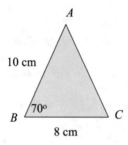

Solution

$$b^2 = a^2 + c^2 - 2ac\, \text{Cos}\, B$$
$$b^2 = 8^2 + 10^2 - 2.8.10.\text{Cos}\, 70$$
$$= 64 + 100 + 160 * 0.342$$
$$= 164 + 54.72$$
$$= 218.72$$
$$b = 14.78 \,\text{cm}$$

The area of the triangle is then given by:

$$\text{Area} = 1/2\, ac\, \text{Sin} B$$
$$= 0.5 * 10 * 8 * \text{Sin}\, 70$$
$$= 37.59 \,\text{cm}^2.$$

3.6.3 Trigonometric Identities

There are several useful trigonometric identities including:

(i) $\text{Sin}\,(-A) = -\text{Sin}\, A$

(ii) $\text{Cos}\,(-A) = \text{Cos}\,A$

(iii) $\text{Cos}^2\,A + \text{Sin}^2\,A = 1$

(iv) $\text{Sin}\,(A + B) = \text{Sin}\,A\,\text{Cos}\,B + \text{Cos}\,A\,\text{Sin}\,B$

(v) $\text{Sin}\,(A - B) = \text{Sin}\,A\,\text{Cos}\,B - \text{Cos}\,A\,\text{Sin}\,B$

(vi) $\text{Cos}\,(A + B) = \text{Cos}\,A\,\text{Cos}\,B - \text{Sin}\,A\,\text{Sin}\,B$

(vii) $\text{Cos}\,(A - B) = \text{Cos}\,A\,\text{Cos}\,B + \text{Sin}\,A\,\text{Sin}\,B$

(viii) $\text{Sin}\,2A = 2\,\text{Sin}\,A\,\text{Cos}\,A$

(ix) $\text{Cos}\,2A = \text{Cos}^2\,A - \text{Sin}^2\,A$

(x) $\text{Sin}\,A + \text{Sin}\,B = 2\,\text{Sin}\,1/2\,(A + B)\,\text{Cos}\,1/2\,(A - B)$

(xi) $\text{Cos}\,A + \text{Cos}\,B = 2\,\text{Cos}\,1/2\,(A + B)\,\text{Cos}\,1/2\,(A - B)$

(xii) $\text{Sec}^2\,A = 1 + \text{Tan}^2\,A$.

3.6.4 Degrees and Radians

The sum of the angles in a triangle is $180°$

The measure (in radians) of the angle at the centre of a circle with radius r is given by the ratio of the arc s to the radius r, i.e.,

$$\theta = \frac{s}{r}$$

That is, the length of the arc s is given by $s = r\theta$, and clearly when the radius is 1 (i.e., the unit circle) then the radian measure of the angle θ is the length of the arc s.

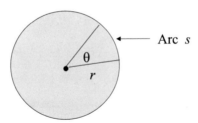

The length of the circumference of a circle with radius r is given by $2\pi r$ and so the measure in radians of the angle at the centre of the circle is

$$\frac{2\pi r}{r} = 2\pi\,(\text{radians}) = 360°$$

That is,

$$2\pi\ \text{radians} = 360°$$

$$\Rightarrow 1\ \text{radian} = 360°/2\pi$$

$$= 57°17'$$

We have the following identities:

$$\pi(\text{rads}) = 180°$$
$$\pi/2(\text{rads}) = 90°$$
$$\pi/4(\text{rads}) = 45°$$
$$\pi/6(\text{rads}) = 30°.$$

3.6.5 Periodic Functions and Sketch of Sine and Cosine Functions

A function $f(x)$ is periodic with period k if $f(x + k) = f(x)$ for all values of x. That is, the curve of $f(x)$ between 0 and k repeats endlessly to the left and to the right, i.e., $f(x + nk) = f(x)$ for $n = 0, \pm 1, \pm 2, \dots$

The sine and cosine functions are periodic functions with period 2π, and this may be seen by:

$$\text{Sin}(x + 2\pi) = \text{Sin}x\,\text{Cos}2\pi + \text{Cos}x\,\text{Sin}2\pi = \text{Sin}x * 1 + \text{Cos}x * 0 = \text{Sin}x$$
$$\text{Cos}(x + 2\pi) = \text{Cos}x\,\text{Cos}2\pi - \text{Sin}x\,\text{Sin}2\pi = \text{Cos}x * 1 - \text{Sin}x * 0 = \text{Cos}x$$

This can be extended to

$$\text{Sin}(x + 2n\pi) = \text{Sin}x \quad n = 0, \pm 1, \pm 2, \dots$$
$$\text{Cos}(x + 2n\pi) = \text{Cos}x \quad n = 0, \pm 1, \pm 2, \dots$$

The graph of the sine function between 0 and 2π is given by:

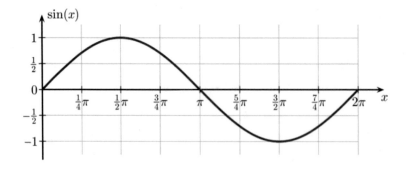

Similarly, the graph of the cosine function between 0 and 2π is given by:

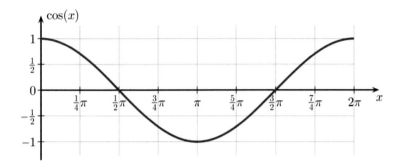

3.6.6 Power Series for Sine and Cosine

The power series for Sin x is given by:

$$\text{Sin}x = x - \frac{x^3}{3!} + \frac{x^5}{5!} - \cdots + (-1)^n \frac{x^{2n+1}}{(2n+1)!} + \cdots$$

$$n = 0, 1, 2, \ldots$$

The power series for Cos x is given by:

$$\text{Cos}x = 1 - \frac{x^2}{2!} + \frac{x^4}{4!} - \cdots + (-1)^n \frac{x^{2n}}{(2n)!} + \cdots$$

$$n = 0, 1, 2, \ldots$$

3.7 Cartesian Coordinates

The Cartesian coordinate system specifies each point uniquely in the plane by a set of numerical coordinates, which are the signed distance of the point to two fixed coordinate axes. The point where the two axes meet is termed the origin of the coordinate system and has the coordinates (0, 0). The Cartesian system was invented by Rene Descartes in the seventeenth century, and it allows geometric shapes such as curves to be described by algebraic equations (Fig. 3.2).

There are three perpendicular axes in the three dimensional Cartesian coordinate system, and the Cartesian coordinates (x, y, z) of a point P in place are the numbers where the planes through P perpendicular to the three axes cut the axes (Fig. 3.3).

The points that lie on the x-axis have their y and z coordinate equal to zero, i.e., they are of the form $(x, 0, 0)$, and similarly points in the y-axis have coordinates $(0, y, 0)$ and points in the z-axis have coordinates $(0, 0, z)$.

The points in a plane perpendicular to the x-axis all have the same x coordinate, and similarly the points in a plane perpendicular to the y-axis all have the same y

Fig. 3.2 Cartesian
coordinate system

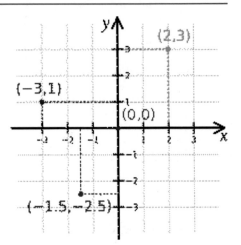

Fig. 3.3 Cartesian
three-dimensional coordinate
system

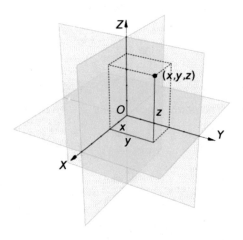

coordinate and the points in a plane perpendicular to the z-axis all have the same z coordinate. It is easy to write equations for these planes as for example, the equation $x = 3$ is an equation to the plane perpendicular to the x-axis at $x = 3$, and similarly the equation $y = 2$ is an equation to the plane perpendicular to the y-axis at $y = 3$ and the equation $z = 4$ is an equation to the plane perpendicular to the z-axis at $z = 4$. The intersection of the three planes is the point (3, 2, 4).

The intersection of the planes $x = 3$ and $y = 2$ is a line that runs parallel to the z-axis, and is given by the equations $x = 3$ and $y = 2$. The equation of the xy-plane is $z = 0$; the equation of the yz-plane is $x = 0$; and the equation of the xz-plane is $y = 0$.

The distance between any two points (x_1, y_1) and (x_2, y_2) in the two dimensional Cartesian plane is given by:

$$d = \sqrt{(x_2 - x_1)^2 + (y_2 - y_1)^2}$$

The distance between any two points (x_1, y_1, z_1) and (x_2, y_2, z_2) in the three dimensional Cartesian plane is given by:

$$d = \sqrt{(x_2 - x_1)^2 + (y_2 - y_1)^2 + (z_2 - z_1)^2}.$$

3.8   Review Questions

1. What is a set? A relation? A function?
2. Explain the difference between a partial and a total function.
3. Determine $A \triangle B$ where $A = \{a, b, c, d\}$ and $B = \{c, d, e\}$.
4. What is the domain and range of $R = \{(a, p), (a, r), (b, q)\}$.
5. Determine the inverse relation R^{-1} where $R = \{(a, 2), (a, 5), (b, 3), (b, 4), (c, 1)\}$.
6. Determine the inverse of the function $f \colon \mathbb{R} \times \mathbb{R} \to \mathbb{R}$ defined by

$$f(x) = \frac{x - 2}{x - 3} \quad (x \neq 3) \quad \text{and } f(3) = 1$$

7. Compute 7/8 * 5/12.
8. Find the prime factorization of 18 and 24.
9. Find the least common multiple of 18 and 24.
10. Find the greatest common divisor of 18 and 24.
11. A company makes a profit of €120,000 which is to be divided between its three partners A, B, and C in the ratio 2:7:6. Find the amount that each partner gets.
12. Solve the proportion 2:7 = 4:x.
13. What number is 15% greater than 140?
14. The length of the hypotenuse and one of the sides of a right-angled triangle is 17 cm and 8 cm respectively. Find the length of the other side.
15. Use the cosine and sine rules to find the angles of a triangle where the sides are 17, 15 and 8. Determine its area.
16. Calculate the number of degrees of:
    (a) $\pi/6$ radians
    (b) 3 radians
17. What is the period of the following functions:
    (a) Sin $2x$
    (b) Cos $3x$
18. Find the distance between (0, 1, 4) and (2, 3, 1).

3.9   Summary

This chapter introduced essential mathematics for computing including set theory, relations, and functions. Sets are collections of well-defined objects; a relation between A and B indicates relationships between members of the sets A and B; and functions are a special type of relation where there is at most one relationship for each element $a \in A$ with an element in B.

A binary relation $R\ (A, B)$ is a subset of the Cartesian product $(A \times B)$ of A and B where A and B are sets. The domain of the relation is A and the codomain of the relation is B. An n-ary relation $R\ (A_1, A_2, \ldots A_n)$ is a subset of $(A_1 \times A_2 \times \ldots \times A_n)$.

A total function $f: A \to B$ is a special relation such that for each element $a \in A$ there is exactly one element $b \in B$. This is written as $f(a) = b$. A function is a relation but not every relation is a function.

Arithmetic is the branch of mathematics that is concerned with the study of numbers and their properties. It includes the study of the integer numbers, and operations on them, such as addition, subtraction, multiplication, and division.

The natural numbers consist of the numbers $\{1, 2, 3, \ldots\}$. The integer numbers are a superset of the set of natural numbers, and the rational numbers are a superset of the set of integer numbers, which consist of all numbers of the form $\{p/q$ where p and q are integers and $q \neq 0\}$. A simple fraction is of the form a/b where a and b are integers, with the number a above the bar termed the numerator and the number b below the bar termed the denominator.

A positive integer $n > 1$ is called prime if its only divisors are n and 1, and a number that is not a prime is called composite. There are an infinite number of prime numbers, and prime numbers are the key building blocks in number theory, and the fundamental theorem of arithmetic states that every number may be written uniquely as the product of factors of prime numbers.

Euclid's algorithm is used for finding the greatest common divisor of two positive integers a and b. The least common multiple (LCM) of two numbers is the smallest number that can be divided by both numbers.

Ratios and proportions are used to solve business problems where a ratio is a comparison of numbers where the quantities are expressed in the same units. The numbers appearing in a ratio are called the terms of the ratio, and the ratios are generally reduced to the lowest terms. One application of ratios is to allocate a quantity into parts by a given ratio (i.e., allocating a portion of a whole into parts).

Percent means "per hundred", and the symbol % indicates parts per hundred (i.e., a percentage is a fraction where the denominator is 100 which provides an easy way to compare two quantities).

Trigonometry is the branch of mathematics that deals with the measurement of sides and angles of triangles and the relationship between them. It has many practical applications in science and engineering.

Pythagoras's expresses the relationship between the hypotenuse and other sides of a right-angled triangle, and the sine, cosine, and tangent of an angle can be expressed in terms of the sides of a right-angled triangle.

The cosine rule and sine rule are used to solve a triangle with the cosine rule applied when given two sides and the included angled or when given three sides. The sine rule is applied when given one side and any two angles or two sides are given and one angle where the angle is opposite one of the sides.

Angles may be measured in degrees or in radians although radians is more common. The sine and cosine functions are trigonometric functions of an angle and are periodic with period of 2π. They are used to model sound and light waves in physics.

References

1. Euclid (1956) The thirteen books of the elements. Vol.1 (Trans: Sir Thomas Heath). Dover Publications (First published in 1925)
2. O'Regan G (2021) Guide to discrete mathematics, 2nd edn. Springer

Introduction to Algorithms

4

Key Topics

Euclid's Algorithm

Sieve of Eratosthenes Algorithm

Early Ciphers

Sorting Algorithms

Insertion Sort and Merge Sort

Analysis of Algorithms

Complexity of Algorithms

NP Complete

4.1 Introduction

An *algorithm* is a well-defined procedure for solving a problem, and it consists of a sequence of steps that takes a set of values as input, and produces a set of values as output. It is an exact specification of how to solve the problem, and it explicitly defines the procedure so that a computer program may implement the solution. The origin of the word "*algorithm*" is from the name of the 9th Persian mathematician, Mohammed Al Khwarizmi.

It is essential that the algorithm is correct, and that it terminates in a reasonable amount of time. This may require mathematical analysis of the algorithm to demonstrate its correctness and efficiency, and to show that termination is within an acceptable timeframe. There may be several algorithms to solve a problem, and

G. O'Regan, *Mathematical Foundations of Software Engineering*,
Texts in Computer Science, https://doi.org/10.1007/978-3-031-26212-8_4

so the choice of the best algorithm (e.g., fastest/most efficient) needs to be considered. For example, there are several well-known sorting algorithms (e.g., *merge sort* and *insertion sort*), and the merge sort algorithm is more efficient $[o(n \lg n)]$ than the insertion sort algorithm $[o(n^2)]$.

An algorithm may be implemented by a computer program written in some programming language (e.g., C++ or Java). The speed of the program depends on the algorithm employed, the input value(s), how the algorithm has been implemented in the programming language, the compiler, the operating system and the computer hardware.

An algorithm may be described in natural language (care is needed to avoid ambiguity), but it is more common to use a more precise formalism for its description. These include *pseudo code* (an informal high-level language description); flowcharts; a programming language such as C or Java; or a formal specification language such as VDM or Z. We shall mainly use natural language and pseudocode to describe an algorithm. One of the earliest algorithms developed was Euclid's algorithm (discussed briefly in Chap. 3) for determining the greatest common divisor of two natural numbers, and it is described in the next section.

4.2   Early Algorithms

Euclid lived in Alexandria during the early Hellenistic period,[1] and he is considered the father of geometry and the deductive method in mathematics. His systematic treatment of geometry and number theory is published in the thirteen books of the Elements [1]. It starts with five axioms, five postulates and twenty-three definitions to logically derive a comprehensive set of theorems in geometry.

His method of proof was generally constructive, in that as well as demonstrating the truth of the theorem, a construction of the required entity was provided. He employed some indirect proofs, and one example was his proof that there are an infinite number of prime numbers. The procedure is to assume the opposite of what one wishes to prove, and to show that a contradiction results. This means that the original assumption must be false, and the theorem is established.

1. Suppose there are a finite number of primes (say n primes).
2. Multiply all n primes together and add 1 to form N.

$$(N = p_1 * p_2 * \ldots * p_n + 1)$$

3. N is not divisible by $p_1, p_2, \ldots, p_n$ as dividing by any of these gives a remainder of one.

---

[1] This refers to the period following the conquests of Alexander the Great, which led to the spread of Greek culture throughout the Middle East and Egypt.

4. Therefore, N must either be prime or divisible by some other prime that was not included in the original list.
5. Therefore, there must be at least $n + 1$ primes.
6. This is a contradiction (it was assumed that there are exactly n primes).
7. Therefore, the assumption that there is a finite number of primes is false.
8. Therefore, there are an infinite number of primes.

His proof that there are an infinite number of primes is indirect, and he does not present an algorithm to as such to construct the set of prime numbers. We present the well-known Sieve of Eratosthenes algorithm for determining the prime numbers up to a given number n later in the chapter.

The material in Euclid's elements is a systematic development of geometry starting from the small set of axioms, postulates, and definitions. It leads to many well-known mathematical results such as Pythagoras's theorem, Thales theorem, sum of angles in a triangle, prime numbers, greatest common divisor and least common multiple, Euclidean algorithm, areas and volumes, tangents to a point, and algebra.

4.2.1 Greatest Common Divisors (GCD)

Let a and b be integers not both zero. The *greatest common divisor d* of a and b is a divisor of a and b (i.e., $d|a$ and $d|b$), and it is the largest such divisor (i.e., if $k|a$ and $k|b$ then $k|d$). It is denoted by gcd (a, b).

Properties of Greatest Common Divisors

(i) Let a and b be integers not both zero then exists integers x and y such that:

$$d = \gcd(a, b) = ax + by$$

(ii) Let a and b be integers not both zero then the set $S = \{ax + by$ where $x, y \in \mathbb{Z}\}$ is the set of all multiples of $d = $ gcd (a, b).

4.2.2 Euclid's Greatest Common Divisor Algorithm

Euclid's algorithm is one of the oldest known algorithms, and it provides the procedure for finding the greatest common divisor of two numbers a and b. It appears in Book VII of Euclid's elements (Fig. 4.1).

The inputs for the gcd algorithm consists of two natural numbers a and b, and the output of the algorithm is d (the greatest common divisor of a and b). It is computed as follows:

$$\gcd(a, b) = \begin{cases} \text{Check if } b \text{ is zero. If so, then } a \text{ is the gcd.} \\ \text{Otherwise, the } \gcd(a, b) \text{ is given by gcd } (b, a \bmod b). \end{cases}$$

Fig. 4.1 Euclid of
Alexandria

It is also possible to determine integers p and q such that $ap + bq = gcd(a, b)$.

The (informal) proof of the Euclidean algorithm is as follows. Suppose a and b are two positive numbers whose greatest common divisor is to be determined, and let r be the remainder when a is divided by b.

1. Clearly $a = qb + r$ where q is the quotient of the division.
2. Any common divisor of a and b is also a divider or r (since $r = a - qb$).
3. Similarly, any common divisor of b and r will also divide a.
4. Therefore, the greatest common divisor of a and b is the same as the greatest common divisor of b and r.
5. The number r is smaller than b and we will reach $r = 0$ in finitely many steps.
6. The process continues until $r = 0$.

Comment 4.1

Algorithms are fundamental in computing as they define the procedure by which a problem is solved. A computer program implements the algorithm in some programming language.

Next, we deal with the Euclidean algorithm more formally, and we start with a basic lemma.

Lemma *Let a, b, q, and r be integers with $b > 0$ and $0 \leq r < b$ such that $a = bq + r$. Then $gcd(a, b) = gcd(b, r)$.*

Proof Let $K = \gcd(a, b)$ and let $L = \gcd(b, r)$ then we need to show that $K = L$. Suppose m is a divisor of a and b, then as $a = bq + r$ we have m is a divisor of r and so any common divisor of a and b is a divisor of r. Therefore, the greatest common divisor K of a and b is a divisor of r. Similarly, any common divisor n of b and r is a divisor of a. Therefore, the greatest common divisor L of b and r is a divisor of a. That is, K divides L and L divides K and so $L = K$, and so the greatest common divisor of a and b is equal to the greatest common divisor of b and r.

Euclid's Algorithm (more formal proof)
Euclid's algorithm for finding the greatest common divisor of two positive integers a and b involves a repeated application of the division algorithm as follows:

$$a = bq_0 + r_1 \quad 0 < r_1 < b$$
$$b = r_1 q_1 + r_2 \quad 0 < r_2 < r_1$$
$$r_1 = r_2 q_2 + r_3 \quad 0 < r_3 < r_2$$
$$\cdots$$
$$r_{n-2} = r_{n-1} q_{n-1} + r_n \quad 0 < r_n < r_{n-1}$$
$$r_{n-1} = r_n q_n$$

Then r_n (i.e., the last non-zero remainder) is the greatest common divisor of a and b: i.e., $\gcd(a, b) = r_n$.

Proof It is clear from the construction that r_n is a divisor of $r_{n-1}, r_{n-2}, \ldots, r_3, r_2, r_1$ and of a and b. Clearly, any common divisor of a and b will also divide r_n. Using the results from the lemma above we have:

$$\begin{aligned} \gcd(a, b) &= \gcd(b, r_1) \\ &= \gcd(r_1 r_2) \\ &= \cdots \\ &= \gcd(r_{n-2} r_{n-1}) \\ &= \gcd(r_{n-1}, r_n) \\ &= r_n \end{aligned}$$

4.2.3 Sieve of Eratosthenes Algorithm

Eratosthenes was a Hellenistic mathematician and scientist who worked in the famous library in Alexandria. He devised a system of latitude and longitude, and he was the first person to estimate of the size of the circumference of the earth. He developed a famous algorithm (the well-known *Sieve of Eratosthenes algorithm*) for determining the prime numbers up to a given number n.

	2	3	4	5	6	7	8	9	10
11	12	13	14	15	16	17	18	19	20
21	22	23	24	25	26	27	28	29	30
31	32	33	34	35	36	37	38	39	40
41	42	43	44	45	46	47	48	49	50

Fig. 4.2 Primes between 1 and 50

The algorithm involves listing all numbers from 2 up to n. The first step is to remove all multiples of 2 up to n; the second step is to remove all multiples of 3 up to n; and so on (Fig. 4.2).

The kth step involves removing multiples of the kth prime p_k up to n and the steps in the algorithm continue while $p \leq \sqrt{n}$. The numbers remaining in the list are the prime numbers from 2 to n.

1. List the integers from 2 to n.
2. For each prime p_k up to $\sqrt{n}$ remove all multiples of p_k.
3. The numbers remaining are the prime numbers between 2 and n.

The list of primes between 1 and 50 is then given by 2, 3, 5, 7, 11, 13, 17, 19, 23, 29, 31, 37, 41, 43, and 47.

The steps in the algorithm may also be described as follows (in terms of two lists):

1. Write a list of the numbers from 2 to the largest number to be tested. This first list is called A.
2. A second list B is created to list the primes. It is initially empty.
3. The number 2 is the first prime number, and it is added to List B.
4. Strike off (or remove) all multiples of 2 from List A.
5. The first remaining number in List A is a prime number, and this prime number is added to List B.
6. Strike off (or remove) this number and all multiples of it from List A.
7. Repeat steps 5 through 7 until no more numbers are left in List A.

4.2.4   Early Cipher Algorithms

Julius Caesar employed a *substitution cipher* on his military campaigns to ensure that important messages were communicated safely (Fig. 4.3). The Caesar cipher is a very simple encryption algorithm, and it involves the substitution of each letter in the *plaintext* (i.e., the original message) by a letter a fixed number of positions down in the alphabet. The Caesar encryption algorithm involves a shift of 3 positions, and causes the letter B to be replaced by E, the letter C by F, and

Alphabet Symbol	abcde fghij klmno pqrst uvwxyz
Cipher Symbol	dfegh ijklm nopqr stuvw xyzabc

Fig. 4.3 Caesar cipher

so on. The Caesar cipher is easily broken, as the frequency distribution of letters may be employed to determine the mapping. The Caesar cipher is defined as:

The process of enciphering a message (i.e., the plaintext) involves mapping each letter in the plaintext to the corresponding cipher letter. For example, the encryption of "summer solstice" involves:

Plaintext:	summer solstice
Cipher Text	vxpphu vrovwleh

The decryption involves the reverse operation, i.e., for each cipher letter the corresponding plaintext letter is determined from the table.

Cipher Text	vxpphu vrovwleh
Plaintext:	summer solstice

The Caesar encryption algorithm may be expressed formally using modular arithmetic. The numbers 0–25 represent the alphabet letters, and the algorithm is expressed using addition (modula 26) to yield the encrypted cipher. The encoding of the plaintext letter x is given by:

$$c = x + 3 \pmod{26}$$

Similarly, the decoding of a cipher letter represented by the number c is given by:

$$x = c - 3 \pmod{26}$$

The emperor Augustus[2] employed a similar substitution cipher (with a shift key of 1). The Caesar cipher remained in use up to the early twentieth century. However, by then frequency analysis techniques were available to break the cipher. The *Vignère cipher* uses a Caesar cipher with a different shift at each position in the text. The value of the shift to be employed with each plaintext letter is defined using a repeating keyword. We shall discuss cryptography in more detail in Chap. 20.

---

[2] Augustus was the first Roman emperor, and his reign ushered in a period of peace and stability following the bitter civil war that occurred after the assassination of Julius Caesar. He was the adopted son of Julius Caesar (he was called Octavion before he became emperor). The civil war broke out between Mark Anthony and Octavion, and Anthony and Cleopatra were defeated by Octavion and Agrippa at the battle of Actium in 31 B.C.

4.3   Sorting Algorithms

One of the most common tasks to be performed in a computer program is that of sorting (e.g., consider the problem of sorting a list of names or numbers). This has led to the development of many sorting algorithms (e.g., selection sort, bubble sort, insertion sort, merge sort, and quicksort) as sorting is a fundamental task to be performed.

For example, consider the problem of specifying the algorithm for sorting a sequence of n numbers. Then, the input to the algorithm is $\langle x_1, x_2, \ldots x_n \rangle$, and the output is $\langle x'_1, x'_2, \ldots x'_n \rangle$, where $x'_1 \leq x'_2 \leq \ldots \leq x'_n$. Further, $\langle x'_1, x'_2, \ldots x'_n \rangle$ is a permutation of $\langle x_1, x_2, \ldots x_n \rangle$, i.e., the same numbers are in both sequences except that the sorted sequence is in ascending order, whereas no order is imposed on the original sequence.

Insertion sort is an efficient algorithm for sorting a small list of elements. It iterates over the input sequence; examines the next input element during the iteration; and builds up the sorted output sequence. During the iteration, insertion sort removes the next element from the input data, and it then finds and inserts it into the location where it belongs in the sorted list. This continues until there are no more input elements to process.

We first give an example of insertion sort and then give a more formal definition of the algorithm (Fig. 4.4). The example considered is that of the algorithm applied to the sequence $A = \langle 5, 3 \ 1, 4 \rangle$. The current input element for each iteration is highlighted, and the arrow points to the location where it is inserted in the sorted sequence. For each iteration, the elements to the left of the current element are already in increasing order, and the operation of insertion sort is to move the current element to the appropriate position in the ordered sequence.

We shall assume that we have an unsorted array A with n elements that we wish to sort. The operation of insertion sort is to rearrange the elements of A within the array, and the output is that the array A contains the sorted output sequence.

```
Insertion sort
for i from 2 to n do
    C ← A[i]
    j ← i-1
    while j > 0 and A[j] > C do
            A[j+1] ← A[j]
            j ← j-1
    A[j+1] ← C
```

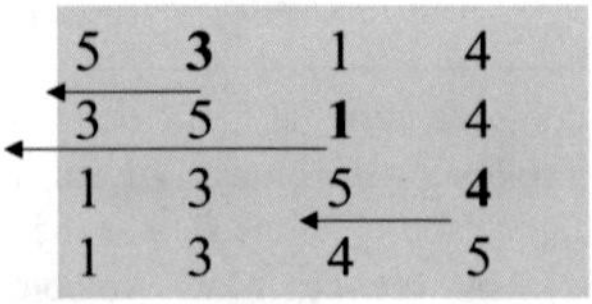

Fig. 4.4 Insertion sort example

The analysis of an algorithm involves determining its efficiency and establishing the resources that it requires (e.g., memory and bandwidth), as well as determining the computational time required. The time taken by the insertion sort algorithm depends on the size of the input sequence (clearly a large sequence will take longer to sort than a short sequence), and on the extent to which the sequences are already sorted. The worst-case running time for the insertion sort algorithm is of order n^2—i.e., $o(n^2)$, where n is the size of the sequence to be sorted (the average case is also of order n^2 with the best case linear).

There are a number of ways to design sorting algorithms, and the insertion sort algorithm uses an incremental approach, with the subarray $A[1 \ldots i - 1]$ already sorted and the element $A[i]$ is then inserted into its correct place to yield the sorted array $A[1 \ldots i]$.

Another approach is to employ divide and conquer techniques, and this technique is used in the merge sort algorithm. This is a more efficient algorithm, and it involves breaking a problem down into several subproblems, and then solving each problem separately. The problem solving may involve recursion or directly solving the subproblem (if it is small enough), and then combining the solutions to the subproblems into the solution for the original problem. The merge sort algorithm involves three steps (Divide, Conquer, and Combine):

1. *Divide* the list A (with n elements) to be sorted into two subsequences (each with $n/2$ elements).
2. Sort each of the subsequences by calling merge sort recursively (*Conquer*)
3. Merge the two sorted subsequences to produce a single sorted list (*Combine*).

The recursive part of the merge sort algorithm bottoms out when the sequence to be sorted is of length 1, as for this case the sequence is of length 1 which is already (trivially) sorted. The key operation then (where all the work is done) is the combine step that merges two sorted sequences to produce a single sorted sequence. The merge sort algorithm may also be described as follows:

1. Divide the sequence (with n elements) to be sorted into n subsequences each with 1 element (a sequence with 1 element is sorted).
2. Repeatedly merge subsequences to form new subsequences (each new sub sequence is sorted), until there is only one remaining subsequence (the sorted sequence).

First, we consider an example (Fig. 4.5) to illustrate how the merge sort algorithm operates, and we then give a formal definition.

It may be seen from the example that the list is repeatedly divided into equal halves with each iteration, until we get to the atomic values that can no longer be divided. The lists are then combined in the order in which they were broken down, and this involves comparing the elements of both lists and combining them to form a sorted list. The merging continues in this way until there are no more

Fig. 4.5   Merge sort example

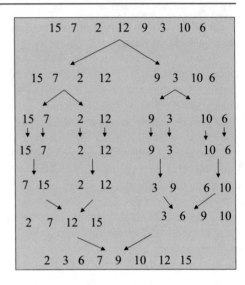

lists to merge, and the list remaining is the sorted list. The formal definition of merge sort is as follows:

```
Merge sort (A, m, n)
If m < n then
    r ← (m + n) div 2
    Merge Sort (A, m, r)
    Merge Sort (A, r+1, n)
    Merge (A, m, r, n)
```

The worst-case and average case running time for the merge sort algorithm is of order n lg n—i.e., $o(n$ lg $n)$, where n is the size of the sequence to be sorted (the average case and best case is also of order $o(n$ lg $n)$).

The merge procedure merges two sorted lists to produce a single sorted list. Merge (A, p, q, r) merges $A[p \dots q]$ with $A[q + 1 \dots r]$ to yield the sorted list $A[p \dots r]$. We use a temporary working array $B[p \dots r]$ with the same index range as A. The indices i and j point to the current element of each subarray, and we move the smaller element into the next position in B (indicated by index k) and then increment either i or j. When we run out of entries in one array then we copy the rest of the other array into B. Finally, we copy the entire contents of B back to A.

```
Merge (A, p, q, r)
    Array B[p … r]
    i ← p
    k ← p
    j ← q+1
    while (i ≤ q ∧ j ≤ r)        i.e., while both subarrays are non-empty
```

```
              if A[i] ≤ A[j].
                 B[k] ← A[i]
                 i ← i+1
              else
                 B[k] ← A[j]
                 j ← j+1
                 k ← k+1
     while (i ≤ q)        … copy any leftover to B
       B[k] ← A[i]
       i ← i+1
       k ← k+1
     while (j ≤ r)        … copy any leftover to B
       B[k] ← A[j]
       j ← j+1
       k ← k+1
     for i = p to r do       … copy B back to A
       A[i] = B[i]
```

4.4 **Binary Trees and Graph Theory**

A *binary tree* (Fig. 4.6) is a tree in which each node has at most two child nodes (termed left and right child node). A node with children is termed a *parent node*, and the top node of the tree is termed the root node. Any node in the tree can be reached by starting from the root node, and by repeatedly taking either the left branch (left child) or right branch (right child) until the node is reached. Binary trees are often used in computing to implement efficient searching algorithms.

The *depth* of a node is the length of the path (i.e., the number of edges) from the root to the node. The depth of a tree is the length of the path from the root to the deepest node in the tree. A *balanced* binary tree is a binary tree in which the depth of the two subtrees of any node never differs by more than one.

Tree traversal is a systematic way of visiting each node in the tree exactly once, and we distinguish between *breadth first search* algorithms in which every node

Fig. 4.6 Sorted binary tree

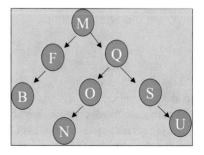

at a particular level is visited before going to a lower level, and *depth first search* algorithms where one starts at the root and explores as far as possible along each branch before backtracking. The traversal in depth first search may be in *preorder*, *inorder,* or *postorder*.

Graph algorithms are employed to solve various problems in graph theory including network cost minimization problems; construction of spanning trees; shortest path algorithms; longest path algorithms; and timetable construction problems. Chapter 7 discusses graph theory in more detail, and the reader may consult texts on graph theory (e.g., [2]) to explore many well-known graph algorithms such as Dijkstra's shortest path and longest path algorithms, Kruskal's minimal spanning tree algorithm, and Prim's minimal spanning tree algorithms.

4.5   Modern Cryptographic Algorithms

A cryptographic system is concerned with the secure transmission of messages. The message is encrypted prior to its transmission, and any unauthorized interception and viewing of the message is meaningless to anyone other than the intended recipient. The recipient uses a key to decrypt the encrypted text to retrieve the original message.

- M represents the message (plaintext)
- C represents the encrypted message (cipher text)
- e_k represents the encryption key
- d_k represents the decryption key
- E represents the encryption
- D represents the decryption

There are essentially two different types of cryptographic systems, namely the public key cryptosystems and secret key cryptosystems. A *public key cryptosystem* is an asymmetric cryptosystem where two different keys are employed: one for encryption and one for decryption. The fact that a person can encrypt a message does not mean that the person is able to decrypt a message.

The same key is used for both encryption and decryption in a *secret key cryptosystem*, and anyone who has knowledge on how to encrypt messages has sufficient knowledge to decrypt messages. The encryption and decryption algorithms satisfy the following equation:

$$Dd_k(C) = Dd_k(Ee_k(M)) = M$$

There are two different keys employed in a public key cryptosystem. These are the encryption key e_k and the decryption key d_k with $e_k. \neq d_k$. It is called asymmetric as the encryption key differs from the decryption key.

A symmetric key cryptosystem (Fig. 20.5) uses the same secret key for encryption and decryption, and so the sender and the receiver first need to agree on a shared key prior to communication. This needs to be done over a secure channel to ensure that the shared key remains secret. Once this has been done they can begin to encrypt and decrypt messages using the secret key.

The encryption of a message is in effect a transformation from the space of messages m to the space of cryptosystems $\mathbb{C}$. That is, the encryption of a message with key k is an invertible transformation f such that:

$$f: m \overset{k}{\to} \mathbb{C}$$

The cipher text is given by $C = E_k(M)$ where $M \in m$ and $C \in \mathbb{C}$. The legitimate receiver of the message knows the secret key k (as it will have transmitted previously over a secure channel), and so the cipher text C can be decrypted by the inverse transformation f^{-1} defined by:

$$f^{-1}: \mathbb{C} \overset{k}{\to} m$$

Therefore, we have that $D_k(C) = D_k (E_k(M)) = M$ the original plaintext message.

A public key cryptosystem (Fig. 20.6) is an asymmetric key system where there is a separate key e_k for encryption and d_k decryption with $e_k \neq d_k$. The fact that a person is able to encrypt a message does not mean that the person has sufficient information to decrypt messages. There is a more detailed account of cryptography in Chap. 20.

4.6   Algorithms in Numerical Analysis

Numerical analysis is concerned with devising methods for approximating solutions to mathematical problems. Often an exact formula is not available for solving a particular equation $f(x) = 0$, and numerical analysis provides techniques to approximate the solution in an efficient manner.

An algorithm is devised to provide the approximate solution, and it consists of a sequence of steps to produce the solution as efficiently as possible within defined accuracy limits. The maximum error due to the application of the numerical methods needs to be determined. The algorithm is implemented in a programming language such as Fortran.

There are several numerical techniques to determine the root of an equation $f(x) = 0$. These include techniques such as the bisection method, which has been used since ancient times, and the Newton–Raphson method developed by Sir Isaac Newton.

The bisection method is employed to find a solution to $f(x) = 0$ for the continuous function f on $[a, b]$ where $f(a)$ and $f(b)$ have opposite signs (Fig. 4.7). The method involves a repeated halving of subintervals of $[a, b]$, with each step locating the half that contains the root. The inputs to the algorithm are the endpoints a

Fig. 4.7 Bisection method

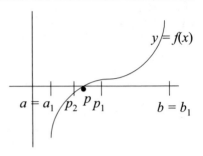

and b, the tolerance (TOL), and the maximum number of iterations N. The steps are as follows:

```
1. Initialize i to 1
2. while i ≤ N
        i. Compute midpoint p
                (p → a + (b − a) / 2)
       ii. If f(p) = 0 or (b − a) / 2 < TOL
                Output p and stop
      iii. If f(a) f(p) > 0
                Set endpoint a → p
                Otherwise set b →p
       iv. i → i + 1
```

The Newton–Raphson method is a well-known technique to determine the roots of a function. It uses tangent lines to approximate the graph of $y = f(x)$ near the points where f is zero. The procedure for Newton's method is:

Newton's Method

(i)   Guess a first approximation to the root of the equation $f(x) = 0$.
(ii) Use the first approximation to get a second, third, and so on.
(iii) To go from the n-th approximation x_n to the next approximation x_{n+1} then use the formula:

where $f'(x_n)$ is the derivative of f at x_n (the derivative is discussed in Chap. 25).

$$x_{n+1} = x_n - \frac{f(x_n)}{f'(x_n)}$$

Newton's method is very efficient for calculating roots as it converges very quickly. However, the method may converge to a different root than expected if the starting value is not close enough to the root sought.

The method involves computing the tangent line at $(x_n, f(x_n))$, and the approximation x_{n+1} is the point where the tangent intersects the x-axis.

4.7   Computational Complexity

An algorithm is of little practical use if it takes millions of years to compute the solution to a problem. That is, the fact that there is an algorithm to solve a problem is not sufficient, as there is also the need to consider the efficiency of the algorithm. The security of the RSA encryption algorithm (see Chap. 20) relies on the fact that there is no known efficient algorithm to determine the prime factors of a large number.

There are often slow and fast algorithms for the same problem, and a measure of the complexity of an algorithm is the number of steps in its computation. An algorithm is of *time complexity* $f(n)$ if for all n and all inputs of length n the execution of the algorithm takes at most $f(n)$ steps.

An algorithm is said to be *polynomially bounded* if there is a polynomial $p(n)$ such that for all n and all inputs of length n the execution of the algorithm takes at most $p(n)$ steps. The notation P is used for all problems that can be solved in polynomial time. A problem is said to be *computationally intractable* if it may not be solved in polynomial time: that is, there is no known algorithm to solve the problem in polynomial time.

A problem L is said to be in the set NP (non-deterministic polynomial time problems) if any given solution to L can be verified quickly in polynomial time. A problem is *NP complete* if it is in the set NP of non-deterministic polynomial time problems, and it is also in the class of *NP hard* problems. A key characteristic to NP complete problems is that there is no known fast solution to them, and the time required to solve the problem using known algorithms increases quickly as the size of the problem grows. Often, the time required to solve the problem is in billions of years. That is, although any given solution may be verified quickly there is no known efficient way to find a solution.

4.8   Review Questions

1. What is an algorithm?
2. Explain why the efficiency of an algorithm is important.
3. What factors should be considered in the choice of algorithm where several algorithms exist for solving the particular problem?
4. Explain the difference between the insertion sort and merge sort algorithms.
5. Investigate famous computer algorithms such as Dijkstra's shortest path, Prim's algorithm, and Kruskal's algorithm.

4.9   Summary

An algorithm is a well-defined procedure for solving a problem, and it consists of a sequence of steps that takes a set of input values and produces a set of output values. It is an exact specification of how to solve the problem, and a computer program implements the algorithm in some programming language. It is essential that the algorithm is correct, and that it terminates in a reasonable period of time. There may be several algorithms for a problem, and so the choice of the best algorithm (e.g., fastest/most efficient) needs to be considered.

This may require mathematical analysis of the algorithm to demonstrate its correctness and efficiency and to show that it terminates in a finite period of time. An algorithm may be implemented by a computer program, and the speed of the program depends on the algorithm employed, the input value(s), how the algorithm has been implemented in the programming language, the compiler, the operating system, and the computer hardware.

References

1. Euclid (1956) The thirteen books of the elements. Vol. 1. (Trans: Sir Thomas Heath). Dover Publications (First published in 1925)
2. Piff M (1991) Discrete mathematics. An introduction for software engineers. Cambridge University Press

Algebra

5

Key Topics

Simultaneous equations

Quadratic equations

Polynomials

Indices

Logs

Abstract Algebra

Groups

Rings

Fields

Vector Spaces

5.1 Introduction

Algebra is the branch of mathematics that uses letters in the place of numbers, where the letters stand for variables or constants that are used in mathematical expressions. It is the study of such mathematical symbols and the rules for manipulating them, and it is a powerful tool for problem solving in science, engineering and business.

The origins of algebra are in work done by Islamic mathematicians during the Golden age in Islamic civilization, and the word "*algebra*" comes from the Arabic

© The Author(s), under exclusive license to Springer Nature Switzerland AG 2023
G. O'Regan, *Mathematical Foundations of Software Engineering*,
Texts in Computer Science, https://doi.org/10.1007/978-3-031-26212-8_5

"*al-jabr*", which appears as part of the title of a book by the Islamic mathematician, Al Khwarizmi, in the ninth century A.D. The third century A.D. Hellenistic mathematician, Diophantus, also did early work on algebra.

Algebra covers many areas such as elementary algebra, linear algebra, and abstract algebra. Elementary algebra includes the study of symbols and rules for manipulating them to form valid mathematical expressions; solving simple equations, simultaneous equations and quadratic equations; polynomials; indices and logarithms. Linear algebra is concerned with the solution of a set of linear equations, and the study of matrices and vectors.

We show how to solve simple equations by bringing the unknown variable to one side of the equation and the values to the other side. We show how simultaneous equations are solved by the method of substitution, the method of elimination, and graphical techniques. We show how to solve quadratic equations by factorization, completing the square, the quadratic formula, and graphical techniques. We show how simultaneous equations and quadratic equations may be used to solve practical problems.

We present the laws of indices and show the relationship between indices and logarithms. We discuss the exponential function e^x and the natural logarithm $\log_e x$ or $\ln x$.

5.2   Simplification of Algebraic Expressions

An algebraic expression is a combination of letters and symbols connected through various operations such as $+$, $-$, $/$, $\times$, $($, and $)$. Arithmetic expressions are formed from terms, and *like terms* (i.e., terms with the same variables and exponents) may be added or subtracted. There are two terms in the algebraic expression below with the terms separated by a '$-$'.

$$\underbrace{5x\left(2x^2 + y\right)}_{\text{term 1}} - \underbrace{4x(x + 2y - 1)}_{\text{term 2}}$$

The terms may include powers of some number (e.g., x^3 represents x raised to the power of 3, and 5^4 represents 5 raised to the power of 4).

In algebra, the like terms may be added or subtracted by adding or subtracting the numerical coefficients of the like terms. For example,

$$4x - 2x = 2x$$
$$5x^2 - 2x^2 = 3x^2$$
$$5x - 2y + 3x - 2y = 8x - 4y$$
$$4x^2 - 2y^3 - 3x^2 + 3y^3 = x^2 + y^3$$
$$- (4x - 3y) = -4x + 3y$$
$$5(3x) = (5 \times 3)x = 15x$$

$$5x \cdot 2x = 10x^2$$

Algebraic expressions may be simplified.

$$(ax + by)(ax - by) = aax^2 + (-ab + ba)xy - bby^2$$
$$= a^2x^2 - b^2y^2$$

Therefore,

$$(2x - 3y)(2x + 3y) = 2^2x^2 - 3^2y^2 = 4x^2 - 9y^2$$
$$(a + b)(a - b) = a^2 - b^2$$

Let $P(x) = (ax + b)$ and $Q(x) = (cx + d)$. Then $P(x)Q(x) =$

That is,

$$(ax + b)(cx + d) = ax(cx + d) + b(cx + d)$$
$$= axcx + axd + bcx + bd$$
$$= acx^2 + (ad + bc)x + bd$$

A polynomial $P(x)$ of degree n is defined as $P(x) = a_nx^n + a_{n-s}^{-1} + a_{n-2}x^{n-2} + \cdots + a_1x + a_0$. A polynomial may be multiplied by another, and when we multiply two polynomials $P(x)$ of degree n and $Q(x)$ of degree m together, the resulting polynomials is of degree $n + m$.

Example
Multiply $(2a + 3b)$ by $(a + b)$.

Solution
This is given by $2a^2 + 2ab + 3ab + 3b^2 = 2a^2 + 5ab + 3b^2$.

5.3   Simple and Simultaneous Equations

A *simple equation* is an equation with one unknown, and the unknown may be on both the left-hand side and right-hand side of the equation. The method of solving such equations is to bring the unknowns to one side of the equation, and the values to the other side.

Simultaneous equations are equations with two (or more) unknowns. There are a number of methods to finding a solution to two simultaneous equations such as elimination, substitution, and graphical techniques. The solution of n linear equations with n unknowns may be done using Gaussian elimination and matrix theory.

Example 5.1 (*Simple Equation*) Solve the simple equation $4 - 3x = 2x - 11$.

Solution (Simple Equation)

$$4 - 3x = 2x - 11$$
$$4 - (-11) = 2x - (3x)$$
$$4 + 11 = 2x + 3x$$
$$15 = 5x$$
$$3 = x$$

Example 5.2 (*Simple Equation*) Solve the simple equation

$$\frac{2y}{5} + \frac{3}{4} + 5 = \frac{1}{20} - \frac{3y}{2}$$

Solution (Simple Equation)

The LCM of 4, 5 and 20 is 20. We multiply both sides by 20 to get:

$$20 * \frac{2y}{5} + 20 * \frac{3}{4} + 20 * 5 = 20 * \frac{1}{20} - 20 * \frac{3y}{2}$$
$$8y + 15 + 100 = 1 - 30y$$
$$38y = -114$$
$$y = -3$$

Simple equations may be used to solve practical problems where there is one unknown value to be determined. The information in the problem is converted into a simple equation with one unknown, and the equation is then solved.

Example 5.3 (*Practical Problem—Simple Equations*) The distance (in metres) travelled in time t seconds is given by the formula $s = ut + 1/2 \, at^2$, where u is the initial velocity in m/s and a is the acceleration in m/s^2. Find the acceleration of the body if it travels 168 m in 6 s, with an initial velocity of 10 m/s.

Solution

Using the formula $s = ut + 1/2 \, at^2$ we get:

$$168 = 10 * 6 + 1/2a * 6^2$$
$$168 = 60 + 18a \qquad \text{(simple equation)}$$
$$108 = 18a$$
$$a = 6\text{m/s}^2$$

5.3.1   Simultaneous Equations

Simultaneous equations are equations with two (or more) unknowns. There are several methods available to find a solution to the simultaneous equations including the method of substitution, the method of elimination, and graphical techniques. We start with the substitution method where we express one of the unknowns in terms of the other. The method of substitution involves expressing x in terms of y and substituting it in the other equation (or vice versa expressing y in terms of x and substituting it in the other equation).

Example 5.4 (*Simultaneous Equation—Substitution Method*) Solve the following simultaneous equations by the method of substitution.

$$x + 2y = -1$$
$$4x - 3y = 18$$

Solution (Simultaneous Equation—Substitution Method)
For this example, we use the first equation to *express x in terms of y*.

$$x + 2y = -1$$
$$x = -1 - 2y$$

We then substitute for x, i.e., instead of writing x we write $(-1 - 2y)$ for x in the second equation, and we get a simple equation involving just the unknown y.

$$4(-1 - 2y) - 3y = 18$$
$$\Rightarrow -4 - 8y - 3y = 18$$
$$\Rightarrow -11y = 18 + 4$$
$$\Rightarrow -11y = 22$$
$$\Rightarrow y = -2$$

We then obtain the value of x from the substitution:

$$x = -1 - 2y$$
$$\Rightarrow x = -1 - 2(-2)$$
$$\Rightarrow x = -1 + 4$$
$$\Rightarrow x = 3$$

We can then verify that our solution is correct by checking our answer for both equations.

$$3 + 2(-2) = -1 \qquad \checkmark$$

$$4(3) - 3(-2) = 18 \quad \checkmark$$

The approach of the method of elimination is to manipulate both equations so that we may eliminate either x or y, and so reduce the equations to a simple equation of one unknown value of x or y. This is best seen by an example.

Example 5.5 (*Simultaneous Equation—Method of Elimination*) Solve the following simultaneous equations by the method of elimination.

$$3x + 4y = 5$$
$$2x - 5y = -12$$

Solution (Simultaneous Equation—Method of Elimination)

We will use the method of elimination in this example to eliminate x, and so we multiply Eq. (5.1) by 2 and Eq. (5.2) by -3, and this yields two equations which have equal but opposite coefficients of x.

$$6x + 8y = 10$$
$$-6x + 15y = 36$$
$$-\;-\;-\;-\;-\;-\;-\;-\;-$$
$$0x + 23y = 46$$
$$y = 2$$

We then add both equations together and conclude that $y = 2$. We then determine the value of x by replacing y with 2 in the first equation.

$$3x + 4(2) = 5$$
$$3x + 8 = 5$$
$$3x = 5 - 8$$
$$3x = -3$$
$$x = -1$$

We can then verify that our solution is correct as before by checking our answer for both equations.

Each simultaneous equation represents a straight line, and so the solution to the two simultaneous equations satisfies both equations and so is on both lines, i.e., the solution is the point of intersection of both lines (if there is such a point). Therefore, the solution involves drawing each line and finding the point of intersection of both lines (Fig. 5.1).

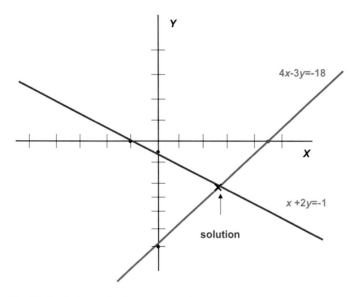

Fig. 5.1   Graphical solution to simultaneous equations

Example 5.6 (*Simultaneous Equation—Graphical Techniques*) Find the solution to the following simultaneous equations using graphical techniques:

$$x + 2y = -1$$
$$4x - 3y = 18$$

Solution (Simultaneous Equation—Graphical Techniques)

First we find two points on line 1, e.g., $(0, -0.5)$ and $(-1, 0)$ are on line 1, since when $x = 0$ we have $2y = -1$ and so $y = -0.5$. Similarly, when $y = 0$ we have $x = -1$. Next we find two points on line 2, e.g., when x is 0 y is -6 and when y is 0 we have $x = 4.5$ and so the points $(0, -6)$ and $(4.5, 0)$ are on line 2.

We then draw the X-axis and the Y-axis, draw the scales on the axes, label the axes, plot the points and draw both lines. Finally, we find the point of intersection of both lines (if there is such a point), and this is our solution to the simultaneous equations.

The graph shows that the two lines intersect, and so we need to determine the point of intersection, and this involves determining the x and y coordinates of the solution which is given by $x = 3$ and $y = -2$. The solution using graphical techniques requires care (as inaccuracies may be introduced from poor drawing) and graph paper is required for accuracy.

The solution to practical problems often involves solving two simultaneous equations.

Example 5.7 (*Simultaneous Equation—Problem Solving*) Three new cars and four new vans supplied to a dealer together cost £97,000, and five new cars and two new vans of the same models cost £103,100. Find the cost of a car and a van.

Solution (Simultaneous Equation—Problem Solving)

We let C represent the cost of a car and V represent the cost of a van. We convert the information provided into two equations with two unknowns and then solve for V and C.

$$3C + 4V = 97,000 \tag{5.1}$$

$$5C + 2V = 103,100 \tag{5.2}$$

We multiply Eq. (5.2) by -2 and add to Eq. (5.1) to eliminate V

$$3C + 4V = 97,000$$
$$-10C - 4V = -206,200$$
$$\text{-----------------}$$
$$-7C = -109,200$$
$$C = £15,600$$

We then calculate the cost of a van by substituting the value of C in Eq. (5.2) to reduce it to an equation of one unknown.

$$5C + 2V = 103,100$$
$$78,000 + 2V = 103,100$$
$$2V = 25,100$$
$$V = £12,550$$

Therefore, the cost of a car is £15,600, and the cost of a van is £12,550.

5.4   Quadratic Equations

A quadratic equation is an equation of the form $ax^2 + bx + c = 0$, and solving the quadratic equation is concerned with finding the unknown value x (roots of the quadratic equation). There may be no solution, one solution (a double root), or two solutions. There are several techniques for solving quadratic equations such as factorization; completing the square; the quadratic formula; and graphical techniques.

Example 5.8 (*Quadratic Equations—Factorization*) Solve the quadratic Eq. $3x^2 - 11x - 4 = 0$ by factorization.

Solution (Quadratic Equations—Factorization)

The approach taken is to find the factors of the quadratic equation. Sometimes this is easy, but often other techniques will need to be employed. For the above quadratic equation we note immediately that its factors are $(3x + 1)(x - 4)$ since

$$(3x + 1)(x - 4)$$
$$= 3x^2 - 12x + x - 4$$
$$= 3x^2 - 11x - 4$$

Next, we note the property that if the product of two numbers A and B is 0 then either A is 0 or B is 0. Another words, $AB = 0 \Rightarrow A = 0$ or $B = 0$. We conclude from this property that as:

$$3x^2 - 11x - 4 = 0$$
$$\Rightarrow (3x + 1)(x - 4) = 0$$
$$\Rightarrow (3x + 1) = 0 \quad \text{or} \quad (x - 4) = 0$$
$$\Rightarrow 3x = -1 \quad \text{or} \quad x = 4$$
$$\Rightarrow x = -0.33 \quad \text{or} \quad x = 4$$

Therefore, the solution (or roots) of the quadratic equation $3x^2 - 11x - 4 = 0$ is $x = -0.33$ or $x = 4$.

Example 5.9 (*Quadratic Equations—Completing the Square*) Solve the quadratic equation $2x^2 + 5x - 3 = 0$ by completing the square.

Solution (Quadratic Equations—Completing the Square)

First we convert the quadratic equation to an equivalent quadratic with a unary coefficient of x^2. This involves division by 2. Next, we examine the coefficient of x (in this case $5/2$), and we add the square of half the coefficient of x to both sides. This allows us to complete the square, and we then to take the square root of both sides. Finally, we solve for x.

$$2x^2 + 5x - 3 = 0$$
$$\Rightarrow x^2 + 5/2x - 3/2 = 0$$
$$\Rightarrow x^2 + 5/2x = 3/2$$
$$\Rightarrow x^2 + 5/2x + (5/4)^2 = 3/2 + (5/4)^2$$
$$\Rightarrow (x + 5/4)^2 = 3/2 + (25/16)$$
$$\Rightarrow (x + 5/4)^2 = 24/16 + (25/16)$$

$$\Rightarrow (x + 5/4)^2 = 49/16$$
$$\Rightarrow (x + 5/4) = \pm 7/4$$
$$\Rightarrow x = -5/4 \pm 7/4$$
$$\Rightarrow x = -5/4 - 7/4 \text{ or } x = -5/4 + 7/4$$
$$\Rightarrow x = -12/4 \text{ or } x = 2/4$$
$$\Rightarrow x = -3 \text{ or } x = 0.5$$

Example 5.10 (*Quadratic Equations—Quadratic Formula*) Establish the quadratic formula for solving quadratic equations.

Solution (Quadratic Equations—Quadratic Formula)

We complete the square, and the result will follow.

$$ax^2 + bx + c = 0$$
$$\Rightarrow x^2 + b/a\,x + c/a = 0$$
$$\Rightarrow x^2 + b/a\,x = -c/a$$
$$\Rightarrow x^2 + b/a\,x + (b/2a)^2 = -c/a + (b/2a)^2$$
$$\Rightarrow (x + b/2a)^2 = -c/a + (b/2a)^2$$
$$\Rightarrow (x + b/2a)^2 = \frac{-4ac}{4a^2} + \frac{b^2}{4a^2}$$
$$\Rightarrow (x + b/2a)^2 = \frac{b^2 - 4ac}{4a^2}$$
$$\Rightarrow (x + b/2a) = \pm \frac{\sqrt{b^2 - 4ac}}{2a}$$
$$\Rightarrow x = \frac{-b \pm \sqrt{b^2 - 4ac}}{2a}.$$

Example 5.11 (*Quadratic Equations—Quadratic Formula*) Solve the quadratic equation $2x^2 + 5x - 3 = 0$ using the quadratic formula.

Solution (Quadratic Equations—Quadratic Formula)

For this example a $= 2$; $b = 5$; and $c = -3$, and we put these values into the quadratic formula.

$$x = \frac{-5 \pm \sqrt{5^2 - 4.2.(-3)}}{2.2} = \frac{-5 \pm \sqrt{25 + 24}}{4}$$
$$x = \frac{-5 \pm \sqrt{49}}{4} = \frac{-5 \pm 7}{4}$$
$$x = 0.5 \text{ or } x = -3.$$

Table 5.1 Table of values for quadratic equation

x	-3	-2	-1	0	1	2	3
$y = 2x^2 - x - 6$	15	4	-3	-6	-5	0	9

Fig. 5.2 Graphical solution
to quadratic equation

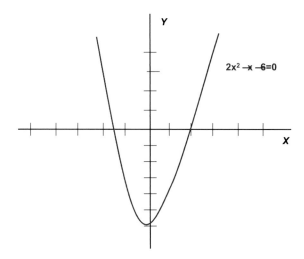

Example 5.12 (*Quadratic Equations—Graphical Techniques*) Solve the quadratic equation $2x^2 - x - 6 = 0$ using graphical techniques given that the roots of the quadratic equation lie between $x = -3$ and $x = 3$.

Solution (Quadratic Equations—Graphical Techniques)

The approach is first to create a table of values for the curve $y = 2x^2 - x - 6$ (Table 5.1), and to draw the X- and Y-axis and scales, and then to plot the points from the table of values, and to join the points together to form the curve (Fig. 5.2).

The graphical solution is to the quadratic equation is then given by the points where the curve intersects the X-axis (i.e., $y = 0$ on the X-axis). There may be no solution (i.e., the curve does not intersect the X-axis), one solution (a double root), or two solutions.

The graph for the curve $y = 2x^2 - x - 6$ is given in Table 5.1, and so the points where the curve intersects the X-axis are determined. We note from the graph that the curve intersects the X-axis at two distinct points, and we see from the graph that the roots of the quadratic equation are given by $x = -1.5$ and $x = 2$.

The solution to quadratic equations using graphical techniques requires care in the plotting the points (as in determining the solution to simultaneous equations using graphical techniques), and graph paper is required for accuracy.

Quadratic equations often arise in solving practical problems as the following example shows.

Example 5.13 (*Quadratic Equations—Practical Problem*) A shed is 7.0 m long and 5.0 m wide. A concrete path of constant width x is laid all the way around the shed, and the area of the path is 30 m^2. Calculate its width x to the nearest centimeter (and use the quadratic formula).

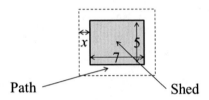

Path   Shed

Solution (Quadratic Equations—Practical Problem)

Let x be the width of the path. Then the area of the path is the difference in area between the area of the large rectangle and the shed. We let A_S denote the area of the shed and let A_{S+P} denote the area of the large rectangle (i.e., the area of the shed + the area of the path).

$$A_S = 7*5 = 35$$
$$A_{S+P} = (7+2x)(5+2x)$$
$$= 35 + 14x + 10x + 4x^2$$
$$= 35 + 24x + 4x^2$$
$$A_P = 30$$
$$= A_{S+P} - A_S$$
$$\Rightarrow 35 + 24x + 4x^2 - 35 = 30$$
$$\Rightarrow -30 + 24x + 4x^2 = 0$$
$$\Rightarrow 4x^2 + 24x - 30 = 0$$
$$\Rightarrow x = 0.91 \text{ m} \quad \text{(from the quadratic formula).}$$

5.5   Indices and Logarithms

The product $a.a.a.a...a$ (n times) is denoted by a^n, and the number n is the index of a. The following are properties of indices.

$$a^0 = 1$$
$$a^{m+n} = a^m \cdot a^n$$
$$a^{mn} = (a^m)^n$$

$$a^{-n} = \frac{1}{a^n}$$

$$a^{\frac{1}{n}} = \sqrt[n]{a}$$

Logarithms are closely related to indices, and if the number b can be written in the form $b = a^x$, then we say that log to the base a of b is x, i.e., $\log_a b = x \Leftrightarrow a^x = b$. Clearly, $\log_{10} 100 = 2$ since $10^2 = 100$. The following are properties of logarithms

$$\log_a AB = \log_a A + \log_a B$$
$$\log_a A^n = n \log_a A$$
$$\log \frac{A}{B} = \log A - \log B$$

We will prove the first property of logarithms. Suppose $\log_a A = x$ and $\log_a B = y$. Then $A = a^x$ and $B = a^y$ and so $AB = a^x a^y = a^{x+y}$ and so $\log_a AB = x + y = \log_a A + \log_a B$.

Example 5.14 Solve $\log_2 1/64$ without a calculator

Solution

$$\log_2 1/64 = x$$
$$\Rightarrow 2^x = 1/64$$
$$\Rightarrow 2^x = 1/4 \times 4 \times 4$$
$$\Rightarrow 2^x = 1/2 \times 2 \times 2 \times 2 \times 2 \times 2$$
$$\Rightarrow 2^x = 1/2^6$$
$$\Rightarrow 2^x = 2^{-6}$$
$$\Rightarrow x = -6.$$

Example 5.15 Write $\log\left\{\frac{16 \times \sqrt[3]{5}}{81}\right\}$ in terms of log 2, log 3, and log 5 to any base.

Solution

$$\log\left\{\frac{16 \times \sqrt[3]{5}}{81}\right\}$$
$$= \log 16 + 1/3 \log 5 - \log 81$$
$$= \log 2^4 + 1/3 \log 5 - \log 3^4$$
$$= 4 \log 2 + 1/3 \log 5 - 4 \log 3$$

The law of logarithms may be used to solve certain indicial equations, and we illustrate this with two examples.

Example 5.16 (*Indicial Equations*) Solve the equation $\log(x - 4) + \log(x + 2) = 2 \log(x - 2)$.

Solution

$$\log(x - 4) + \log(x + 2) = 2 \log(x - 2)$$
$$\Rightarrow \log(x - 4)(x + 2) = \log(x - 2)^2$$
$$\Rightarrow \log(x^2 - 2x - 8) = \log(x^2 - 4x + 4)$$
$$\Rightarrow x^2 - 2x - 8 = x^2 - 4x + 4$$
$$\Rightarrow -2x - 8 = -4x + 4$$
$$\Rightarrow 2x = 12$$
$$\Rightarrow x = 6.$$

Example 5.17 (*Indicial Equations*) Solve the equation $2^x = 3$, correct to 4 significant places.

Solution

$$2^x = 3$$
$$\Rightarrow \log_{10} 2^x = \log_{10} 3$$
$$\Rightarrow x \log_{10} 2 = \log_{10} 3$$
$$\Rightarrow x = \frac{\log_{10} 3}{\log_{10} 2}$$
$$= \frac{0.4771}{0.3010}$$
$$\Rightarrow x = 1.585.$$

5.6   Exponentials and Natural Logarithms

The number e is a mathematical constant that occurs frequently in mathematics and its value is approximately equal to 2.7183 (it is an irrational number). The exponential function e^x (where e is the base and x is the exponent) is widely used in mathematics and science, and especially in problems involving growth and decay. The exponential function has the property that it is the unique function that is equal to its own derivative (i.e., $d/dx \, e^x = e^x$). The number e is the base of the natural logarithm, and e is sometimes called Euler's number or Euler's constant.

The value of the exponential function may be determined from its power series, and the value of $e^{0.1}$ may be determined by substituting 0.1 for x in the power series. However, it is more common to determine its value by using a scientific calculator which contains the e^x function.

$$e^x = 1 + x + \frac{x^2}{2!} + \cdots + \frac{x^n + \cdots}{n!}$$

Logarithms to the base e are termed natural logarithms (or Napierian logs). The natural logarithm of x is written as $\log_e x$ or more commonly as $\ln x$.

Example 5.18 (*Natural Logs*) Solve the equation $7 = 4e^{-3x}$ to find x, correct to 4 decimal places.

Solution (Natural Logs)

$$7 = 4e^{-3x}$$
$$\Rightarrow 7/4 = e^{-3x}$$
$$\Rightarrow \ln(1.75) = \ln\left(e^{-3x}\right) = -3x$$
$$0.55966 = -3x$$
$$x = -0.1865.$$

Example 5.19 (*Practical Problem*) The length of a bar, l, at a temperature θ is given by $l = l_0 \, e^{\alpha\theta}$, where l_0 and α are constants. Evaluate l, correct to four significant figures, when $l_0 = 2.587$, $\theta = 321.7$ and $\alpha = 1.771 \times 10^{-4}$.

Solution (Practical Problem)

$$l = l_0 e^{\alpha\theta},$$
$$= 2.587 e^{1.771 \times 10^{-4} * 321.7}$$
$$= 2.587 * 0.56973$$
$$= 1.4739.$$

5.7   Horner's Method for Polynomials

Horner's method is a computationally efficient way to evaluate a polynomial function. It is named after William Horner who was a nineteenth century British mathematician and schoolmaster. Chinese mathematicians were familiar with the method in the third century A.D.

The normal method for the evaluation of a polynomial involves computing exponentials, and this is computationally expensive. Horner's method has the advantage that fewer calculations are required, and it eliminates all exponentials by using nested multiplication and addition. It also provides a computationally efficient way to determine the derivative of the polynomial.

Horner's Method and Algorithm

Consider a polynomial $P(x)$ of degree n defined by:

$$P(x) = a_n x^n + a_{n-1} x^{n-1} + a_{n-2} x^{n-2} + \cdots + a_1 x + a_0$$

The Horner method to evaluate $P(x_0)$ essentially involves writing $P(x)$ as:

$$P(x) = ((((((a_n x + a_{n-1})x + a_{n-2})x + \cdots + a_1)x + a_0$$

The computation of $P(x_0)$ involves defining a set of coefficients b_k such that:

$$b_m = a_n$$
$$b_{n-1} = a_{n-1} + b_n x_0$$
$$\cdots$$
$$b_k = a_k + b_{k+1} x_0$$
$$\cdots$$
$$b_1 = a_1 + b_2 x_0$$
$$b_0 = a_0 + b_1 x_0$$

Then the computation of $P(x_0)$ is given by:

$$P(x_0) = b_0$$

Further, if $Q(x) = b_n x^{n-1} + b_{n-1} x^{n-2} + b_{n-2} x^{n-3} + \cdots + b_1$ then it is easy to verify that:

$$P(x) = (x - x_0) Q(x) + b_0$$

This also allows the derivative of $P(x)$ to be easily computed for x_0 since:

$$P'(x) = Q(x) + (x - x_0) Q'(x)$$
$$P'(x_0) = Q(x_0)$$

Algorithm (To evaluate polynomial and its derivative)

(i)   Initialize y to a_n and z to a_n (Compute b_n for P and b_{n-1} for Q)
(ii) For each j from $n - 1, n - 2$ to 1 compute b_j for P and b_{j-1} for Q by
         Set y to $x_0 y + a_j$ (i.e., b_j for P) and z to $x_0 z + y$ (i.e., b_{j-1} for Q)
(iii) Compute b_0 by setting y to $x_0 y + a_0$

Then $P(x_0) = y$ and $P'(x_0) = z$.

5.8   Abstract Algebra

One of the important features of modern mathematics is the power of abstraction. This has opened up whole new areas of mathematics, and it has led to a large body of new results and problems. The term *"abstract"* is subjective, as what is abstract to one person may be quite concrete to another. We shall introduce some important algebraic structures in this section including monoids, groups, rings, fields, and vector spaces. Chapter 21 will show how abstract structures such as vector spaces may be used for error correcting codes in coding theory.

5.8.1   Monoids and Groups

A non-empty set M together with a binary operation '$*$' is called a *monoid* if for all elements $a, b, c \in M$ the following properties hold:

(1) $a * b \in M$ (Closure property)
(2) $a * (b * c) = (a * b) * c$ (Associative property)
(3) $\exists u \in M$ such that: $a * u = u * a = a (\forall a \in M)$ (Identity Element)

A monoid is commutative if $a * b = b * a$ for all $a, b \in M$. A *semi-group* $(M, *)$ is a set with a binary operation '$*$' such that the closure and associativity properties hold (but it may not have an identity element).

Example 5.20 (*Monoids*)

(i) The set of sequences Σ^* under concatenation with the empty sequence Λ the identity element.
(ii) The set of integers under addition forms an infinite monoid in which 0 is the identity element.

A non-empty set G together with a binary operation '$*$' is called a *group* if for all elements $a, b, c \in G$ the following properties hold

(1) $a * b \in G$ (Closure property)
(2) $a * (b * c) = (a * b) * c$ (Associative property)
(3) $\exists e \in G$ such that: $a * e = e * a = a (\forall a \in G)$ (Identity Element)
(4) For every $a \in G, \exists a^{-1} \in G$, such that: $a * a^{-1} = a^{-1} * a = e$ (Inverse Element)

The identity element is unique, and the inverse a^{-1} of an element a is unique (see Exercise 5). A *commutative group* has the additional property that $a * b = b * a$ for all $a, b \in G$. The order of a group G is the number of elements in G and is denoted by $o(G)$. If the order of G is finite then G is said to be a finite group.

Example 5.21 (*Groups*)

(i)   The set of integers under addition $(\mathbb{Z}, +)$ forms an infinite group in which 0 is
      the identity element.
(ii) The set of integer 2×2 matrices under addition, where the identity element is
      $$\begin{pmatrix} 0 & 0 \\ 0 & 0 \end{pmatrix}$$
(iii) The set of integers under multiplication $(\mathbb{Z}, \times)$ forms an infinite monoid with 1
      as the identity element.

A *cyclic group* is a group where all elements $g \in G$ are obtained from the powers a^i
of one element $a \in G$, with $a^0 = e$. The element 'a' is termed the generator of the
cyclic group G. A finite cyclic group with n elements is of the form $\{a^0, a^1, a^2, ...,$
$a^{n-1}\}$.

A non-empty subset H of a group G is said to be a *subgroup* of G if for all
$a, b \in H$ then $a * b \in H$, and for any $a \in H$ then $a^{-1} \in H$. A subgroup N is termed
a *normal subgroup* of G if $gng^{-1} \in N$ for all $g \in G$ and all $n \in N$. Further, if G is
a group and N is a normal subgroup of G, then the *quotient group* G/N may be
formed.

Lagrange's theorem states the relationship between the order of a subgroup H
of G, and the order of G. The theorem states that if G is a finite group, and H is a
subgroup of G, then $o(H)$ is a divisor of $o(G)$.

We may also define mapping between similar algebraic structures termed *homomorphism*, and these mapping preserve structure. If the homomorphism is one to
one and onto it is termed an *isomorphism*, which means that the two structures are
identical in some sense (apart from a relabelling of elements).

5.8.2   Rings

A *ring* is a non-empty set R together with two binary operations '+' and '$\times$'
where $(R, +)$ is a commutative group; $(R, \times)$ is a semi-group; and the left and
right distributive laws hold. Specifically, for all elements $a, b, c \in R$ the following
properties hold:

(1) $a + b \in R$ (Closure property)
(2) $a + (b + c) = (a + b) + c$ (Associative property)
(3) $\exists 0 \in R$ such that $\forall a \in R : a + 0 = 0 + a = a$ (Identity property)
(4) $\forall a \in R : \exists (-a) \in R : a + (-a) = (-a) + a = 0$ (Inverse Element)
(5) $a + b = b + a$ (Commutativity)
(6) $a \times b \in R$ (Closure property)
(7) $a \times (b \times c) = (a \times b) \times c$ (Associative property)
(8) $a \times (b + c) = a \times b + a \times c$ (Distributive Law)
(9) $(b + c) \times a = b \times a + c \times a$ (Distributive Law)

The element 0 is the identity element under addition, and the additive inverse of an element a is given by $-a$. If a ring $(R, \times, +)$ has a multiplicative identity 1 where $a \times 1 = 1 \times a = a$ for all $a \in R$ then R is termed a ring with a unit element. If $a \times b = b \times a$ for all $a, b \in R$ then R is termed a *commutative ring*.

An element $a \neq 0$ in a ring R is said to be a *zero divisor* if there exists $b \in R$, with $b \neq 0$ such that $ab = 0$. A commutative ring is an *integral domain* if it has no zero divisors. A ring is said to be a *division ring* if its nonzero elements form a group under multiplication.

Example 5.22 (*Rings*)

(i) The set of integers $(\mathbb{Z}, +, \times)$ forms an infinite commutative ring with multiplicative unit element 1. Further, since it has no zero divisors it is an integral domain.

(ii) The set of integers mod 4 (i.e., $\mathbb{Z}_4$ where addition and multiplication is performed modulo 4)[1] is a finite commutative ring with unit element $[1]_4$. Its elements are $\{[0]_4, [1]_4, [2]_4, [3]_4\}$. It has zero divisors since $[2]_4[2]_4 = [0]_4$ and so it is not an integral domain.

(iii) The quaternions (discussed in Chap. 28) are an example of a non-commutative ring (they form a division ring).

(iv) The set of integers mod 5 (i.e., $\mathbb{Z}_5$ where addition and multiplication is performed modulo 5) is a finite commutative division ring,[2] and it has no zero divisors.

5.8.3 Fields

A *field* is a non-empty set F together with two binary operation '+' and '$\times$' where $(F, +)$ is a commutative group; $(F \setminus \{0\}, \times)$ is a commutative group; and the distributive properties hold. The properties of a field are:

(1) $a + b \in F$ (Closure property)
(2) $a + (b + c) = (a + b) + c$ (Associative property)
(3) $\exists 0 \in F$ such that $\forall a \in F : a + 0 = 0 + a = a$ (Identity Element)
(4) $\forall a \in F : \exists (-a) \in F \, a + (-a) = (-a) + a = 0$ (Inverse Element)
(5) $a + b = b + a$ (Commutativity)
(6) $a \times b \in F$ (Closure property)
(7) $a \times (b \times c) = (a \times b) \times c$ (Associative property)
(8) $\exists 1 \in F$ such that $\forall a \in F \, a \times 1 = 1 \times a = a$ (Identity Element)

---

[1] Recall that $\mathbb{Z}/n\mathbb{Z} = \mathbb{Z}_n = \{[a]_n : 0 \leq a \leq n - 1\} = \{[0]_n, [1]_n, \ldots, [n - 1]_n\}$.
[2] A finite division ring is actually a field (i.e., it is commutative under multiplication), and this classic result was proved by Wedderburn.

(9) $\forall a \in F\backslash\{0\} \exists a^{-1} \in F a \times a^{-1} = a^{-1} \times a = 1$ (Inverse Element)
(10) $a \times b = b \times a$ (Commutativity)
(11) $a \times (b + c) = a \times b + a \times c$ (Distributive Law)
(12) $(b + c) \times a = b \times a + c \times a$ (Distributive Law)

The following are examples of fields:

Example 5.23 *(Fields)*

(i)   The set of rational numbers $(\mathbb{Q}, +, \times)$ forms an infinite commutative field. The additive identity is 0, and the multiplicative identity is 1.

(ii) The set of real numbers $(\mathbb{R}, +, \times)$ forms an infinite commutative field. The additive identity is 0, and the multiplicative identity is 1.

(iii) The set of complex numbers $(\mathbb{C}, +, \times)$ forms an infinite commutative field. The additive identity is 0, and the multiplicative identity is 1.

(iv) The set of integers mod 7 (i.e., $\mathbb{Z}_7$ where addition and multiplication is performed mod 7) is a finite field.

(v)   The set of integers mod p where p is a prime (i.e., $\mathbb{Z}_p$ where addition and multiplication is performed mod p) is a finite field with p elements. The additive identity is [0], and the multiplicative identity is [1].

A field is a commutative division ring but not every division ring is a field. For example, the quaternions (discovered by Hamilton) are an example of a division ring, which is not a field (quaternion multiplication is not commutative). If the number of elements in the field F is finite then F is called a finite field, and F is written as F_q where q is the number of elements in F. In fact, every finite field has $q = p^k$ elements for some prime p, and some $k \in \mathbb{N}$ and $k > 0$.

5.8.4   Vector Spaces

A non-empty set V is said to be a *vector space* over a field F if V is a commutative group under vector addition $+$, and if for every $\alpha \in F$, $v \in V$ there is an element αv in V such that the following properties hold for $v, w \in V$ and $\alpha, \beta \in F$:

1. $u + v \in V$ (Closure property)
2. $u + (v + w) = (u + v) + w$ (Associative)
3. $\exists 0 \in V$ such that $\forall v \in V + 0 = 0 + v = v$ (Identity element)
4. $\forall v \in V : \exists (-v) \in V$ such that $v + (-v) = (-v) + v = 0$ (Inverse)
5. $v + w = w + v$ (Commutative)
6. $\alpha(v + w) = \alpha v + \alpha w$
7. $(\alpha + \beta)v = \alpha v + \beta v$
8. $\alpha(\beta v) = (\alpha\beta)v$
9. $1v = v$

The elements in V are referred to as *vectors* and the elements in F are referred to as *scalars*. The element 1 refers to the identity element of the field F under multiplication. A *normed vector space* is a vector space where a norm is defined (the norm is similar to the distance function).

Application of Vector Spaces to Coding Theory
The representation of codewords in coding theory (discussed in Chap. 21) is by n-dimensional vectors over the finite field F_q. A codeword vector v is represented as the n-tuple:

$$v = (a_0, a_1, \ldots, a_{n-1})$$

where each $a_i \in F_q$. The set of all n-dimensional vectors is the n-dimensional vector space $\mathbf{F}^n_q$ with q^n elements. The addition of two vectors v and w, where $v = (a_0, a_1, \ldots a_{n-1})$ and $w = (b_0, b_1, \ldots b_{n-1})$ is given by:

$$v + w = (a_0 + b_0, a_1 + b_1, \ldots, a_{n-1}b_{n-1})$$

The scalar multiplication of a vector $v = (a_0, a_1, \ldots a_{n-1}) \in \mathbf{F}^n_q$ by a scalar $\beta \in F_q$ is given by:

$$\beta v = (\beta a_0, \beta a_1, \ldots \beta a_{n-1})$$

The set $\mathbf{F}^n_q$ is called the vector space over the finite field F_q if the vector space properties above hold. A finite set of vectors $v_1, v_2, \ldots v_k$ is said to be *linearly independent* if

$$\beta_1 v_1 + \beta_2 v_2 + \cdots + \beta_k v_k = 0 \Rightarrow \beta_1 = \beta_2 = \ldots \beta_k = 0$$

Otherwise, the set of vectors $v_1, v_2, \ldots v_k$ is said to be *linearly dependent*.

A non-empty subset W of a vector space V ($W \subseteq V$) is said to be a *subspace* of V, if W forms a vector space over F under the operations of V. This is equivalent to W being closed under vector addition and scalar multiplication, i.e., $w_1, w_2 \in W$, $\alpha, \beta \in F$ then $\alpha w_1 + \beta w_2 \in W$.

The *dimension* (dim W) of a subspace $W \subseteq V$ is k if there are k linearly independent vectors in W but every $k + 1$ vectors are linearly dependent. A subset of a vector space is a *basis* for V if it consists of linearly independent vectors, and its linear span is V (i.e., the basis generates V). We shall employ the basis of the vector space of codewords (see Chap. 21) to create the generator matrix to simplify the encoding of the information words. The linear span of a set of vectors $v_1, v_2, \ldots, v_k$ is defined as $\beta_1 v_1 + \beta_2 v_2 + \cdots + \beta_k v_k$.

Example 5.24 (*Vector Spaces*)

(i) The real coordinate space $\mathbb{R}^n$ forms an n-dimensional vector space over $\mathbb{R}$. The elements of $\mathbb{R}^n$ are the set of all n tuples of elements of $\mathbb{R}$, where an element x in $\mathbb{R}^n$ is written as:

$$x = (x_1, x_2, \ldots x_n)$$

where each $x_i \in \mathbb{R}$ and vector addition and scalar multiplication are given by:

$$\alpha x = (\alpha x_1, \alpha x_2, \ldots, \alpha x_n)$$
$$x + y = (x_1 + y_1, x_2 + y_2 \ldots x_n + y_n)$$

(ii) The set of $m \times n$ matrices over the real numbers forms a vector space, with vector addition given by matrix addition, and the multiplication of a matrix by a scalar given by the multiplication of each entry in the matrix by the scalar.

5.9   Review Questions

1. Solve the simple equation: $4(3x + 1) = 7(x + 4) - 2(x + 5)$
2. Solve the following simultaneous equations by

$$x + 2y = -1$$
$$4x - 3y = 18$$

3. Solve the quadratic Eq. $3x^2 + 5x - 2 = 0$ given that the solution is between $x = -3$ and $x = 3$ by:
   (a) Graphical techniques
   (b) Factorization
   (c) Quadratic Formula
4. Solve the following indicial equation using logarithms

$$2^{x=1} = 3^{2x-1}$$

5. Explain the differences between semigroups, monoids and groups.
6. Show that the following properties are true for groups.
   (i) The identity element is unique in a group.
   (ii) The inverse of an element is unique in a group.
7. Explain the differences between rings, commutative rings, integral domains, division rings and fields.
8. What is a vector space?

9. Explain how vector spaces may be applied to coding theory (see Chap. 11 for more details).

5.10 Summary

This chapter provided a brief introduction to algebra, which is the branch of mathematics that studies mathematical symbols and the rules for manipulating them. Algebra is a powerful tool for problem solving in science and engineering.

Elementary algebra includes the study of simultaneous equations (i.e., two or more equations with two or more unknowns); the solution of quadratic equations $ax^2 + bx + c = 0$; and the study of polynomials, indices and logarithms. Linear algebra is concerned with the solution of a set of linear equations, and the study of matrices and vector spaces.

Abstract algebra is concerned with the study of abstract algebraic structures such as monoids, groups, rings, integral domains, fields, and vector spaces. The abstract approach in modern mathematics has opened up whole new areas of mathematics as well as applications in areas such as coding theory in the computing field.

Mathematical Induction and Recursion

6

6.1   Introduction

Mathematical induction is an important proof technique used in mathematics, and it is often used to establish the truth of a statement for all natural numbers. There are two parts to a proof by induction, and these are the base step and the inductive step. The *base case* involves showing that the statement is true for some natural number (usually the number 1). The second step is termed the *inductive step*, and it involves showing that if the statement is true for some natural number $n = k$, then the statement is true for its successor $n = k + 1$. This is often written as $P(k) \rightarrow P(k + 1)$.

The statement $P(k)$ that is assumed to be true when $n = k$ is termed the *inductive hypothesis*. From the base step and the inductive step, we infer that the

© The Author(s), under exclusive license to Springer Nature Switzerland AG 2023   109
G. O'Regan, *Mathematical Foundations of Software Engineering*,
Texts in Computer Science, https://doi.org/10.1007/978-3-031-26212-8_6

statement is true for all natural numbers (that are greater than or equal to the number specified in the base case). Formally, the proof technique used in mathematical induction is of the form[1]:

$$(P(1) \wedge (P(k) \rightarrow P(k+1))) \rightarrow \forall n\, P(n).$$

Mathematical induction (weak induction) may be used to prove a wide variety of theorems and especially theorems of the form $\forall n\, P(n)$. It may be used to provide a proof of theorems about summation formulae, inequalities, set theory, and the correctness of algorithms and computer programs. One of the earliest inductive proofs was the sixteenth-century proof that the sum of the first n odd integers is n^2, which was proved by Francesco Maurolico in 1575. Later mathematicians made the method of mathematical induction more precise.

We distinguish between *strong induction* and *weak induction*, where strong induction also has a base case and an inductive step, but the inductive step is a little different. It involves showing that if the statement is true for **all** natural numbers less than or equal to an arbitrary number k, then the statement is true for its successor $k + 1$. Weak induction involves showing that if the statement is true for **some** natural number $n = k$, then the statement is true for its successor $n = k + 1$. *Structural induction* is another form of induction, and this mathematical technique is used to prove properties about recursively defined sets and structures.

Recursion is often used in mathematics to define functions, sequences, and sets. However, care is required with a recursive definition to ensure that it actually defines something, and that what is defined makes sense. Recursion defines a concept in terms of itself, and we need to ensure that the definition is not circular (i.e., that it does not lead to a vicious circle).

Recursion and induction are closely related and are often used together. Recursion is extremely useful in developing algorithms for solving complex problems, and induction is a useful technique in verifying the correctness of such algorithms.

Example 6.1 Show that the sum of the first n natural numbers is given by the formula:

$$1 + 2 + 3 + \cdots + n = \frac{n(n+1)}{2}$$

Proof
Base Case

We consider the case where $n = 1$ and clearly $1 = \frac{1(1+1)}{2}$ and so the base case $P(1)$ is true.

---

[1] This definition of mathematical induction covers the base case of $n = 1$ and would need to be adjusted if the number specified in the base case is higher.

Inductive Step

Suppose the result is true for some number k, then we have $P(k)$

$$1 + 2 + 3 + \cdots + k = \frac{k(k+1)}{2}$$

Then consider the sum of the first $k+1$ natural numbers, and we use the inductive hypothesis to show that its sum is given by the formula.

$$1 + 2 + 3 + \cdots + k + (k+1)$$
$$= \frac{k(k+1)}{2} + (k+1) \quad \text{(by inductive hypothesis)}$$
$$= \frac{k^2 + k}{2} + \frac{(2k+2)}{2}$$
$$= \frac{k^2 + 3k + 2}{2}$$
$$= \frac{(k+1)(k+2)}{2}$$

Thus, we have shown that if the formula is true for an arbitrary natural number k, then it is true for its successor $k + 1$. That is, $P(k) \rightarrow P(k+1)$. We have shown that $P(1)$ is true, and so it follows from mathematical induction that $P(2)$, $P(3)$, ... are true, and so $P(n)$ is true, for all natural numbers and the theorem is established.

Note 6.1 There are opportunities to make errors in proofs with induction, and the most common mistakes are not to complete the base case or inductive step correctly. These errors can lead to strange results, and so care is required. It is important to be precise in the statements of the base case and inductive step.

Example 6.2 (*Binomial Theorem*) Prove the binomial theorem using induction (permutations and combinations are discussed in Chap. 8). That is,

$$(1+x)^n = 1 + \binom{n}{1}x + \binom{n}{2}x^2 + \cdots + \binom{n}{r}x^r + \cdots + \binom{n}{n}x^n$$

Proof
Base Case

We consider the case where $n = 1$ and clearly $(1+x)^1 = (1+x) = 1 + \binom{1}{1}x^1$, and so the base case $P(1)$ is true.

Inductive Step

Suppose the result is true for some number k, then we have $P(k)$

$$(1+x)^k = 1 + \binom{k}{1}x + \binom{k}{2}x^2 + \cdots + \binom{k}{r}x^r + \cdots + \binom{k}{k}x^k$$

Then consider $(1+x)^{k+1}$ and we use the inductive hypothesis to show that it is given by the formula.

$$(1+x)^{k+1} = (1+x)^k(1+x)$$

$$= \left(1 + \binom{k}{1}x + \binom{k}{2}x^2 + \cdots + \binom{k}{r}x^r + \cdots + \binom{k}{k}x^k\right)(1+x)$$

$$= \left(1 + \binom{k}{1}x + \binom{k}{2}x^2 + \cdots + \binom{k}{r}x^r + \cdots + \binom{k}{k}x^k\right)$$

$$+ x + \binom{k}{1}x^2 + \cdots + \binom{k}{r}x^{r+1} + \cdots + \binom{k}{k}x^{k+1}$$

$$= 1 + \binom{k}{1}x + \binom{k}{2}x^2 + \cdots + \binom{k}{r}x^r + \cdots + \binom{k}{k}x^k$$

$$+ \binom{k}{0}x + \binom{k}{1}x^2 + \cdots + \binom{k}{r-1}x^r + \cdots + \binom{k}{k-1}x^k + \binom{k}{k}x^{k+1}$$

$$= 1 + \binom{k+1}{1}x + \cdots + \binom{k+1}{r}x^r + \cdots + \binom{k+1}{k}x^k + \binom{k+1}{k+1}x^{k+1}$$

(which follows from Exercise 7 below).

Thus, we have shown that if the binomial theorem is true for an arbitrary natural number k, then it is true for its successor $k+1$. That is, $P(k) \rightarrow P(k+1)$. We have shown that $P(1)$ is true, and so it follows from mathematical induction that $P(n)$ is true, for all natural numbers, and so the theorem is established.

The standard formula of the binomial theorem $(x + y)^n$ follows immediately from the formula for $(1 + x)^n$, by noting that $(x + y)^n = \{x(1 + y/x)\}^n = x^n(1 + y/x)^n$.

6.2   Strong Induction

Strong induction is another form of mathematical induction, which is often employed when we cannot prove a result with (weak) mathematical induction. It is similar to weak induction in that there is a base step and an inductive step. The base step is identical to weak mathematical induction, and it involves show-ing that the statement is true for some natural number (usually the number 1). The inductive step is a little different, and it involves showing that if the statement is true for all natural numbers less than or equal to an arbitrary number k, then the statement is true for its successor $k + 1$. This is often written as $(P(1) \land P(2) \land \ldots \land P(k)) \to P(k + 1)$.

From the base step and the inductive step, we infer that the statement is true for all natural numbers (that are greater than or equal to the number specified in the base case). Formally, the proof technique used in mathematical induction is of the form[2]:

$$(P(1) \land [(P(1) \land P(2) \land \cdots \land P(k)) \to P(k + 1)]) \to \forall n P(n).$$

Strong and weak mathematical induction are equivalent in that any proof done by weak mathematical induction may also be considered a proof using strong induction, and a proof conducted with strong induction may also be converted into a proof using weak induction.

Weak mathematical induction is generally employed when it is reasonably clear how to prove $P(k + 1)$ from $P(k)$, with strong mathematical typically employed where it is not so obvious. The validity of both forms of mathematical induction follows from the *well-ordering property* of the natural numbers, which states that every non-empty set has a least element.

Well-Ordering Principle
Every non-empty set of natural numbers has a least element. The well-ordering principle is equivalent to the principle of mathematical induction.

Example 6.3 Show that every natural number greater than one is divisible by a prime number.

Proof
Base Case

We consider the case of $n = 2$ which is trivially true, since 2 is a prime number and is divisible by itself.

Inductive Step (strong induction)

Suppose that the result is true for every number less than or equal to k. Then we consider $k + 1$, and there are there are two cases to consider. If $k + 1$ is prime, then it is divisible by itself. Otherwise it is composite, and it may be factored as the product of two numbers each of which is less than or equal to k. Each of these numbers is divisible by a prime number by the strong inductive hypothesis, and so $k + 1$ is divisible by a prime number.

Thus, we have shown that if all natural numbers less than or equal to k are divisible by a prime number, then $k + 1$ is divisible by a prime number. We have shown that

---

[2] As before this definition covers the base case of $n = 1$ and would need to be adjusted if the number specified in the base case is higher.

the base case $P(2)$ is true, and so it follows from strong mathematical induction that every natural numbers greater than one is divisible by some prime number.

6.3   Recursion

Some functions (or objects) used in mathematics (e.g., the Fibonacci sequence) are difficult to define explicitly and are best defined by a *recurrence relation*: (i.e., an equation that recursively defines a sequence of values, once one or more initial values are defined). Recursion may be employed to define functions, sequences, and sets.

There are two parts to a recursive definition, namely the *base case* and the *recursive step*. The base case usually defines the value of the function at $n = 0$ or $n = 1$, whereas the recursive step specifies how the application of the function to a number may be obtained from its application to one or more smaller numbers.

It is important that care is taken with the recursive definition, to ensure that that it is not circular, and does not lead to an infinite regress. The argument of the function on the right-hand side of the definition in the recursive step is usually smaller than the argument on the left-hand side to ensure termination (there are some unusual recursively defined functions such as the *McCarthy* 91 *function* where this is not the case).

It is natural to ask when presented with a recursive definition whether it means anything at all, and in some cases, the answer is negative. Fixed-point theory provides the mathematical foundations for recursion and ensures that the functions/objects are well defined.

Chapter 12 (see Sect. 12.6) discusses various mathematical structures such as partial orders, complete partial orders, and lattices, which may be employed to give a secure foundation for recursion. A precise mathematical meaning is given to recursively defined functions in terms of domains and fixed-point theory, and it is essential that the conditions in which recursion may be used safely be understood. The reader is referred to [1] for more detailed information.

A recursive definition will include at least one non-recursive branch with every recursive branch occurring in a context that is different from the original and brings it closer to the non-recursive case. Recursive definitions are a powerful and elegant way of giving the denotational semantics of language constructs.

Next, we present examples of the recursive definition of the factorial function and Fibonacci numbers.

Example 6.4 (*Recursive Definition of Functions*) The factorial function $n!$ is very common in mathematics, and its well-known definition is $n! = n(n - 1)(n - 2) \ldots 3.2.1$ and $0! = 1$. The formal definition in terms of a base case and inductive step is given as follows:

Base Step   fac $(0) = 1$

Recursive Step   fac $(n) = n * \text{fac}(n - 1)$

This recursive definition defines the procedure by which the factorial of a number is determined from the base case, or by the product of the number by the factorial of its predecessor. The definition of the factorial function is built up in a sequence: fac(0), fac(1), fac(2), ...

The Fibonacci sequence[3] is named after the Italian mathematician Fibonacci, who introduced it in the 13th century. It had been previously described in Indian mathematics, and the Fibonacci numbers are the numbers in the following integer sequence:

$$1, 1, 2, 3, 5, 8, 13, 21, 34$$

Each Fibonacci number (apart from the first two in the sequence) is obtained by adding the two previous Fibonacci numbers in the sequence together. Formally, the definition is given by

Base Step $F_1 = 1, F_2 = 1$
Recursive Step $F_n = F_{n-1} + F_{n-2}$ (Definition for when $n > 2$)

Example 6.5 (*Recursive Definition of Sets and Structures*) Sets and sequences may also be defined recursively, and there are two parts to the recursive definition (as before). The base case specifies an initial collection of elements in the set, whereas the recursive step provides rules for adding new elements to the set based on those already there. Properties of recursively defined sets may often be proved by a technique called structural induction.

Consider the subset S of the natural numbers defined by

Base Step $5 \in S$
Recursive Step   For $x \in S$ then $x + 5 \in S$

Then the elements in S are given by the set of all multiples of 5, as clearly $5 \in S$; therefore by the recursive step $5 + 5 = 10 \in S$; $5 + 10 = 15 \in S$, and so on.

The recursive definition of the set of strings $\Sigma*$ over an alphabet Σ is given by

Base Step $\Lambda \in \Sigma*$ (Λ is the empty string)
Recursive Step   For $\sigma \in \Sigma*$ and $v \in \Sigma$ then $\sigma v \in \Sigma*$

Clearly, the empty string is created from the base step. The recursive step states that a new string is obtained by adding a letter from the alphabet to the end of an existing string in $\Sigma*$. Each application of the inductive step produces a new string

---

[3] We are taking the Fibonacci sequence as starting at 1, whereas others take it as starting at 0.

that contains one additional character. For example, if $\Sigma = \{0, 1\}$, then the strings in $\Sigma*$ are the set of bit strings Λ, 0, 1, 00, 01, 10, 11, 000, 001, 010, etc.

We can define an operation to determine the length of a string (len: $\Sigma* \to \mathbb{N}$) recursively.

Base Step   len $(\Lambda) = 0$
Recursive Step   len $(\sigma v) = $ len$(\sigma) + 1$ (where $\sigma \in \Sigma*$ and $v \in \Sigma$)

A binary tree[4] is a well-known data structure in computer science, and it consists of a root node together with a left and right binary tree. It is defined as a finite set of nodes (starting with the root node), where each node consists of a data value and a link to a left subtree and a right subtree. Recursion is often used to define the structure of a binary tree.

Base Step   A single node is a binary tree (root)
Recursive Step   (i) Suppose X and Y are binary trees and x is a node then XxY
                     is a binary tree, where X is the left subtree, Y the right subtree,
                     and x is the new root node.

                     (ii) Suppose X is a binary tree and x is a node, then xX and Xx
                     are binary trees, which consist of the root node x and a single
                     child left or right subtree.

That is, a binary tree has a root node, and it may have no subtrees; it may consist of a root node with a left subtree only; a root node with a right subtree only; or a root node with both a left and right subtree.

6.4   Structural Induction

Structural induction is a mathematical technique that is used to prove properties about recursively defined sets and structures. It may be used to show that all members of a recursively defined set have a certain property, and there two parts to the proof (as before), namely the base case and the recursive (inductive) step.

The first part of the proof is to show that the property holds for all elements specified in the base case of the recursive definition. The second part of the proof involves showing that if the property is true for all elements used to construct the new elements in the recursive definition, then the property holds for the new elements. From the base case and the recursive step we deduce that the property holds for all elements of the set (structure).

---

[4] We will give an alternate definition of a tree in terms of a connected acyclic graph in Chap. 7 on graph theory.

Example 6.6 (*Structural Induction*) We gave a recursive definition of the subset S of the natural numbers that consists of all multiples of 5. We did not prove that all elements of the set S are divisible by 5, and we use structural induction to prove this.

Base Step	$5 \in S$ (and clearly the base case is divisible by 5)
Inductive Step	Suppose $q \in S$ then $q = 5\,k$ for some k. From the inductive hypothesis $q + 5 \in S$ and $q + 5 = 5\,k + 5 = 5\,(k + 1)$ and so $q + 5$ is divisible by 5
	Therefore, all elements of S are divisible by 5.

6.5   Review Questions

1. Show that $9^n + 7$ is always divisible by 8.
2. Show that the sum of $1^2 + 2^2 + \cdots + n^2 = n(n + 1)(2n + 1)/6$.
3. Explain the difference between strong and weak induction.
4. What is structural induction?
5. Explain how recursion is used in mathematics.
6. Investigate the recursive definition of the McCarthy 91 function, and explain how it differs from usual recursive definitions.
7. Show that $\binom{r}{r} + \binom{n}{r - 1} = \binom{n + 1}{r}$
8. Determine the standard formula for the binomial theorem $(x + y)^n$ from the formula for $(1 + x)^n$.

6.6   Summary

Mathematical induction is an important proof technique that is used to establish the truth of a statement for all natural numbers. There are two parts to a proof by induction, and these are the base case and the inductive step. The base case involves showing that the statement is true for some natural number (usually for the number $n = 1$). The inductive step involves showing that if the statement is true for some natural number $n = k$, then the statement is true for its successor $n = k + 1$.

From the base step and the inductive step, we infer that the statement is true for all natural numbers (that are greater than or equal to the number specified in the base case). Mathematical induction may be used to prove a wide variety of theorems, such as theorems about summation formulae, inequalities, set theory, and the correctness of algorithms and computer programs.

Strong induction is often employed when we cannot prove a result with (weak) mathematical induction. It also has a base case and an inductive step, where the inductive step is a little different, and it involves showing that if the statement is true for all natural numbers less than or equal to an arbitrary number k, then the statement is true for its successor $k + 1$.

Recursion may be employed to define functions, sequences, and sets in mathematics, and there are two parts to a recursive definition, namely the base case and the recursive step. The base case usually defines the value of the function at $n = 0$ or $n = 1$, whereas the recursive step specifies how the application of the function to a number may be obtained from its application to one or more smaller numbers. It is important that care is taken with the recursive definition, to ensure that that it is not circular, and does not lead to an infinite regress.

Structural induction is a mathematical technique that is used to prove properties about recursively defined sets and structures. It may be used to show that all members of a recursively defined set have a certain property, and there are two parts to the proof, namely the base case and the recursive (inductive) step.

Reference

1. Meyer B (1990) Introduction to the theory of programming languages. Prentice Hall

Graph Theory

7

Key Topics

Directed Graphs

Adirected Graphs

Incidence Matrix

Degree of Vertex

Walks and Paths

Hamiltonian Path

Graph Algorithms

7.1 Introduction

Graph theory is a practical branch of mathematics that deals with the arrangements of certain objects known as vertices (or nodes) and the relationships between them. It has been applied to practical problems such as the modelling of computer networks, determining the shortest driving route between two cities, the link structure of a website, the travelling salesman problem, and the four-colour problem.[1]

Consider a map of the London underground, which is issued to users of the underground transport system in London. Then this map does not represent every feature of the city of London, as it includes only material that is relevant to the

---

[1] The four-colour theorem states that given any map it is possible to colour the regions of the map with no more than four colours such that no two adjacent regions have the same colour. This result was finally proved in the mid-1970s.

© The Author(s), under exclusive license to Springer Nature Switzerland AG 2023
G. O'Regan, *Mathematical Foundations of Software Engineering*,
Texts in Computer Science, https://doi.org/10.1007/978-3-031-26212-8_7

Fig. 7.1 Königsberg Seven
Bridges Problem

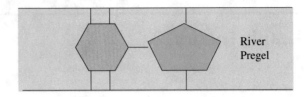

Fig. 7.2 Königsberg graph

users of the London underground transport system. In this map the exact geographical location of the stations is unimportant, and the essential information is how the stations are interconnected to one another, as this allows a passenger to plan a route from one station to another. That is, the map of the London underground is essentially a model of the transport system that shows how the stations are interconnected.

The Seven Bridges of Königsberg[2] (Fig. 7.1) is one of the earliest problems in graph theory. The city was set on both sides of the Pregel River in the early eighteenth century, and it consisted of two large islands that were connected to each other and the mainland by seven bridges. The problem was to find a walk through the city that would cross each bridge once and once only.

Euler showed that the problem had no solution, and his analysis helped to lay the foundations for graph theory as a discipline. The Königsberg problem in graph theory is concerned with the question as to whether it is possible to travel along the edges of a graph starting from a vertex and returning to it and travelling along each edge exactly once. An Euler Path in a graph G is a simple path containing every edge of G.

Euler noted that a walk through a graph traversing each edge exactly once depends on the *degree* of the nodes (i.e., the number of edges touching it). He showed that a necessary and sufficient condition for the walk is that the graph is connected and has zero or two nodes of odd degree. For the Konigsberg graph, the four nodes (i.e., the land masses) have odd degree (Fig. 7.2).

A *graph* is a collection of objects that are interconnected in some way. The objects are typically represented by vertices (or nodes), and the interconnections between them are represented by edges (or lines). We distinguish between directed and adirected graphs, where a *directed graph* is mathematically equivalent to a

---

[2] Königsberg was founded in the thirteenth century by Teutonic knights and was one of the cities of the Hanseatic League. It was the historical capital of East Prussia (historical part of Germany), and it was annexed by Russia at the end of the Second World War. The German population either fled the advancing Red army or were expelled by the Russians in 1949. The city is now called Kaliningrad. The famous German philosopher, Immanuel Kant, spent all his life in the city and is buried there.

binary relation, and an *adirected (undirected) graph* is equivalent to a symmetric binary relations.

7.2 Undirected Graphs

An *undirected graph* (*adirected graph*) (Fig. 7.3) G is a pair of finite sets (V, E) such that E is a binary symmetric relation on V. The set of vertices (or nodes) is denoted by $V(G)$, and the set of edges is denoted by $E(G)$.

A *directed graph* (Fig. 7.4) is a pair of finite sets (V, E) where E is a binary relation (that may not be symmetric) on V. A *directed acyclic graph* (*dag*) is a directed graph that has no cycles. The example below is of a directed graph with three edges and four vertices.

An edge $e \in E$ consists of a pair $< x, y >$ where x, y are adjacent nodes in the graph. The *degree* of x is the number of nodes that are adjacent to x. The set of edges is denoted by $E(G)$, and the set of vertices is denoted by $V(G)$.

A *weighted graph* is a graph $G = (V, E)$ together with a weighting function w: $E \to \mathbb{N}$, which associates a weight with every edge in the graph. A weighting function may be employed in modelling computer networks: for example, the weight of an edge may be applied to model the bandwidth of a telecommunications link between two nodes. Another application of the weighting function is in determining the distance (or shortest path) between two nodes in the graph (where such a path exists).

For an adirected graph the weight of the edge is the same in both directions: i.e., $w(v_i, v_j) = w(v_j, v_i)$ for all edges $< v_i, v_j >$ in the graph G, whereas the weights may be different for a directed graph.

Two vertices x, y are adjacent if $xy \in E$, and x and y are said to be incident to the edge xy. A matrix may be employed to represent the adjacency relationship.

Fig. 7.3 Undirected graph

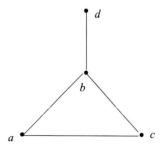

Fig. 7.4 Directed graph

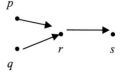

Example 7.1

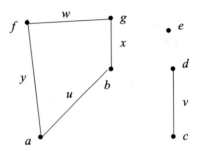

Consider the graph $G = (V, E)$ where $E = \{u = ab, v = cd, w = fg, x = bg, y = af\}$.

An adjacency matrix (Fig. 7.5) may be employed to represent the relationship of adjacency in the graph. Its construction involves listing the vertices in the rows and columns, and an entry of 1 is made in the table if the two vertices are adjacent and 0 otherwise.

Similarly, we can construct a table describing the incidence of edges and vertices by constructing an incidence matrix (Fig. 7.6). This matrix lists the vertices and edges in the rows and columns, and an entry of 1 is made if the vertex is one of the nodes of the edge and 0 otherwise.

Two graphs $G = (V, E)$ and $G' = (V', E')$ are said to be isomorphic if there exists a bijection $f: V \rightarrow V'$ such that for any $u, v \in V$, $uv \in E$, $f(u)f(v) \in E'$. The mapping f is called an isomorphism. Two graphs that are isomorphic are essentially equivalent apart from a relabelling of the nodes and edges.

Let $G = (V, E)$ and $G' = (V', E')$ be two graphs then G' is a *subgraph* of G if $V' \subseteq V$ and $E' \subseteq E$. Given $G = (V, E)$ and $V' \subseteq V$ then we can induce a subgraph $G' = (V', E')$ by restricting G to V' (denoted by $G|_{V'}|$). The set of edges in E' is defined as

$$E' = \{e \in E : e = uv \text{ and } u, v \in V'\}.$$

Fig. 7.5 Adjacency matrix

	a	b	c	d	e	f	g
a	0	1	0	0	0	1	0
b	1	0	0	0	0	0	1
c	0	0	0	1	0	0	0
d	0	0	1	0	0	0	0
e	0	0	0	0	0	0	0
f	1	0	0	0	0	0	1
g	0	1	0	0	0	1	0

Fig. 7.6 Incidence matrix

	u	v	w	x	y
a	1	0	0	0	1
b	1	0	0	1	0
c	0	1	0	0	0
d	0	1	0	0	0
e	0	0	0	0	0
f	0	0	1	0	1
g	0	0	1	1	0

The *degree* of a vertex v is the number of distinct edges incident to v. It is denoted by deg v where

$$\deg v = |\{e \in E : e = vx \text{ for some } x \in V\}|$$
$$= |\{x \in V : vx \in E\}|.$$

A vertex of degree 0 is called an isolated vertex.

Theorem 7.1 Let $G = (V, E)$ be a graph then

$$\sum_{v \in V} \deg v = 2|E|.$$

Proof This result is clear since each edge contributes one to each of the vertex degrees. The formal proof is by induction based on the number of edges in the graph, the basis case is for a graph with no edges (i.e., where every vertex is isolated), and the result is immediate for this case.

The inductive step (strong induction) is to assume that the result is true for all graphs with k or fewer edges. We then consider a graph $G = (V, E)$ with $k + 1$ edges.

Choose an edge $e = xy \in E$ and consider the graph $G' = (V, E')$ where $E' = E \backslash \{e\}$. Then G' is a graph with k edges and therefore letting $\deg' v$ represent the degree of a vertex in G' we have

$$\sum_{v \in V} \deg' v = 2|E'| = 2(|E| - 1) = 2|E| - 2.$$

The degree of x and y are one less in G' than they are in G. That is,

$$\sum_{v \in V} \deg v - 2 = \sum_{v \in V} \deg' v = 2|E| - 2$$
$$\Rightarrow \sum_{v \in V} \deg v = 2|E|.$$

A graph $G = (V, E)$ is said to be *complete* if all the vertices are adjacent : i.e., $E = V \times V$. A graph $G = (V, E)$ is said to be *simple graph* if each edge connects two different vertices, and no two edges connect the same pair of vertices. Similarly, a graph that may have multiple edges between two vertices is termed a *multigraph*.

A common problem encountered in graph theory is determining whether or not there is a route from one vertex to another. Often, once a route has been identified the problem then becomes that of finding the shortest or most efficient route to the destination vertex. A graph is said to be *connected* if for any two given vertices v_1, v_2 in V, there is a path from v_1 to v_2.

Consider a person walking in a forest from A to B where the person does not know the way to B. Often, the route taken will involve the person wandering around aimlessly, and often retracing parts of the route until eventually the destination B is reached. This is an example of a *walk* from v_1 to v_k where in a walk there may be repetition of edges.

If all of the edges of a walk are distinct, then it is called a *trail*. A *path* v_1, v_2, ..., v_k from vertex v_1 to v_k is of length $k-1$ and consists of the sequence of edges $< v_1, v_2 >$, $< v_2, v_3 >$,..., $< v_{k-1}, v_k >$ where each $< v_i, v_{i+1} >$ is an edge in E. The vertices in the path are all distinct apart from possibly v_1 and v_k. The path is said to be a cycle if $v_1 = v_k$. A graph is said to be *acyclic* if it contains no cycles.

Theorem 7.2 Let $G = (V, E)$ be a graph and $W = v_1, v_2, ..., v_k$ be a walk from v_1 to v_k. Then there is a path from v_1 to v_k using only edges of W.

Proof The walk W may be reduced to a path by successively replacing redundant parts in the walk of the form $v_i, v_{i+1} ..., v_j$ where $v_i = v_j$ with v_i. That is, we successively remove cycles from the walk, and this clearly leads to a path (not necessarily the shortest path) from v_1 to v_k.

Theorem 7.3 Let $G = (V, E)$ be a graph and let $u, v \in V$ with $u \neq v$. Suppose that there exists two different paths from u to v in G, then G contains a cycle.

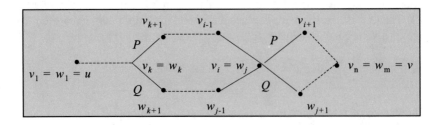

Suppose that $P = v_1, v_2, ..., v_n$ and $Q = w_1, w_2, ..., w_m$ are two distinct paths from u to v (where $u \neq v$), and $u = v_1 = w_1$ and $v = v_n = w_m$.

Fig. 7.7 Travelling salesman problem

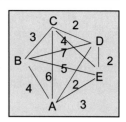

Suppose P and Q are identical for the first k vertices (k could be 1) and then differ (i.e., $v_{k+1} \neq w_{k+1}$). Then Q crosses P again at $v_n = w_m$, and possibly several times before then. Suppose the first occurrence is at $v_i = w_j$ with $k < i \leq n$. Then $w_k, w_{k+1}, w_{k+2}, \ldots, w_j, v_{i-1}, v_{i-2}, \ldots, v_k$ is a closed path (i.e., a cycle) since the vertices are all distinct.

If there is a path from v_1 to v_2, then it is possible to define the *distance* between v_1 and v_2. This is defined to be the total length (number of edges) of the shortest path between v_1 and v_2.

7.2.1 Hamiltonian Paths

A *Hamiltonian path*[3] in a graph $G = (V, E)$ is a path that visits every vertex once and once only. Other words, the length of a Hamiltonian path is $|V|-1$. A graph is Hamiltonian connected if for every pair of vertices there is a Hamiltonian path between the two vertices.

Hamiltonian paths are applicable to the travelling salesman problem, where a salesman[4] wishes to travel to k cities in the country without visiting any city more than once. In principle, this problem may be solved by looking at all of the possible routes between the various cities and choosing the route with the minimal distance.

For example, Fig. 7.7 shows five cities and the connections (including distance) between them. Then, a travelling salesman starting at A would visit the cities in the order AEDCBA (or in reverse order ABCDEA) covering a total distance of 14.

However, the problem becomes much more difficult to solve as the number of cities increases, and there is no general algorithm for its solution. For example, for the case of ten cities, the total number of possible routes is given by $9! = 362,880$, and an exhaustive search by a computer is feasible and the solution may be determined quite quickly. However, for 20 cities, the total number of routes is given by $19! = 1.2 \times 10^{17}$, and in this case it is no longer feasible to do an exhaustive search by a computer.

---

[3] These are named after Sir William Rowan Hamilton, a nineteenth-century Irish mathematician and astronomer, who is famous for discovering quaternions discussed in a later chapter.
[4] We use the term "salesman" to stand for "salesman" or "saleswoman".

There are several sufficient conditions for the existence of a Hamiltonian path, and Theorem 7.4 describes one condition that is sufficient for its existence.

Theorem 7.4 Let $G = (V, E)$ be a graph with $|V| = n$ and such that deg v + deg $w \geq n-1$ for all non-adjacent vertices v and w. Then G possesses a Hamiltonian path.

Proof The first part of the proof involves showing that G is connected, and the second part involves considering the largest path in G of length $k-1$ and assuming that $k < n$. A contradiction is then derived, and it is deduced that $k = n$.

We assume that $G' = (V', E')$ and $G'' = (V'', E'')$ are two connected components of G, then $|V'| + |V''| \leq n$ and so if $v \in V'$ and $w \in V''$ then $n-1 \leq$ deg v + deg $w \leq |V'|-1 + |V''|-1 = |V'| + |V''|-2 \leq n-2$ which is a contradiction, and so G must be connected.

Let $P = v_1, v_2, \ldots, v_k$ be the largest path in G and suppose $k < n$. From this, a contradiction is derived, and the details for are in [1].

7.3   Trees

An acyclic graph is termed a *forest*, and a connected forest is termed a *tree*. A graph G is a tree if and only if for each pair of vertices in G there exists a unique path in G joining these vertices. This is since G is connected and acyclic with the connected property giving the existence of at least one path and the acyclic property giving uniqueness.

A *spanning tree* $T = (V, E')$ for the connected graph $G = (V, E)$ is a tree with the same vertex set V. It is formed from the graph by removing edges from it until it is acyclic (while ensuring that the graph remains connected).

Theorem 7.5 Let $G = (V, E)$ be a tree and let $e \in E$ then $G' = (V, E\backslash\{e\})$ is disconnected and has two components.

Proof Let $e = uv$ then since G is connected and acyclic uv is the unique path from u to v, and thus G' is disconnected since there is no path from u to v in G'.

It is thus clear that there are at least two components in G' with u and v in different components. We show that any other vertex w is connected to u or to v in G'. Since G is connected, there is a path from w to u in G, if this path does not use e, then it is in G' as well, and therefore u and w are in the same component of G'.

If it does use e, then e is the last edge of the graph since u cannot appear twice in the path, and so the path is of the form $w, \ldots, v, u$ in G. Therefore, there is a path from w to v in G', and so w and v are in the same component in G'. Therefore, there are only two components in G'.

Theorem 7.6 Any connected graph $G = (V, E)$ possesses a spanning tree.

Proof This result is proved by considering all connected subgraphs of $(G = V, E)$ and choosing a subgraph T with $|E'|$ as small as possible. The final step is to show that T is the desired spanning tree, and this involves showing that T is acyclic. The details of the proof are left to the reader.

Theorem 7.7 Let $G = (V, E)$ be a connected graph, then G is a tree if and only if $|E| = |V| - 1$.

Proof This result may be proved by induction on the number of vertices $|V|$ and the applications of Theorems 7.5 and 7.6.

7.3.1 Binary Trees

A *binary tree* (Fig. 7.8) is a tree in which each node has at most two child nodes (termed left and right child nodes). A node with children is termed a *parent node*, and the top node of the tree is termed the root node. Any node in the tree can be reached by starting from the root node and by repeatedly taking either the left branch (left child) or right branch (right child) until the node is reached. Binary trees are used in computing to implement efficient searching algorithms (we gave an alternative recursive definition of a binary tree in Chap. 6).

The *depth* of a node is the length of the path (i.e., the number of edges) from the root to the node. The depth of a tree is the length of the path from the root to the deepest node in the tree. A *balanced* binary tree is a binary tree in which the depth of the two subtrees of any node never differs by more than one. The root of the binary tree in Fig. 7.8 is A, and its depth is 4. The tree is unbalanced and unsorted.

Tree traversal is a systematic way of visiting each node in the tree exactly once, and we distinguish between *breadth first search* in which every node on a particular level is visited before going to a lower level, and *depth first search* where one starts at the root and explores as far as possible along each branch before backtracking. The traversal in depth first search may be in preorder, inorder, or postorder.

Fig. 7.8 Binary tree

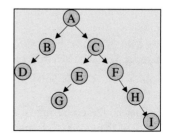

7.4   Graph Algorithms

Graph algorithms are employed to solve various problems in graph theory including network cost minimization problems; construction of spanning trees; shortest path algorithms; longest path algorithms; and timetable construction problems.

A length function $l: E \rightarrow \mathbb{R}$ may be defined on the edges of a connected graph $G = (V, E)$, and a shortest path from u to v in G is a path P with edge set E' such that $l(E')$ is minimal.

The reader may consult the many texts on graph theory to explore many well-known graph algorithms. These include Dijkstra's shortest path algorithm and longest path algorithm, and Kruskal's minimal spanning tree algorithm and Prim's minimal spanning tree algorithms are all described in [1]. Next, we briefly discuss graph colouring.

7.5   Graph Colouring and Four-Colour Problem

It is very common for maps to be coloured in such a way that neighbouring states or countries are coloured differently. This allows different states or countries to be easily distinguished as well as the borders between them. The question naturally arises as to how many colours are needed (or determining the least number of colours needed) to colour the entire map, as it might be expected that a large number of colours would be needed to colour a large complicated map.

However, it may come as a surprise that in fact very few colours are required to colour any map. A former student of the British logician, Augustus De Morgan, had noticed this in the mid-1800s, and he proposed the conjecture of the four-colour theorem. There were various attempts to prove that four colours were sufficient from the mid-1800s onwards, and it remained a famous unsolved problem in mathematics until the late twentieth century.

Kempe gave an erroneous proof of the four-colour problem in 1879, but his attempt led to the proof that five colours are sufficient (which was proved by Heawod in the late 1800s). Appel and Haken of the University of Illinois finally provided the proof that four colours are sufficient in the mid-1970s (using over 1000 h of computer time in their proof).

Each map in the plane can be represented by a graph, with each region of the graph represented by a vertex. Edges connect two vertices if the regions have a common border. The colouring of a graph is the assignment of a colour to each vertex of the graph so that no two adjacent vertices in this graph have the same colour.

Definition Let $G = (V, E)$ be a graph and let C be a finite set called the colours. Then, a colouring of G is a mapping $\kappa: V \rightarrow C$ such that if $uv \in E$ then $\kappa(u) \neq \kappa(v)$.

That is, the colouring of a simple graph is the assignment of a colour to each vertex of the graph such that if two vertices are adjacent, then they are assigned a

Fig. 7.9 Determining the
chromatic colour of G

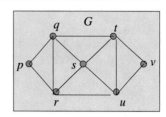

Fig. 7.10 Chromatic
colouring of G

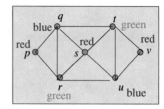

different colour. The chromatic number of a graph is the least number of colours
needed for a colouring of the graph. It is denoted by $\chi(G)$.

Example 7.2 Show that the chromatic colour of the graph G in Fig. 7.9 is 3.

Solution

The chromatic colour of G must be at least 3 since vertices p, q, and r must have
different colours, and so we need to show that three colours are in fact sufficient
to colour G. We assign the colours red, blue, and green to p, q, and r, respectively.
We immediately deduce that the colour of s must be red (as adjacent to q and r).
From this, we deduce that t is coloured green (as adjacent to q and s) and u is
coloured blue (as adjacent to s and t). Finally, v must be coloured red (as adjacent
to u and t). This leads to the colouring of the graph G in Fig. 7.10.

Theorem 7.8 (Four-Colour Theorem) *The chromatic number of a planar graph G
is less than or equal to 4.*

7.6   Review Questions

1. What is a graph and explain the difference between an adirected graph
   and a directed graph.
2. Determine the adjacency and incidence matrices of the following graph
   where $V = \{a, b, c, d, e\}$ and $E = \{ab, bc, ae, cd, bd\}$.

3. Determine if the two graphs G and G' defined below are isomorphic.

$$G = (V, E), \ V = \{a, b, c, d, e, f, g\} \text{ and } E$$
$$= \{ab, ad, ae, bd, ce, cf, dg, fg, bf\}$$

$$G' = (V', E'), \ V' = \{a, b, c, d, e, f, g\} \text{ and } E'$$
$$= \{ab, bc, cd, de, ef, fg, ga, ac, be\}$$

4. What is a binary tree? Describe applications of binary trees.
5. Describe the travelling salesman problem and its applications.
6. Explain the difference between a walk, trail, and path.
7. What is a connected graph?
8. Explain the difference between an incidence matrix and an adjacency matrix.
9. Describe the four-colour problem and its applications.

7.7   Summary

Graph theory is a practical branch of mathematics that deals with the arrangements of vertices and the edges between them. It has been applied to practical problems such as the modelling of computer networks, determining the shortest driving route between two cities, and the travelling salesman problem.

An undirected graph G is a pair of finite sets (V, E) such that E is a binary symmetric relation on V, whereas a directed graph is a binary relation that is not symmetric. An adjacency matrix is used to represent whether two vertices are adjacent to each other, whereas an incidence matrix indicates whether a vertex is part of a particular edge.

A Hamiltonian path in a graph is a path that visits every vertex once and once only. Hamiltonian paths are applicable to the travelling salesman problem, where a salesman wishes to travel to k cities in the country without visiting any city more than once.

Graph colouring arose to answer the question as to how many colours are needed to colour an entire map. It may be expected that many colours would be required, but the four-colour theorem demonstrates that in fact four colours are sufficient to colour a planar graph.

A tree is a connected and acyclic graph, and a binary tree is a tree in which each node has at most two child nodes.

Reference

1. Piff M (1991) Discrete mathematics. An introduction for software engineers. Cambridge University Press

Sequences, Series, and Permutations and Combinations

8

Key Topics

Arithmetic Sequence

Arithmetic Series

Geometric Sequence

Geometric Series

Permutations and Combinations

Counting Principle

Sum and Product Rule

Pigeonhole Principle

8.1 Introduction

The goal of this chapter is to give an introduction to sequences and series, including arithmetic and geometric sequences, and arithmetic and geometric series. We derive formulae for the sum of an arithmetic series and geometric series, and we discuss the convergence of a geometric series when $|r| < 1$, and the limit of its sum as n gets larger and larger.

We consider the counting principle where one operation has m possible outcomes and a second operation has n possible outcomes. We determine that the total number of outcomes after performing the first operation followed by the second operation to be $m \times n$, whereas the total number of outcomes from performing the first operation or the second operation is given by $m + n$. We discuss

© The Author(s), under exclusive license to Springer Nature Switzerland AG 2023
G. O'Regan, *Mathematical Foundations of Software Engineering*,
Texts in Computer Science, https://doi.org/10.1007/978-3-031-26212-8_8

the pigeonhole principle which states that if n items are placed into m containers (with $n > m$), then at least one container must contain more than one item.

A permutation is an arrangement of a given number of objects, by taking some or all of them at a time. The order of the arrangement is important, as the arrangement "abc" is different from "cba".

A combination is a selection of a number of objects in any order, where the order of the selection is unimportant. That is, the selection "abc" is the same as the selection "cba".

8.2   Sequences and Series

A sequence $a_1, a_2, \ldots a_n \ldots$ is any succession of terms (usually numbers). For example, each term in the Fibonacci sequence (apart from the first two terms) is obtained from the sum of the previous two terms in the sequence (see Sect. 6.3 for a formal definition of the Fibonacci sequence).

$$1, 1, 2, 3, 5, 8, 13, 21, \ldots.$$

A sequence may be finite (with a fixed number of terms) or infinite. The Fibonacci sequence is infinite, whereas the sequence 2, 4, 6, 8, 10 is finite. We distinguish between convergent and divergent sequences, where a *convergent* sequence approaches a certain value as n gets larger and larger. That is, we say that $\lim_{n \to \infty} a_n$ exists (i.e., the limit of a_n exists), and otherwise, the sequence is said to be *divergent*.

Often, there is a mathematical expression for the nth term of a sequence (e.g., for the sequence of even integers 2, 4, 6, 8, ... the general expression for a_n is given by $a_n = 2n$). Clearly, the sequence of the even integers is divergent, as it does not approach a particular value, as n gets larger and larger. Consider the following sequence

$$1, -1, 1, -1, 1, -1 \ldots$$

Then this sequence is divergent since it does not approach a certain value, as n gets larger and larger, since it continues to alternate between 1 and -1. The formula for the nth term in the sequence may be given by

$$(-1)^{n+1}$$

The sequence 1, 1/2, 1/3, 1/4, ... 1/n ... is convergent, and it converges to 0. The nth term in the sequence is given by $1/n$, and as n gets larger and larger, it gets closer and closer to 0.

A series is the sum of the terms in a sequence, and the sum of the first n terms of the sequence $a_1, a_2, \ldots a_n \ldots$ is given by $a_1 + a_2 + \cdots + a_n$ which is denoted by

$$\sum_{k=1}^{n} a_k$$

A series is convergent if its sum approaches a certain value S as n gets larger and larger, and this is written formally as

$$\lim_{n \to \infty} \sum_{k=1}^{n} a_k = S$$

Otherwise, the series is said to be divergent.

8.3 Arithmetic and Geometric Sequences

Consider the sèquence 1, 4, 7, 10, ..., where each term is obtained from the previous term by adding the constant value 3. This is an example of an arithmetic sequence, and there is a difference of 3 between any term and the previous one. The general form of a term in this sequence is $a_n = 3n - 2$.

The general form of an *arithmetic sequence* is given by

$$a, a + d, a + 2d, a + 3d, \ldots a + (n-1)d, \ldots.$$

The value a is the initial term in the sequence, and the value d is the constant difference between a term and its successor. For the sequence, 1, 4, 7, ..., we have $a = 1$ and $d = 3$, and the sequence is not convergent. In fact, all arithmetic sequences (apart from the constant sequence a, a, ... a which converges to a) are divergent.

Consider the sequence 1, 3, 9, 27, 81, ..., where each term is achieved from the previous term by multiplying by the constant value 3. This is an example of a geometric sequence, and the general form of a geometric sequence is given by

$$a, ar, ar^2, ar^3, \ldots, ar^{n-1}.$$

The first term in the geometric sequence is a and r is the common ratio. Each term is obtained from the previous one by multiplying by the common ratio r. For the sequence 1, 3, 9, 27 the value of a is 1 and r is 3.

A geometric sequence is convergent if $r < 1$, and for this case it converges to 0. It is also convergent if $r = 1$, as for this case it is simply the constant sequence a, a, a, ..., which converges to a. For the case where $r > 1$ the sequence is divergent.

8.4   Arithmetic and Geometric Series

An arithmetic series is the sum of the terms in an arithmetic sequence, and a geometric sequence is the sum of the terms in a geometric sequence. It is possible to derive a simple formula for the sum of the first n terms in an arithmetic and geometric series.

Arithmetic Series

We write the series two ways: first the normal left to right addition and then the reverse, and then we add both series together.

$$S_n = a \qquad\qquad + (a + d) + (a + 2d) + (a + 3d) + \cdots + (a + (n - 1))d$$
$$S_n = a + (n - 1)d + a + (n - 2)d + \cdots + \qquad +(a + d) + a$$

- -

$$2S_n = [2a + (n - 1)d] + [2a + (n - 1)d] + \cdots + [2a + (n - 1)d] \quad (n\,\text{times})$$
$$2S_n = n \times [2a + (n - 1)d]$$

Therefore, we conclude that

$$S_n = \frac{n}{2}[2a + (n - 1)d]$$

Example 8.1 (*Arithmetic Series*) Find the sum of the first n terms in the following arithmetic series $1, 3, 5, 7, 9, \ldots$.

Solution

Clearly, $a = 1$ and $d = 2$. Therefore, applying the formula we get

$$S_n = \frac{n}{2}[2.1 + (n - 1)2] = \frac{2n^2}{2} = n^2$$

Geometric Series

For a geometric series we have

$$S_n = a + ar + ar^2 + ar^3 + \cdots + ar^{n-1}$$
$$\Rightarrow r S_n = \qquad ar + ar^2 + ar^3 + \cdots + ar^{n-1} + ar^n$$

- -

$$\Rightarrow r S_n - S_n \; = ar^n - a$$
$$= a(r^n - 1)$$
$$\Rightarrow (r - 1)S_n \; = a(r^n - 1).$$

Therefore, we conclude that (where $r \neq 1$) that

$$S_n = \frac{a(r^n - 1)}{r - 1} = \frac{a(1 - r^n)}{1 - r}$$

The case of when $r = 1$ corresponds to the arithmetic series $a + a + \cdots + a$, and the sum of this series is simply na. The geometric series converges when $|r| < 1$ as $r^n \to 0$ as $n \to \infty$, and so

$$S_n \to \frac{a}{1 - r} \quad \text{as } n \to \infty$$

Example 8.2 (*Geometric Series*) Find the sum of the first n terms in the following geometric series $1, {}^1/_2, {}^1/_4, {}^1/_8, \ldots$. What is the sum of the series?

Solution

Clearly, $a = 1$ and $r = {}^1/_2$. Therefore, applying the formula we get

$$S_n = \frac{1(1 - 1/2^n)}{1 - 1/2} = \frac{(1 - 1/2^n)}{1 - 1/2} = 2(1 - 1/2^n)$$

The sum of the series is the limit of the sum of the first n terms as n approaches infinity. This is given by

$$\lim_{n \to \infty} S_n = \lim_{n \to \infty} 2(1 - 1/2^n) = 2$$

8.5   Permutations and Combinations

A permutation is an arrangement of a given number of objects, by taking some or all of them at a time. A combination is a selection of a number of objects where the order of the selection is unimportant. Permutations and combinations are defined in terms of the factorial function, which is defined as:

$$n! = n(n - 1) \ldots 3.2.1.$$

Principles of Counting

(a) Suppose one operation has m possible outcomes and a second operation has n possible outcomes, then the total number of possible outcomes when performing the first operation **followed by** the second operation is $m \times n$. (***Product Rule***).

(b) Suppose one operation has m possible outcomes and a second operation has n possible outcomes then the possible outcomes of the first operation **or** the second operation is given by $m + n$. (***Sum Rule***).

Example 8.3 (*Counting Principle (a)*) Suppose a dice is thrown and a coin is then tossed. How many different outcomes are there and what are they?

Solution
There are six possible outcomes from a throw of the dice: 1, 2, 3, 4, 5, or 6, and
two possible outcomes from the toss of a coin: H or T. Therefore, the total number
of outcomes is determined from the product rule as $6 \times 2 = 12$. The outcomes are
given by

$$(1, H), (2, H), (3, H), (4, H), (5, H), (6, H), (1, T), (2, T),$$
$$(3, T), (4, T), (5, T), (6, T).$$

Example 8.4 (*Counting Principle (b)*) Suppose a dice is thrown and if the number
is even a coin is tossed and if it is odd then there is a second throw of the dice. How
many different outcomes are there?

Solution
   There are two experiments involved with the first experiment involving an even
number and a toss of a coin. There are three possible outcomes that result in an even
number and two outcomes from the toss of a coin. Therefore, there are $3 \times 2 = 6$
outcomes from the first experiment.
   The second experiment involves an odd number from the throw of a dice and the
further throw of the dice. There are three possible outcomes that result in an odd
number and six outcomes from the throw of a dice. Therefore, there are $3 \times 6 = 18$
outcomes from the second experiment.
   Finally, there are six outcomes from the first experiment and 18 outcomes from
the second experiment, and so from the sum rule there are a total of $6 + 18 = 24$
outcomes.

Pigeonhole Principle
The pigeonhole principle states that if n items are placed into m containers (with
$n > m$), then at least one container must contain more than one item.

Example 8.5 (*Pigeonhole Principle*)

(a) Suppose there is a group of 367 people, then there must be at least two people
    with the same birthday.
    This is clear as there are 365 days in a year (with 366 days in a leap year), and
    so as there are at most 366 possible birthdays in a year. The group size is 367
    people, and so there must be at least two people with the same birthday.[1]
(b) Suppose that a class of 102 students are assessed in an examination (the outcome
    from the exam is a mark between 0 and 100). Then, there are at least two students
    who receive the same mark.

---

[1] The *birthday paradox* is the unintuitive result that in a group as small as 23 people the probability
that there is a pair of people with the same birthday is above 0.5 (over 50%).

This is clear as there are 101 possible outcomes from the test (as the mark that a student may achieve is between 0 and 100), and as there are 102 students in the class and 101 possible outcomes from the test, then there must be at least two students who receive the same mark.

Permutations

A permutation is an arrangement of a number of objects in a definite order.

Consider the three letters A, B, and C. If these letters are written in a row, then there are six possible arrangements

ABC ACB BAC BCA CAB CBA.

There is a choice of three letters for the first place, then there is a choice of two letters for the second place, and there is only one choice for the third place. Therefore, there are $3 \times 2 \times 1 = 6$ arrangements.

If there are n different objects to arrange, then the total number of arrangements (permutations) of n objects is given by $n! = n(n-1)(n-2) \dots 3.2.1$.

Consider the four letters A, B, C, and D. How many arrangements (taking two letters at a time with no repetition) of these letters can be made?

There are four choices for the first letter and three choices for the second letter, and so there are 12 possible arrangements. These are given by:

AB AC AD BA BC BD CA CB CD DA DB DC.

The total number of arrangements of n different objects taking r at a time ($r \le n$) is given by $^nP_r = n(n-1)(n-2) \dots (n-r+1)$. It may also be written as:

$$^nP_r = \frac{n!}{(n-r)!}$$

Example 8.6 (*Permutations*) Suppose A, B, C, D, E, and F are six students. How many ways can they be seated in a row if:

(i) There is no restriction on the seating.
(ii) A and B must sit next to one another.
(iii) A and B must not sit next to one another.

Solution

For unrestricted seating the number of arrangements is given by $6.5.4.3.2.1 = 6! = 720$.

For the case where A and B must be seated next to one another, then consider A and B as one person, and then the five people may be arranged in $5! = 120$ ways. There are $2! = 2$ ways in which AB may be arranged, and so there are $2! \times 5! = 240$ arrangements.

AB	C	D	E	F

For the case where A and B must not be seated next to one another, then this is given by the difference between the total number of arrangements and the number of arrangements with A and B together: i.e., $720 - 240 = 480$.

Combintations

A combination is a selection of a number of objects in any order, and the order of the selection is unimportant, in that both AB and BA represent the same selection.

The total number of arrangements of n different objects taking r at a time is given by ^nP_r, and the number of ways that r objects can be selected from n different objects may be determined from this, since each selection may be permuted $r!$ times.

That is, the total number of arrangements is $r! \times$ total number of combinations. That is, $^nP_r = r! \times {}^nC_r$, and we may also write this as:

$$\binom{n}{r} = \frac{n!}{r!(n-r)!} = \frac{n(n-1)\ldots(n-r+1)}{r!}$$

It is clear from the definition that

$$\binom{n}{r} = \binom{n}{n-r}$$

Example 8.7 (*Combinations*) How many ways are there to choose a team of 11 players from a panel of 15 players?

Solution

Clearly, the number of ways is given by $\binom{15}{11} = \binom{15}{4}$

That is, $15.14.13.12/4.3.2.1 = 1365$.

Example 8.8 (*Combinations*) How many ways can a committee of four people be chosen from a panel of ten people where

(i)   There is no restriction on membership of the panel.
(ii) A certain person must be a member.
(iii) A certain person must not be a member.

Solution

For (*i*) with no restrictions on membership the number of selections of a committee of four people from a panel of ten people is given by: $\binom{10}{4} = 210$.

For (*ii*) where one person must be a member of the committee then this involves choosing three people from a panel of nine people and is given by: $\binom{9}{3} = 84$.

For (*iii*) where one person must not be a member of the committee then this involves choosing four people from a panel of nine people and is given by: $\binom{9}{4} = 126$.

8.6   Review Questions

1. Determine the formula for the general term and the sum of the following arithmetic sequence

$$1, 4, 7, 10, \ldots.$$

2. Write down the formula for the *n*th term in the following sequence

$$1/4, 1/12, 1/36, 1/108, \ldots.$$

3. Find the sum of the following geometric sequence

$$1/3, 1/6, 1/12, 1/24, \ldots.$$

4. How many different five-digit numbers can be formed from the digits 1, 2, 3, 4, 5 where:
   (i)   No restrictions on digits and repetitions allowed.
   (ii) The number is odd and no repetitions are allowed.
   (iii) The number is even and repetitions are allowed.
5.
   (i)   How many ways can a group of five people be selected from nine people?
   (ii) How many ways can a group be selected if two particular people are always included?
   (iii) How many ways can a group be selected if two particular people are always excluded?

8.7   Summary

This chapter provided a brief introduction to sequences and series, including arithmetic and geometric sequences, and arithmetic series and geometric series. We derived formulae for the sum of an arithmetic series and geometric series, and we discussed the convergence of a geometric series when $|r| < 1$.

We considered counting principles including the product and sum rules. The product rule is concerned with where one operation has m possible outcomes and a second operation has n possible outcomes then the total number of possible outcomes when performing the first operation followed by the second operation is $m \times n$.

We discussed the pigeonhole principle, which states that if n items are placed into m containers (with $n > m$), then at least one container must contain more than one item. We discussed permutations and combinations where permutations are an arrangement of a given number of objects, by taking some or all of them at a time. A combination is a selection of a number of objects in any order, and the order of the selection is unimportant.

A Short History of Logic

9

Key Topics

Syllogistic Logic

Fallacies

Paradoxes

Stoic Logic

Boole's Symbolic Logic

Digital Computing

Propositional Logic

Predicate Logic

Universal and Existential Quantifiers

9.1 Introduction

Logic is concerned with reasoning and with establishing the validity of arguments. It allows conclusions to be deduced from premises according to logical rules, and the logical argument establishes the truth of the conclusion provided that the premises are true.

The origins of logic are with the Greeks who were interested in the nature of truth. The sophists (e.g., Protagoras and Gorgias) were teachers of rhetoric, who taught their pupils techniques in winning an argument and convincing an audience. Plato explores the nature of truth in some of his dialogues, and he is critical of the position of the sophists who argue that there is no absolute truth and that truth

© The Author(s), under exclusive license to Springer Nature Switzerland AG 2023
G. O'Regan, *Mathematical Foundations of Software Engineering*,
Texts in Computer Science, https://doi.org/10.1007/978-3-031-26212-8_9

instead is always relative to some frame of reference. The classic sophist position is stated by Protagoras "*Man is the measure of all things: of things which are, that they are, and of things which are not, that they are not.*" Other words, what is true for you is true for you, and what is true for me is true for me.

Socrates had a reputation for demolishing an opponents position, and the Socratean enquiry consisted of questions and answers in which the opponent would be led to a conclusion incompatible with his original position. The approach was similar to a *reductio ad absurdum* argument, although Socrates was a moral philosopher who did no theoretical work on logic.

Aristotle did important work on logic, and he developed a system of logic, called *syllogistic logic*, that remained in use up to the nineteenth century. Syllogistic logic is a 'term-logic', with letters used to stand for the individual terms. A syllogism consists of two premises and a conclusion, where the conclusion is a valid deduction from the two premises. Aristotle also did some early work on modal logic.

The Stoics developed an early form of propositional logic, where the assertable (propositions) have a truth-value such that at any time they are either true or false. The assertable may be simple or non-simple, and various connectives such as conjunctions, disjunctions, and implication are used in forming more complex assertables.

George Boole developed his symbolic logic in the mid-1800s, and it later formed the foundation for digital computing. Boole argued that logic should be considered as a separate branch of mathematics, rather than a part of philosophy. He argued that there are mathematical laws to express the operation of reasoning in the human mind, and he showed how Aristotle's syllogistic logic could be reduced to a set of algebraic equations.

Frege is considered (along with Boole) to be one of the founders of modern logic. He also made important contributions to the foundations of mathematics, and he attempted to show that all of the basic truths of mathematics (or at least of arithmetic) could be derived from a limited set of logical axioms.

Logic plays a key role in reasoning and deduction in mathematics, but it is considered a separate discipline to mathematics. There were attempts in the early twentieth century to show that all mathematics can be derived from formal logic, and that the formal system of mathematics would be complete, with all the truths of mathematics provable in the system (see Chap. 14). However, this program failed when the Austrian logician, Kurt Goedel, showed that the first-order arithmetic is incomplete.

9.2   Syllogistic Logic

Early work on logic was done by Aristotle in the fourth century B.C. in the *Organon* [1]. Aristotle regarded logic as a useful tool of enquiry into any subject, and his *syllogistic logic* provides more rigour in reasoning. This is a form of reasoning in which a conclusion is drawn from two premises, where each premise is in a subject-predicate form. A common or middle term is present in each of the two premises but not in the conclusion. For example:

All Greeks are mortal.

Socrates is a Greek

— — — — — — — —

Therefore Socrates is mortal

The common (or middle) term in this example is 'Greek'. It occurs in both premises but not in the conclusion. The above argument is valid, and Aristotle studied and classified the various types of syllogistic arguments to determine those that were valid or invalid. Each premise contains a subject and a predicate, and the middle term may act as subject or a predicate. Each premise is a positive or negative affirmation, and an affirmation may be universal or particular. The universal and particular affirmations and negatives are described in Table 9.1.

This leads to four basic forms of syllogistic arguments (Table 9.2) where the middle is the subject of both premises; the predicate of both premises; and the subject of one premise and the predicate of the other premise.

There are four types of premises (A, E, I, O) and therefore sixteen sets of premise pairs for each of the forms above. However, only some of these premise pairs will yield a valid conclusion. Aristotle went through every possible premise pair to determine if a valid argument may be derived. The syllogistic argument above is of form (iv) and is valid

G A M

S I G

— — —

S I M.

Table 9.1 Types of syllogistic premises

Type	Symbol	Example
Universal affirmative	G A M	All Greeks are mortal
Universal negative	G E M	No Greek is mortal
Particular affirmative	G I M	Some Greek is mortal
Particular negative	G O M	Some Greek is not mortal

Table 9.2 Forms of syllogistic premises

	Form (i)	Form (ii)	Form (iii)	Form (iv)
Premise 1	M P	P M	P M	M P
Premise 2	M S	S M	M S	S M
Conclusion	S P	S P	S P	S P

Syllogistic logic is a 'term-logic' with letters used to stand for the individual terms. Syllogistic logic was the first attempt at a science of logic, and it remained in use up to the nineteenth century. There are many limitations to what it may express and on its suitability as a representation of how the mind works.

9.3   Paradoxes and Fallacies

A paradox is a statement that apparently contradicts itself, and it presents a situation that appears to defy logic. Some logical paradoxes have a solution, whereas others are contradictions or invalid arguments. There are many examples of paradoxes, and they often arise due to self-reference in which one or more statements refer to each other. We discuss several paradoxes such as the *liar paradox* and the *sorites paradox*, which were invented by Eubulides of Miletus, and the *barber paradox*, which was introduced by Russell to explain the contradictions in naïve set theory.

An example of the *liar paradox* is the statement "Everything that I say is false", which is made by the liar. This looks like a normal sentence, but it is also saying something about itself as a sentence. If the statement is true, then the statement must be false, since the meaning of the sentence is that every statement (including the current statement) made by the liar is false. If the current statement is false, then the statement that everything that I say is false is false, and so this must be a true statement.

The *Epimenides paradox* is a variant of the liar paradox. Epimenides was a Cretan who allegedly stated *"All Cretans are liars"*. If the statement is true, then since Epimenides is Cretan, he must be a liar, and so the statement is false and we have a contradiction. However, if we assume that the statement is false and that Epimenides is lying about all Cretan being liars, then we may deduce (without contradiction) that there is at least one Cretan who is truthful. So in this case the paradox can be avoided.

The *sorites paradox* (paradox of the heap) involves a heap of sand in which grains are individually removed. It is assumed that removing a single grain of sand does not turn a heap into a non-heap, and the paradox is to consider what happens after when the process is repeated often enough. Is a single remaining grain a heap? When does it change from being a heap to a non-heap? This paradox may be avoided by specifying a fixed boundary of the number of grains of sand required to form a heap, or to define a heap as a collection of multiple grains ($\geq$ 2 grains). Then any collection of grains of sand less than this boundary is not a heap.

The *barber paradox* is a variant of Russell's paradox (a contradiction in naïve set theory). In a village there is a barber who shaves everyone who does not shave himself, and no one else. Who shaves the barber? The answer to this question results in a contradiction, as the barber cannot shave himself, since he shaves only those who do not shave themselves. Further, as the barber does not shave himself, then he falls into the group of people who would be shaved by the barber (himself). Therefore, we conclude that there is no such barber (or that the barber has a beard).

The purpose of a debate is to convince an audience of the correctness of your position and to challenge and undermine your opponent's position. Often, the arguments made are factual, but occasionally individuals skilled in rhetoric and persuasion introduce bad arguments as a way to persuade the audience. Aristotle studied and classified bad arguments (known as *fallacies*), and these include fallacies such as the *ad hominem* argument; the *appeal to authority* argument; and the *straw man* argument. The fallacies are described in more detail in Table 9.3.

9.4 Stoic Logic

The Stoic school[1] was founded in the Hellenistic period by Zeno of Citium (in Cyprus) in the late 4th/early third century B.C. The school presented its philosophy as a way of life, and it emphasized ethics as the main focus of human knowledge. The Stoics stressed the importance of living a good life in harmony with nature (Fig. 9.1).

The Stoics recognized the importance of reason and logic, and Chrysippus, the head of the Stoics in the third century B.C., developed an early version of propositional logic. This was a system of deduction in which the smallest unanalyzed expressions are assertables (the Stoic equivalent of propositions). The assertables have a truth-value such that at any moment of time they are either true or false. True assertables are viewed as facts in the Stoic system of logic, and false assertables are defined as the contradictories of true ones.

Truth is temporal, and assertions may change their truth-value over time. The assertables may be simple or non-simple (more than one assertible), and there may be present tense, past tense, and future tense assertables. Chrysippus distinguished between simple and compound propositions, and he introduced a set of logical connectives for conjunction, disjunction, and implication that are used to form non-simple assertables from existing assertables.

The conjunction connective is of the form '*both .. and ..*', and it has two conjuncts. The disjunction connective is of the form '*either .. or .. or ..*', and it consists of two or more disjuncts. Conditionals are formed from the connective '*if.., ..*' and they consist of an antecedent and a consequence.

His deductive system included various logical argument forms such as *modus ponens* and *modus tollens*.[2] His propositional logic differed from syllogistic logic, in that the Stoic logic was based on propositions (or statements) as distinct from

---

[1] The origin of the word Stoic is from the *Stoa Poikile* (Στοα Ποιλικη), which was a covered walkway in the Agora of Athens. Zeno taught his philosophy in a public space at this location, and his followers became known as Stoics.

[2] Modus ponens is a rule of inference where from P and $P \rightarrow Q$ we can deduce Q, whereas modus tollens is a rule of inference where from $P \rightarrow Q$ and $\neg Q$ we can deduce $\neg P$.

Table 9.3   Fallacies in arguments

Fallacy	Description/Example
Hasty/accident generalization	This is a bad argument that involves a generalization that disregards exceptions
Slippery slope	This argument outlines a chain reaction leading to a highly undesirable situation that will occur if a certain situation is allowed. The claim is that even if one step is taken onto the slippery slope, then we will fall all the way down to the bottom
Against the person *Ad Hominem*	The focus of this argument is to attack the person rather than the argument that the person has made
Appeal to people *Ad Populum*	This argument involves an appeal to popular belief to support an argument, with a claim that the majority of the population supports this argument. However, popular opinion is not always correct
Appeal to authority (*Ad Verecundiam*)	This argument is when an appeal is made to an authoritative figure to support an argument and where the authority is not an expert in this area
Appeal to pity (*Ad Misericordiam*)	This is where the arguer tries to get people to accept a conclusion by making them feel sorry for someone
Appeal to ignorance	The arguer makes the case that there is no conclusive evidence on the issue at hand and that therefore his conclusion should be accepted
Straw man argument	The arguer sets up a version of an opponent's position of his argument and defeats this watered down version of his opponent's position rather than the real subject of the argument
Begging the question (Petitio Principii)	This is a circular argument where the arguer relies on a premise that says the same thing as the conclusion and without providing any real evidence for the conclusion
Red herring	The arguer goes off on a tangent that has nothing to do with the argument in question
False dichotomy	The arguer presents the case that there are only two possible outcomes (often there are more). One of the possible outcomes is then eliminated leading to the desired outcome. The argument suggests that there is only one outcome

Aristotle's term-logic. However, he could express the universal affirmation in syllogistic logic (e.g., All *A*s are *B*) by rephrasing it as a conditional statement that if something is *A* then it is *B*.

Fig. 9.1 Zeno of Citium

Chrysippus's propositional logic did not replace Aristotle's syllogistic logic, and syllogistic logic remained in use up to the mid-nineteenth century. George Boole developed his symbolic logic in the mid-1800s, and this logic is discussed in the next section.

9.5 Boole's Symbolic Logic

George Boole was born in Lincoln, England, in 1815. His father (a cobbler who was interested in mathematics and optical instruments) taught him mathematics and showed him how to make optical instruments. Boole inherited his father's interest in knowledge, and he was self-taught in mathematics and Greek. He taught at various schools near Lincoln, and he developed his mathematical knowledge by working his way through Newton's Principia, as well as applying himself to the work of mathematicians such as Laplace and Lagrange.

He developed his symbolic algebra, which is the foundation for modern computing, and he is considered (along with Babbage) to be one of the grandfathers of computing. His work was theoretical, and he never actually built a computer or calculating machine. *However, Boole's symbolic logic was the perfect mathematical model for switching theory and for the design of digital circuits.*

Boole published a pamphlet titled "Mathematical Analysis of Logic" in 1847 [2]. This short book developed novel ideas on a logical method, and he argued that logic should be considered as a separate branch of mathematics, rather than a part of philosophy. He argued that there are mathematical laws to express the operation of reasoning in the human mind, and he showed how Aristotle's syllogistic logic could be reduced to a set of algebraic equations. He corresponded regularly on logic with Augustus De Morgan.[3]

---

[3] De Morgan was a 19th British mathematician based at University College London. De Morgan's laws in Set Theory and Logic state that: $(A \cup B)^c = A^c \cap B^c$ and $\neg (A \vee B) \equiv \neg A \wedge \neg B$.

He introduced two quantities "0" and "1" with the quantity 1 used to represent the universe of thinkable objects (i.e., the universal set), and the quantity 0 represents the absence of any objects (i.e., the empty set). He then employed symbols such as x, y, z, etc., to represent collections or classes of objects given by the meaning attached to adjectives and nouns. Next, he introduced three operators $(+, -$, and $\times)$ that combined classes of objects.

He showed that these symbols obeyed a rich collection of algebraic laws and could be added, multiplied, etc., in a manner that is similar to real numbers. These symbols may be used to reduce propositions to equations, and algebraic rules may be employed to solve the equations.

Boole applied the symbols to encode Aristotle's syllogistic logic, and he showed how the syllogisms could be reduced to equations. This allowed conclusions to be derived from premises by eliminating the middle term in the syllogism. He refined his ideas on logic further in his book *"An Investigation of the Laws of Thought"* [3]. This book aimed to identify the fundamental laws underlying reasoning in the human mind and to give expression to these laws in the symbolic language of a calculus.

He considered the equation $x^2 = x$ to be a fundamental laws of thought. It allows the principle of contradiction to be expressed (i.e., for an entity to possess an attribute and at the same time not to possess it: i.e., $x - x^2 = 0$ or equivalently $x(1 - x) = 0$).

Boole's logic appeared to have no practical use, but this changed with Claude Shannon's 1937 Master's Thesis, which showed its applicability to switching theory and to the design of digital circuits.

9.5.1   Switching Circuits and Boolean Algebra

Claude Shannon showed in his famous Master's Thesis (*"A Symbolic Analysis of Relay and Switching Circuits)"* [4] that Boole's symbolic algebra provided the perfect mathematical model for switching theory and for the design of digital circuits. He realized that you could combine switches in circuits in such a manner as to carry out symbolic logic operations. This allowed binary arithmetic and more complex mathematical operations to be performed by relay circuits. He designed a circuit, which could add binary numbers, and he later designed circuits that could make comparisons and thus be capable of performing a conditional statement. *This was the birth of digital logic and the digital computing age.*

He showed that the binary digits (i.e., 0 and 1) can be represented by electrical switches. The implications of this were enormous, as it allowed binary arithmetic and more complex mathematical operations to be performed by relay circuits. This provided electronics engineers with the mathematical tool they needed to design digital electronic circuits and provided the foundation of digital electronic design.

His *Master's Thesis is a key milestone in computing*, and it shows how to lay out circuits according to Boolean principles. It provides the theoretical foundation of switching circuits, and *his insight of using the properties of electrical switches to do Boolean logic is the basic concept that underlies all electronic digital computers.*

The use of the properties of electrical switches to process logic is the basic concept that underlies all modern electronic digital computers. Digital computers use the binary digits 0 and 1, and Boolean logical operations may be implemented by electronic AND, OR, and NOT gates. More complex circuits (e.g., arithmetic) may be designed from these fundamental building blocks.

9.6 Frege

Gottlob Frege (Fig. 9.2) was a German mathematician and logician who is considered (along with Boole) to be one of the founders of modern logic. He also made important contributions to the foundations of mathematics, and he attempted to show that all of the basic truths of mathematics (or at least of arithmetic) could be derived from a limited set of logical axioms (this approach is known as *logicism*).

He invented predicate logic and the universal and existential quantifiers, and predicate logic was a significant advance on Aristotle's syllogistic logic. Predicate logic is described in more detail in the next chapter.

Frege's first logical system contained nine axioms and one rule of inference. It was the first axiomization of logic, and it was complete in its treatment of propositional logic and first-order predicate logic. He published several important books on logic, including *Begriffsschrift* (term writing) in 1879; *Die Grundlagen der Arithmetik* (The Foundations of Arithmetic) in 1884; and the two-volume work *Grundgesetze der Arithmetik* (Basic Laws of Arithmetic), which were published in 1893 and 1903. These books described his invention of axiomatic predicate logic; the use of quantified variables; and the application of his logic to the foundations of arithmetic.

Fig. 9.2 Gottlob Frege

Frege presented his predicate logic in his books, and he began to use it to define the natural numbers and their properties. He had intended producing three volumes of the Basic Laws of Arithmetic, with the later volumes dealing with the real numbers and their properties. However, Bertrand Russell discovered a contradiction in Frege's system, which he communicated to Frege shortly before the publication of the second volume. Frege was astounded by the contradiction and he struggled to find a satisfactory solution, and Russell later introduced the theory of types in the *Principia Mathematica* as a solution.

9.7   Review Questions

1. What is logic?
2. What is a fallacy?
3. Give examples of fallacies in arguments in natural language (e.g., in politics, marketing, debates).
4. Investigate some of the early paradoxes (e.g., the Tortoise and Achilles paradox or the arrow in flight paradox) and give your interpretation of the paradox.
5. What is syllogistic logic and explain its relevance.
6. What is stoic logic and explain its relevance.
7. Explain the significance of the equation $x^2 = x$ in Boole's symbolic logic.
8. Describe how Boole's symbolic logic provided the foundation for digital computing.
9. Describe Frege's contributions to logic.

9.8   Summary

This chapter gave a short introduction to logic, and logic is concerned with reasoning and with establishing the validity of arguments. It allows conclusions to be deduced from premises according to logical rules, and the logical argument establishes the truth of the conclusion provided that the premises are true.

The origins of logic are with the Greeks who were interested in the nature of truth. Aristotle did important work on logic, and he developed a system of logic, *syllogistic logic*, that remained in use up to the nineteenth century. Syllogistic logic is a 'term-logic', with letters used to stand for the individual terms. A syllogism consists of two premises and a conclusion, where the conclusion is a valid deduction from the two premises. He also did some early work on modal logic.

The Stoics developed an early form of propositional logic, where the assertables (propositions) have a truth-value such that at any time they are either true or false.

George Boole developed his symbolic logic in the mid-1800s, and it later formed the foundation for digital computing. Boole argued that logic should be considered as a separate branch of mathematics, rather than a part of philosophy. He argued that there are mathematical laws to express the operation of reasoning in the human mind, and he showed how Aristotle's syllogistic logic could be reduced to a set of algebraic equations.

Gottlob Frege made important contributions to logic and to the foundations of mathematics. He attempted to show that all of the basic truths of mathematics (or at least of arithmetic) could be derived from a limited set of logical axioms (this approach is known as *logicism*). He invented predicate logic and the universal and existential quantifiers, and predicate logic was a significant advance on Aristotle's syllogistic logic.

References

1. Ackrill JL (1994) Aristotle the philosopher. Clarendon Press Oxford
2. Boole G (1848) The calculus of logic. Camb Dublin Math J 3:183–198
3. Boole G (1958) An investigation into the laws of thought. Dover Publications (First published in 1854)
4. Shannon C (1937) A symbolic analysis of relay and switching circuits. Masters Thesis, Massachusetts Institute of Technology

Propositional and Predicate Logic

10

Key Topics

Propositions

Truth Tables

Semantic Tableaux

Natural Deduction

Proof

Predicates

Universal Quantifiers

Existential Quantifiers

10.1 Introduction

Logic is the study of reasoning and the validity of arguments, and it is concerned with the truth of statements (propositions) and the nature of truth. Formal logic is concerned with the form of arguments and the principles of valid inference. Valid arguments are truth preserving, and for a valid deductive argument the conclusion will always be true if the premises are true.

Propositional logic is the study of propositions, where a proposition is a statement that is either true or false. Propositions may be combined with other propositions (with a logical connective) to form compound propositions. Truth tables are used to give operational definitions of the most important logical connectives, and they provide a mechanism to determine the truth-values of more complicated logical expressions. Truth tables are used to give operational definitions of the most important logical connectives, and they provide a mechanism to determine the truth-values of more complicated logical expressions.

© The Author(s), under exclusive license to Springer Nature Switzerland AG 2023
G. O'Regan, *Mathematical Foundations of Software Engineering*,
Texts in Computer Science, https://doi.org/10.1007/978-3-031-26212-8_10

Propositional logic may be used to encode simple arguments that are expressed in natural language and to determine their validity. The validity of an argument may be determined from truth tables, or using the inference rules such as modus ponens to establish the conclusion via deductive steps.

Predicate logic is richer and more expressive than propositional logic, and it allows complex facts about the world to be represented, with new facts determined via deductive reasoning. Predicate calculus includes predicates, variables, and quantifiers, and a *predicate* is a characteristic or property that the subject of a statement can have. A predicate may include variables, and statements with variables become propositions once the variables are assigned values.

The universal quantifier is used to express a statement such as that all members of the domain of discourse have property P. This is written as $(\forall x)\ P(x)$, and it expresses the statement that the property.

$P(x)$ is true for all x.

The existential quantifier states that there is at least one member of the domain of discourse that has property P. This is written as $(\exists x)P(x)$.

10.2 Propositional Logic

Propositional logic is the study of propositions where a proposition is a statement that is either true or false. There are many examples of propositions such as "1 + 1 = 2" which is a true proposition, and the statement that 'Today is Wednesday' which is true if today is Wednesday and false otherwise. The statement $x > 0$ is not a proposition as it contains a variable x, and it is only meaningful to consider its truth or falsity only when a value is assigned to x. Once the variable x is assigned a value, it becomes a proposition. The statement "This sentence is false" is not a proposition as it contains a self-reference that contradicts itself. Clearly, if the statement is true, it is false, and if is false, it is true.

A propositional variable may be used to stand for a proposition (e.g., let the variable P stand for the proposition '2 + 2 = 4' which is a true proposition). A propositional variable takes the value true or false. The negation of a proposition P (denoted $\neg P$) is the proposition that is true if and only if P is false, and is false if and only if P is true.

A well-formed formula (*wff*) in propositional logic is a syntactically correct formula created according to the syntactic rules of the underlying calculus. A well-formed formula is built up from variables, constants, terms, and logical connectives such as conjunction (and), disjunction (or), implication (if.. then..), equivalence (if and only if), and negation. A distinguished subset of these well-formed formulae is the *axioms* of the calculus, and there are *rules of inference* that allow the truth of new formulae to be derived from the axioms and from formulae that have already demonstrated to be true in the calculus.

A formula in propositional calculus may contain several propositional variables, and the truth or falsity of the individual variables needs to be known prior to determining the truth or falsity of the logical formula.

Table 10.1 Truth table for
formula W

A	B	$W\,(A,B)$
T	T	T
T	F	F
F	T	F
F	F	T

Each propositional variable has two possible values, and a formula with n-propositional variables has 2^n values associated with the n-propositional variables. The set of values associated with the n variables may be used to derive a truth table with 2^n rows and $n + 1$ columns. Each row gives each of the 2^n truth-values that the n variables may take, and column $n + 1$ gives the result of the logical expression for that set of values of the propositional variables. For example, the propositional formula W defined in the truth table above has two propositional variables A and B, with $2^2 = 4$ rows for each of the values that the two propositional variables may take. There are $2 + 1 = 3$ columns with W defined in the third column (Table 10.1).

A rich set of connectives is employed in the calculus to combine propositions and to build up the well-formed formulae. This includes the conjunction of two propositions $(A \wedge B)$; the disjunction of two propositions $(A \vee B)$; and the implication of two propositions $(A \to B)$. These connectives allow compound propositions to be formed, and the truth of the compound propositions is determined from the truth-values of its constituent propositions and the rules associated with the logical connective. The meaning of the logical connectives is given by truth tables.[1]

Logic involves proceeding in a methodical way from the axioms and rules of inference to derive further truths. A valid argument is truth preserving: i.e., for a valid logical argument if the set of premises is true, then the conclusion (i.e., the deduced proposition) will also be true. The rules of inference include rules such as *modus ponens*, which states that given the truth of the proposition A, and the proposition $A \to B$, then the truth of proposition B may be deduced.

10.2.1 Truth Tables

Truth tables give operational definitions of the most important logical connectives, and they provide a mechanism to determine the truth-values of more complicated compound expressions. Compound expressions are formed from propositions and connectives, and the truth-values of a compound expression containing several propositional variables are determined from the underlying propositional variables and the logical connectives.

---

[1] Basic truth tables were first used by Frege and developed further by Post and Wittgenstein.

Table 10.2 Conjunction

A	B	$A \wedge B$
T	T	T
T	F	F
F	T	F
F	F	F

Table 10.3 Disjunction

A	B	$A \vee B$
T	T	T
T	F	T
F	T	T
F	F	F

The conjunction of A and B (denoted $A \wedge B$) is true if and only if both A and B are true and is false in all other cases (Table 10.2). The disjunction of two propositions A and B (denoted $A \vee B$) is true if at least one of A and B are true and false in all other cases (Table 10.3). The disjunction operator is known as the '*inclusive or*' operator, and there is also an *exclusive or* operator that is true exactly when one of A or B is true and is false otherwise.

Example 10.1

Consider proposition A given by "An orange is a fruit" and the proposition B given by "$2 + 2 = 5$" then A is true and B is false. Therefore,

(i) $A \wedge B$ (i.e., An orange is a fruit and $2 + 2 = 5$) is false.
(ii) $A \vee B$ (i.e., An orange is a fruit or $2 + 2 = 5$) is true.

The implication operation $(A \rightarrow B)$ is true if whenever A is true means that B is also true and also whenever A is false (Table 10.4). It is equivalent (as shown by a truth table) to $\neg A \vee B$. The equivalence operation $(A \leftrightarrow B)$ is true whenever both A and B are true or whenever both A and B are false (Table 10.5).

The not operator $(\neg)$ is a unary operator (i.e., it has one argument) such that $\neg A$ is true when A is false and is false when A is true (Table 10.6).

Table 10.4 Implication

A	B	$A \rightarrow B$
T	T	T
T	F	F
F	T	T
F	F	T

Table 10.5 Equivalence

A	B	$A \leftrightarrow B$
T	T	T
T	F	F
F	T	F
F	F	T

Table 10.6 Not operation

A	$\neg A$
T	F
F	T

Example 10.2

Consider proposition A given by "Jaffa cakes are biscuits" and the proposition B given by "$2 + 2 = 5$" then A is true and B is false. Therefore,

(i) $A \rightarrow B$ (i.e., Jaffa cakes are biscuits implies $2 + 2 = 5$) is false.
(ii) $A \leftrightarrow B$ (i.e., Jaffa cakes are biscuits is equivalent to $2 + 2 = 5$) is false.
(iii) $\neg B$ (i.e., $2 + 2 \neq 5$) is true.

Creating a Truth Table

The truth table for a well-formed formula $W(P_1, P_2, \ldots, P_n)$ is a table with 2^n rows and $n + 1$ columns. Each row lists a different combination of truth-values of the propositions $P_1, P_2, \ldots, P_n$ followed by the corresponding truth-value of W.

The example above (Table 10.7) gives the truth table for a formula W with three propositional variables (meaning that there are $2^3 = 8$ rows in the truth table).

Table 10.7 Truth table for $W(P, Q, R)$

P	Q	R	$W(P, Q, R)$
T	T	T	F
T	T	F	F
T	F	T	F
T	F	F	T
F	T	T	T
F	T	F	F
F	F	T	F
F	F	F	T

10.2.2 Properties of Propositional Calculus

There are many well-known properties of the propositional calculus such as the commutative, associative, and distributive properties. These ease the evaluation of complex expressions and allow logical expressions to be simplified.

The *commutative property* holds for the conjunction and disjunction operators, and it states that the order of evaluation of the two propositions may be reversed without affecting the resulting truth-value: i.e.,

$$A \wedge B = B \wedge A$$
$$A \vee B = B \vee A.$$

The *associative property* holds for the conjunction and disjunction operators. This means that order of evaluation of a subexpression does not affect the resulting truth-value: i.e.,

$$(A \wedge B) \wedge C = A \wedge (B \wedge C)$$
$$(A \vee B) \vee C = A \vee (B \vee C).$$

The conjunction operator *distributes* over the disjunction operator and vice versa.

$$A \wedge (B \vee C) = (A \wedge B) \vee (A \wedge C)$$
$$A \vee (B \wedge C) = (A \vee B) \wedge (A \vee C)$$

The result of the logical conjunction of two propositions is false if one of the propositions is false (irrespective of the value of the other proposition).

$$A \wedge F = F \wedge A = F$$

The result of the logical disjunction of two propositions is true if one of the propositions is true (irrespective of the value of the other proposition).

$$A \vee T = T \vee A = T$$

The result of the logical disjunction of two propositions, where one of the propositions is known to be false, is given by the truth-value of the other proposition. That is, the Boolean value 'F' acts as the identity for the disjunction operation.

$$A \vee F = A = F \vee A$$

The result of the logical conjunction of two propositions, where one of the propositions is known to be true, is given by the truth-value of the other proposition. That is, the Boolean value 'T' acts as the identity for the conjunction operation.

$$A \wedge T = A = T \wedge A$$

The $\wedge$ and $\vee$ operators are *idempotent*. That is, when the arguments of the conjunction or disjunction operator are the same proposition A, the result is A. The idempotent property allows expressions to be simplified.

$$A \wedge A = A$$
$$A \vee A = A$$

The *law of the excluded middle* is a fundamental property of the propositional calculus. It states that a proposition A is either true or false: i.e., there is no third logical value.

$$A \vee \neg A$$

We mentioned earlier that $A \rightarrow B$ is logically equivalent to $\neg A \vee B$ (same truth table), and clearly $\neg A \vee B$ is the same as $\neg A \vee \neg\neg B = \neg\neg B \vee \neg A$ which is logically equivalent to $\neg B \rightarrow \neg A$. Another word, $A \rightarrow B$, is logically equivalent to $\neg B \rightarrow \neg A$ (this is known as the *contrapositive*).

De Morgan was a contemporary of Boole in the nineteenth century, and the following law is known as De Morgan's law.

$$\neg(A \wedge B) \equiv \neg A \vee \neg B$$
$$\neg(A \vee B) \equiv \neg A \wedge \neg B$$

Certain well-formed formulae are true for all values of their constituent variables. This can be seen from the truth table when the last column of the truth table consists entirely of true values.

A proposition that is true for all values of its constituent propositional variables is known as a *tautology*. An example of a tautology is the proposition $A \vee \neg A$ (Table 10.8).

A proposition that is false for all values of its constituent propositional variables is known as a *contradiction*. An example of a contradiction is the proposition $A \wedge \neg A$.

Table 10.8 Tautology $B \vee \neg B$

B	$\neg B$	$B \vee \neg B$
T	F	T
F	T	T

10.2.3 Proof in Propositional Calculus

Logic enables further truths to be derived from existing truths by rules of inference that are truth preserving. Propositional calculus is both *complete* and *consistent*. The completeness property means that all true propositions are deducible in the calculus, and the consistency property means that there is no formula A such that both A and $\neg A$ are deducible in the calculus.

An argument in propositional logic consists of a sequence of formulae that are the premises of the argument and a further formula that is the conclusion of the argument. One elementary way to see if the argument is valid is to produce a truth table to determine if the conclusion is true whenever all of the premises are true.

Consider a set of premises $P_1, P_2, \ldots P_n$ and conclusion Q. Then to determine if the argument is valid using a truth table involves adding a column in the truth table for each premise $P_1, P_2, \ldots P_n$, and then to identify the rows in the truth table for which these premises are all true. The truth-value of the conclusion Q is examined in each of these rows, and if Q is true for each case for which $P_1, P_2, \ldots P_n$ are all true then the argument is valid. This is equivalent to $P_1 \wedge P_2 \wedge \ldots \wedge P_n \rightarrow Q$ is a tautology.

An alternate approach to proof with truth tables is to assume the negation of the desired conclusion (i.e., $\neg Q$) and to show that the premises and the negation of the conclusion result in a contradiction (i.e., $P_1 \wedge P_2 \wedge \ldots \wedge P_n \wedge \neg Q$) are a contradiction.

The use of truth tables becomes cumbersome when there are a large number of variables involved, as there are 2^n truth table entries for n-propositional variables.

Procedure for Proof by Truth Table

(i) Consider argument $P_1, P_2, \ldots, P_n$ with conclusion Q.
(ii) Draw truth table with column in truth table for each premise $P_1, P_2, \ldots, P_n$.
(iii) Identify rows in truth table for when these premises are all true.
(iv) Examine truth-value of Q for these rows.
(v) If Q is true for each case that $P_1, P_2, \ldots P_n$ are true, then the argument is valid.
(vi) That is $P_1 \wedge P_2 \wedge \ldots \wedge P_n \rightarrow Q$ is a tautology.

Example 10.3 (*Truth Tables*)

Consider the argument adapted from [1] and determine if it is valid.
If the pianist plays the concerto, then crowds will come if the prices are not too high.
If the pianist plays the concerto, then the prices will not be too high.
Therefore, if the pianist plays the concerto, then crowds will come.

Table 10.9 Proof of argument with a truth table

P	C	H	¬H	¬H → C	P → (¬H → C)	P → ¬H	P → C	¬(P → C)	*
T	T	T	F	T	T	F	T	F	F
T	T	F	T	T	T	T	T	F	F
T	F	T	F	T	T	F	F	T	F
T	F	F	T	F	F	T	F	T	F
F	T	T	F	T	T	T	T	F	F
F	T	F	T	T	T	T	T	F	F
F	F	T	F	T	T	T	T	F	F
F	F	F	T	F	T	T	T	F	F

Solution

We will adopt a common proof technique that involves showing that the negation of the conclusion is incompatible (inconsistent) with the premises, and from this we deduce the conclusion must be true. First, we encode the argument in propositional logic:

Let P stand for "The pianist plays the concerto"; C stands for "Crowds will come"; and H stands for "Prices are too high". Then the argument may be expressed in propositional logic as

$$P \rightarrow (\neg H \rightarrow C)$$
$$P \rightarrow \neg H$$
$$P \rightarrow C.$$

Then we negate the conclusion $P \rightarrow C$ and check the consistency of $P \rightarrow (\neg H \rightarrow C) \wedge (P \rightarrow \neg H) \wedge \neg (P \rightarrow C)$* using a truth table (Table 10.9).

It can be seen from the last column in the truth table that the negation of the conclusion is incompatible with the premises, and therefore it cannot be the case that the premises are true and the conclusion false. Therefore, the conclusion must be true whenever the premises are true, and we conclude that the argument is valid.

Logical Equivalence and Logical Implication

The laws of mathematical reasoning are truth preserving and are concerned with deriving further truths from existing truths. Logical reasoning is concerned with moving from one line in mathematical argument to another and involves deducing the truth of another statement Q from the truth of P.

The statement Q may be in some sense be logically equivalent to P, and this allows the truth of Q to be immediately deduced. In other cases the truth of P is sufficiently strong to deduce the truth of Q; in other words P logically implies Q. This leads naturally to a discussion of the concepts of logical equivalence ($W_1 \equiv W_2$) and logical implication ($W_1 \vdash W_2$).

Table 10.10 Logical equivalence of two WFFs

P	Q	$P \wedge Q$	$\neg P$	$\neg Q$	$\neg P \vee \neg Q$	$\neg(\neg P \vee \neg Q)$
T	T	T	F	F	F	T
T	F	F	F	T	T	F
F	T	F	T	F	T	F
F	F	F	T	T	T	F

Logical Equivalence

Two well-formed formulae W_1 and W_2 with the same propositional variables $(P,Q,R \ldots)$ are logically equivalent $(W_1 \equiv W_2)$ if they are always simultaneously true or false for any given truth-values of the propositional variables.

If two well-formed formulae are logically equivalent, then it does not matter which of W_1 and W_2 is used and $W_1 \leftrightarrow W_2$ is a tautology. In Table 10.10, we see that $P \wedge Q$ is logically equivalent to $\neg(\neg P \vee \neg Q)$.

Logical Implication

For two well-formed formulae W_1 and W_2 with the same propositional variables $(P,Q,R \ldots)$, W_1 logically implies W_2 $(W_1 \vdash W_2)$ if any assignment to the propositional variables which makes W_1 true also makes W_2 true. That is, $W_1 \rightarrow W_2$ is a tautology.

Example 10.4

Show by truth tables that $(P \wedge Q) \vee (Q \wedge \neg R) \vdash (Q \vee R)$.

The formula $(P \wedge Q) \vee (Q \wedge \neg R)$ is true on rows 1, 2, and 6, and formula $(Q \vee R)$ is also true on these rows (Table 10.11). Therefore, $(P \wedge Q) \vee (Q \wedge \neg R) \vdash (Q \vee R)$.

Table 10.11 Logical implication of two WFFs

P	Q	R	$(P \wedge Q) \vee (Q \wedge \neg R)$	$Q \vee R$
T	T	T	T	T
T	T	F	T	T
T	F	T	F	T
T	F	F	F	F
F	T	T	F	T
F	T	F	T	T
F	F	T	F	T
F	F	F	F	F

10.2.4 Semantic Tableaux in Propositional Logic

We showed in Example 10.3 how truth tables may be used to demonstrate the validity of a logical argument. However, the problem with truth tables is that they can get extremely large very quickly (as the size of the table is 2^n where n is the number of propositional variables), and so in this section we will consider an alternate approach known as semantic tableaux.

The basic idea of semantic tableaux is to determine if it is possible for a conclusion to be false when all of the premises are true. If this is not possible, then the conclusion must be true when the premises are true, and so the conclusion is *semantically entailed* by the premises. The method of semantic tableaux is a technique to expose inconsistencies in a set of logical formulae, by identifying conflicting logical expressions.

We present a short summary of the rules of semantic tableaux in Table 10.12, and we then proceed to provide a proof for Example 10.3 using semantic tableaux instead of a truth table.

Table 10.12 Rules of semantic tableaux

Rule No.	Definition	Description
1	$A \wedge B$ A B	If $A \wedge B$ is true, then both A and B are true and may be added to the branch containing $A \wedge B$
2	$A \vee B$ $A \qquad B$	If $A \vee B$ is true, then either A or B is true, and we add two new branches to the tableaux, one containing A and one containing B
3	$A \rightarrow B$ $\neg A \qquad B$	If $A \rightarrow B$ is true, then either $\neg A$ or B is true, and we add two new branches to the tableaux, one containing $\neg A$ and one containing B
4	$A \leftrightarrow B$ $A \wedge B \qquad \neg A \wedge \neg B$	If $A \leftrightarrow B$ is true, then either $A \wedge B$ or $\neg A \wedge \neg B$ is true, and we add two new branches, one containing $A \wedge B$ and one containing $\neg A \wedge \neg B$
5	$\neg\neg A$ A	If $\neg\neg A$ is true, then A may be added to the branch containing $\neg\neg A$
6	$\neg(A \wedge B)$ $\neg A \qquad \neg B$	If $\neg(A \wedge B)$ is true, then either $\neg A$ or $\neg B$ is true, and we add two new branches to the tableaux, one containing $\neg A$ and one containing $\neg B$
7	$\neg(A \vee B)$ $\neg A$ $\neg B$	If $\neg(A \vee B)$ is true, then both $\neg A$ and $\neg B$ are true, and may be added to the branch containing $\neg(A \vee B)$
8	$\neg(A \rightarrow B)$ A $\neg B$	If $\neg(A \rightarrow B)$ is true, then both A and $\neg B$ are true and may be added to the branch containing $\neg(A \rightarrow B)$

Whenever a logical expression A and its negation $\neg A$ appear in a branch of the tableau, then an inconsistency has been identified in that branch, and the branch is said to be *closed*. If all of the branches of the semantic tableaux are closed, then the logical propositions from which the tableau was formed are mutually inconsistent and cannot be true together.

The method of proof with semantic tableaux is to negate the conclusion and to show that all branches in the semantic tableau are closed, and thus it is not possible for the premises of the argument to be true and for the conclusion to be false. Therefore, the argument is valid and the conclusion follows from the premises.

Example 10.5 (*Semantic Tableaux*) Perform the proof for Example 10.3 using semantic tableaux.

Solution

We formalized the argument previously as

$$\text{(Premise 1)} \quad P \rightarrow (\neg H \rightarrow C)$$
$$\text{(Premise 2)} \quad P \rightarrow \neg H$$
$$\text{(Conclusion)} \; P \rightarrow C.$$

We negate the conclusion to get $\neg(P \rightarrow C)$, and we show that all branches in the semantic tableau are closed, and that therefore it is not possible for the premises of the argument to be true and for the conclusion false. Therefore, the argument is valid, and the truth of the conclusion follows from the truth of the premises.

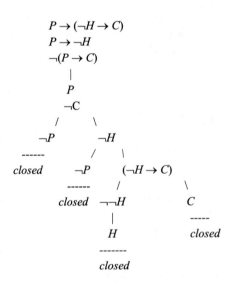

Fig. 10.1 Gerhard Gentzen

We have showed that all branches in the semantic tableau are closed, and that therefore it is not possible for the premises of the argument to be true and for the conclusion to be false. Therefore, the argument is valid as required.

10.2.5 Natural Deduction

The German mathematician, Gerhard Gentzen (Fig. 10.1), developed a method for logical deduction known as '*Natural Deduction*', and this formal approach aims to be as close as possible to natural reasoning. Gentzen worked as an assistant to David Hilbert (Hilbert's program is discussed in Chap. 14) at the University of Göttingen, and he died of malnutrition in Prague at the end of the Second World War.

Natural deduction includes rules for $\wedge$, $\vee$, $\rightarrow$ introduction and elimination and also for *reductio ab absurdum*. There are ten inference rules in the system, and they include two inference rules for each of the five logical operators $\wedge$, $\vee$, $\neg$, $\rightarrow$, and $\leftrightarrow$. There are two inference rules per operator (an introduction rule and an elimination rule), and the rules are defined in Table 10.13.

Natural deduction may be employed in logical reasoning, and it is described in detail in [1, 2].

10.2.6 Sketch of Formalization of Propositional Calculus

Truth tables provide an informal approach to proof, and the proof is provided in terms of the meanings of the propositions and logical connectives. The formalization of propositional logic includes the definition of an alphabet of symbols and well-formed formulae of the calculus, the axioms of the calculus, and rules of inference for logical deduction.

The deduction of a new formulae Q is via a sequence of well-formed formulae $P_1, P_2, \ldots P_n$ (where $P_n = Q$) such that each P_i is either an axiom, a hypothesis, or deducible from an earlier pair of formula P_j, P_k, (where P_k is of the form $P_j \Rightarrow$

Table 10.13 Natural deduction rules

Rule	Definition	Description
$\wedge I$	$\dfrac{P_1, P_2, \dots P_n}{P_1 \wedge P_2 \wedge \dots \wedge P_n}$	Given the truth of propositions P_1, P_2, ... P_n, then the truth of the conjunction $P_1 \wedge P_2 \wedge \dots \wedge P_n$ follows. This rule shows how conjunction can be introduced
$\wedge E$	$\dfrac{P_1 \wedge P_2 \wedge \dots \wedge P_n}{P_i}$ where $i \in \{1,\dots,n\}$	Given the truth the conjunction $P_1 \wedge P_2 \wedge \dots \wedge P_n$, then the truth of proposition P_i ($1 \leq i \leq n$) follows. This rule shows how a conjunction can be eliminated
$\vee I$	$\dfrac{P_i}{P_1 \vee P_2 \vee \dots \vee P_n}$	Given the truth of propositions P_i, then the truth of the disjunction $P_1 \vee P_2 \vee \dots \vee P_n$ follows. This rule shows how a disjunction can be introduced
$\vee E$	$\dfrac{P_1 \vee \dots \vee P_n, P_1 \to E, \dots P_n \to E}{E}$	Given the truth of the disjunction $P_1 \vee P_2 \vee \dots \vee P_n$, and that each disjunct implies E, then the truth of E follows. This rule shows how a disjunction can be eliminated
$\to I$	$\dfrac{\textbf{From } P_1, P_2, \dots P_n \textbf{ infer } P}{(P_1 \wedge P_2 \wedge \dots \wedge P_n) \to P}$	This rule states that if we have a theorem that allows P to be inferred from the truth of premises P_1, P_2, ... P_n (or previously proved), then we can deduce $(P_1 \wedge P_2 \wedge \dots \wedge P_n) \to P$. This is known as the *deduction theorem*
$\to E$	$\dfrac{P_i \to P_j, P_i}{P_j}$	This rule is known as *modus ponens*. The consequence of an implication follows if the antecedent is true (or has been previously proved)
$\equiv I$	$\dfrac{P_i \to P_j, P_j \to P_i}{P_i \leftrightarrow P_j}$	If proposition P_i implies proposition P_j and vice versa, then they are equivalent (i.e., $P_i \leftrightarrow P_j$)
$\equiv E$	$\dfrac{P_i \leftrightarrow P_j}{P_i \to P_j, P_j \to P_i}$	If proposition P_i is equivalent to proposition P_j, then proposition P_i implies proposition P_j and vice versa
$\neg I$	$\dfrac{\textbf{From } P \textbf{ infer } P_1 \wedge \neg P_1}{\neg P}$	If the proposition P allows a contradiction to be derived, then $\neg P$ is deduced. This is an example of a *proof by contradiction*
$\neg E$	$\dfrac{\textbf{From } \neg P \textbf{ infer } P_1 \wedge \neg P_1}{P}$	If the proposition $\neg P$ allows a contradiction to be derived, then P is deduced. This is an example of a proof by contradiction

P_i) and modus ponens. *Modus ponens* is a rule of inference that states that given propositions A, and $A \Rightarrow B$ then proposition B may be deduced. The deduction of a formula Q from a set of hypothesis H is denoted by $H \vdash Q$, and where Q is deducible from the axioms alone this is denoted by $\vdash Q$.

The *deduction theorem* of propositional logic states that if $H \cup \{P\} \vdash Q$, then $H \vdash P \to Q$, and the converse of the theorem is also true: i.e., if $H \vdash P \to Q$, then $H \cup \{P\} \vdash Q$. Formalism (this approach was developed by the German mathematician, David Hilbert) allows reasoning about symbols according to rules and to derive theorems from formulae irrespective of the meanings of the symbols and formulae.

Propositional calculus is *sound*; i.e., any theorem derived using the Hilbert approach is true. Further, the calculus is also *complete*, and every tautology has a proof (i.e., is a theorem in the formal system). The propositional calculus is *consistent*: (i.e., it is not possible that both the well-formed formula A and $\neg A$ are deducible in the calculus).

Propositional calculus is *decidable*: i.e., there is an algorithm (truth table) to determine for any well-formed formula A whether A is a theorem of the formal system. The Hilbert style system is slightly cumbersome in conducting proof and is quite different from the normal use of logic in mathematical deduction.

10.2.7 Applications of Propositional Calculus

Propositional calculus may be employed in reasoning with arguments in natural language. First, the premises and conclusion of the argument are identified and formalized into propositions. Propositional logic is then employed to determine if the conclusion is a valid deduction from the premises.

Consider, for example, the following argument that aims to prove that Superman does not exist.

> *If Superman were able and willing to prevent evil, he would do so. If Superman were unable to prevent evil he would be impotent; if he were unwilling to prevent evil he would be malevolent; Superman does not prevent evil. If superman exists he is neither malevolent nor impotent; therefore Superman does not exist.*

First, letters are employed to represent the propositions as follows:

a: Superman is able to prevent evil
w: Superman is willing to prevent evil
i: Superman is impotent
m: Superman is malevolent
p: Superman prevents evil
e: Superman exists.

Then, the argument above is formalized in propositional logic as follows:

Premises

$\quad P_1 \quad (a \wedge w) \rightarrow p$
$\quad P_2 \quad (\neg a \rightarrow i) \wedge (\neg w \rightarrow m)$
$\quad P_3 \quad \neg p$
$\quad P_4 \quad e \rightarrow \neg i \wedge \neg m$

— — — — — — — — — — — — — — — — —

Conclusion $P_1 \wedge P_2 \wedge P_3 \wedge P_4 \Rightarrow \neg e$

Proof that Superman does not exist

1.	$a \wedge w \rightarrow p$	*Premise 1*
2.	$(\neg a \rightarrow i) \wedge (\neg w \rightarrow m)$	*Premise 2*
3.	$\neg p$	*Premise 3*
4.	$e \rightarrow (\neg i \wedge \neg m)$	*Premise 4*
5.	$\neg p \rightarrow \neg(a \wedge w)$	*1, Contrapositive*
6.	$\neg(a \wedge w)$	*3,5 Modus Ponens*
7.	$\neg a \vee \neg w$	*6, De Morgan's Law*
8.	$\neg(\neg i \wedge \neg m) \rightarrow \neg e$	*4, Contrapositive*
9.	$i \vee m \rightarrow \neg e$	*8, De Morgan's Law*
10.	$(\neg a \rightarrow i)$	*2, $\wedge$ Elimination*
9.	$(\neg w \rightarrow m)$	*2, $\wedge$ Elimination*
12.	$\neg\neg a \vee i$	*10, A $\rightarrow$ B equivalent to $\neg$A$\vee$ B*
13.	$\neg\neg a \vee i \vee m$	*11, $\vee$ Introduction*
14.	$\neg\neg a \vee (i \vee m)$	
15.	$\neg a \rightarrow (i \vee m)$	*14, A $\rightarrow$ B equivalent to $\neg$A$\vee$ B*
16.	$\neg\neg w \vee m$	*11, A $\rightarrow$ B equivalent to $\neg$A$\vee$ B*
17.	$\neg\neg w \vee (i \vee m)$	
18.	$\neg w \rightarrow (i \vee m)$	*17, A $\rightarrow$ B equivalent to $\neg$A$\vee$ B*
19.	$(i \vee m)$	*7, 15, 18 $\vee$ Elimination*
20.	$\neg e$	*9, 19 Modus Ponens*

Second Proof

1.	$\neg p$	P_3
2.	$\neg(a \wedge w) \vee p$	P_1 $(A \rightarrow B \equiv \neg A \vee B)$
3.	$\neg(a \wedge w)$	$1, 2$ $A \vee B, \neg B \vdash A$
4.	$\neg a \vee \neg w$	3, De Morgan's Law
5.	$(\neg a \rightarrow i)$	P_2 ($\wedge$-Elimination)
6.	$\neg a \rightarrow i \vee m$	$5, x \rightarrow y \vdash x \rightarrow y \vee z$
7.	$(\neg w \rightarrow m)$	P_2 ($\wedge$-Elimination)
8.	$\neg w \rightarrow i \vee m$	$7, x \rightarrow y \vdash x \rightarrow y \vee z$
9.	$(\neg a \vee \neg w) \rightarrow (i \vee m)$	$8, x \rightarrow z, y \rightarrow z \vdash x \vee y \rightarrow z$
10.	$(i \vee m)$	4, 9 Modus Ponens
9.	$e \rightarrow \neg(i \vee m)$	P_4 (De Morgan's Law)
12.	$\neg e \vee \neg (i \vee m)$	$11, (A \rightarrow B \equiv \neg A \vee B)$
13.	$\neg e$	$10, 12$ $A \vee B, \neg B \vdash A$

Therefore, the conclusion that Superman does not exist is a valid deduction from the given premises.

10.2.8 Limitations of Propositional Calculus

The propositional calculus deals with propositions only. It is incapable of dealing with the syllogism 'All Greeks are mortal; Socrates is a Greek; therefore Socrates is mortal'. This would be expressed in propositional calculus as three propositions A, B therefore C, where A stands for 'All Greeks are mortal', B stands for 'Socrates is a Greek', and C stands for 'Socrates is mortal'. Propositional logic does not allow the conclusion that all Greeks are mortal to be derived from the two premises.

Predicate calculus deals with these limitations by employing variables and terms and using universal and existential quantification to express that a particular property is true of all (or at least one) value(s) of a variable.

10.3 Predicate Calculus

Predicate logic is a richer system than propositional logic, and it allows complex facts about the world to be represented. It allows new facts about the world to be derived in a way that guarantees that if the initial premises are true, then the conclusions are true. Predicate calculus includes predicates, variables, constants, and quantifiers.

A *predicate* is a characteristic or property that an object can have, and we are predicating some property of the object. For example, "*Socrates is a Greek*" could be expressed as $G(s)$, with capital letters standing for predicates and small letters standing for objects. A predicate may include variables, and a statement with a variable becomes a proposition once the variables are assigned values. For example, $G(x)$ states that the variable x is a Greek, whereas $G(s)$ is an assignment of values to x. The set of values that the variables may take is termed the universe of discourse (the variables take values from this set).

Predicate calculus employs quantifiers to express properties such as all members of the domain have a particular property: e.g., $(\forall x)P(x)$, or that there is at least one member that has a particular property: e.g., $(\exists x)P(x)$. These are referred to as the *universal and existential quantifiers*.

The syllogism 'All Greeks are mortal; Socrates is a Greek; therefore Socrates is mortal' may be easily expressed in predicate calculus by

$$(\forall x)(G(x) \rightarrow M(x))$$
$$G(s)$$

$$- - - - - - -$$

$$M(s).$$

In this example, the predicate $G(x)$ stands for x is a Greek and the predicate $M(x)$ stands for x is mortal. The formula $G(x) \rightarrow M(x)$ states that if x is a Greek, then x is mortal, and the formula $(\forall x)(G(x) \rightarrow M(x))$ states for any x that if x is a

Greek, then x is mortal. The formula $G(s)$ states that Socrates is a Greek, and the formula $M(s)$ states that Socrates is mortal.

Example 10.6 (*Predicates*) A predicate may have one or more variables. A predicate that has only one variable (i.e., a unary or 1-place predicate) is often related to sets; a predicate with two variables (a 2-place predicate) is a relation; and a predicate with n variables (a n-place predicate) is a n-ary relation. Propositions do not contain variables and so they are 0-place predicates. The following are examples of predicates:

i.   The predicate *Prime*(x) states that x is a prime number (with the natural numbers being the universe of discourse).
ii.   *Lawyer*(a) may stand for a is a lawyer.
iii. Mean(m,x,y) states that m is the mean of x and y: i.e., $m = \frac{1}{2}(x + y)$.
iv.   LT($x,6$) states that x is less than 6.
v. $G(x, \pi)$ states that x is greater than π (where π is the constant 3.14159)
vi. $G(x,y)$ states that x is greater than y.
vii. EQ(x, y) states that x is equal to y.
viii. LE(x,y) states that x is less than or equal to y.
ix.   Real(x) states that x is a real number.
x.   Father(x,y) states that x is the father of y.
xi. $\neg(\exists x)(Prime(x) \wedge B(x,32,36))$ states that there is no prime number between 32 and 36.

Universal and Existential Quantification

The universal quantifier is used to express a statement such as that all members of the domain have property P. This is written as $(\forall x)P(x)$ and expresses the statement that the property.

$P(x)$ is true for all x. Similarly, $(\forall x_1, x_2, ..., x_n) P(x_1, x_2, ..., x_n)$ states that property $P(x_1, x_2, ..., x_n)$ is true for all $x_1, x_2, ..., x_n$. Clearly, the predicate $(\forall x) P(a,b)$ is identical to $P(a,b)$ since it contains no variables, and the predicate $(\forall y \in \mathbb{N})$ $(x \leq y)$ is true if $x = 1$ and false otherwise.

The existential quantifier states that there is at least one member in the domain of discourse that has property P. This is written as $(\exists x)P(x)$, and the predicate $(\exists x_1, x_2, ..., x_n) P(x_1, x_2, ..., x_n)$ states that there is at least one value of $(x_1, x_2, ..., x_n)$ such that $P(x_1, x_2, ..., x_n)$ is true.

Example 10.7 (*Quantifiers*)

(i) $(\exists p)$ (Prime(p) $\wedge$ $p > 1{,}000{,}000$) is true

     It expresses the fact that there is at least one prime number greater than a million, which is true as there are an infinite number of primes.

(ii) $(\forall x)$ $(\exists y)$ $x < y$ is true

This predicate expresses the fact that given any number x we can always find a larger number: e.g., take $y = x + 1$.

(iii) $(\exists y)\,(\forall x)\ x < y$ is false

This predicate expresses the statement that there is a natural number y such that all natural numbers are less than y. Clearly, this statement is false since there is no largest natural number, and so the predicate $(\exists y)\,(\forall x)\ x < y$ is false.

Comment 10.1

It is important to be careful with the order in which quantifiers are written, as the meaning of a statement may be completely changed by the simple transposition of two quantifiers.

The well-formed formulae in the predicate calculus are built from terms and predicates, and the rules for building the formulae are described briefly in Sect. 10.3.1. Examples of well-formed formulae include

$$(\forall x)(x > 2)$$
$$(\exists x)x^2 = 2$$
$$(\forall x)(x > 2 \wedge x < 10)$$
$$(\forall x)(\exists y)x^2 = y$$
$$(\forall x)(\exists y)\ Love\ (y, x)\quad \text{(everyone is loved by someone)}$$
$$(\exists y)(\forall x)\ Love\ (y, x)\quad \text{(someone loves everyone)}$$

The formula $(\forall x)(x > 2)$ states that every x is greater than the constant 2; $(\exists x)\ x^2 = 2$ states that there is an x that is the square root of 2; $(\forall x)\,(\exists y)\ x^2 = y$ states that for every x there is a y such that the square of x is y.

10.3.1 Sketch of Formalization of Predicate Calculus

The formalization of predicate calculus includes the definition of an alphabet of symbols (including constants and variables), the definition of function and predicate letters, logical connectives, and quantifiers. This leads to the definitions of the terms and well-formed formulae of the calculus.

The predicate calculus is built from an alphabet of constants, variables, function letters, predicate letters, and logical connectives (including the logical connectives discussed earlier in propositional logic and universal and existential quantifiers).

The definition of terms and well-formed formulae specifies the syntax of the predicate calculus, and the set of well-formed formulae gives the language of the predicate calculus. The terms and well-formed formulae are built from the symbols, and these symbols are not given meaning in the formal definition of the syntax.

The language defined by the calculus needs to be given an *interpretation* in order to give a meaning to the terms and formulae of the calculus. The interpretation needs to define the domain of values of the constants and variables and provide meaning to the function letters, the predicate letters, and the logical connectives.

Terms are built from constants, variables, and function letters. A constant or variable is a term, and if t_1, t_2, ..., t_k are terms, then $f_i^k(t_1, t_2, ..., t_k)$ is a term (where f_i^k is a k-ary function letter). Examples of terms include.

x^2 where x is a variable and square is a 1-ary function letter.

$x^2 + y^2$ where $x^2 + y^2$ is shorthand for the function add (square(x), square(y))
where add is a 2-ary function letter and square is a 1-ary function letter.

The well-formed formulae are built from terms as follows. If P_i^k is a k-ary predicate letter, t_1, t_2, ..., t_k are terms, then $P_i^k(t_1, t_2, ..., t_k)$ is a well-formed formula. If A and B are well-formed formulae, then so are $\neg A$, $A \wedge B$, $A \vee B$, $A \rightarrow B$, $A \leftrightarrow B$, $(\forall x)A$, and $(\exists x)A$.

There is a set of axioms for predicate calculus and two rules of inference used for the deduction of new formulae from the existing axioms and previously deduced formulae. The deduction of a new formula Q is via a sequence of well-formed formulae P_1, P_2, ... P_n (where $P_n = Q$) such that each P_i is either an axiom, a hypothesis, or deducible from one or more of the earlier formulae in the sequence.

The two rules of inference are *modus ponens* and *generalization*. Modus ponens is a rule of inference that states that given predicate formulae A, and $A \rightarrow B$, then the predicate formula B may be deduced. Generalization is a rule of inference that states that given predicate formula A, then the formula $(\forall x)A$ may be deduced where x is any variable.

The deduction of a formula Q from a set of hypothesis H is denoted by $H \vdash Q$, and where Q is deducible from the axioms alone this is denoted by $\vdash Q$. The *deduction theorem* states that if $H \cup \{P\} \vdash Q$, then $H \vdash P \rightarrow Q$[2] and the converse of the theorem is also true: i.e., if $H \vdash P \rightarrow Q$, then $H \cup \{P\} \vdash Q$.

The approach allows reasoning about symbols according to rules and to derive theorems from formulae irrespective of the meanings of the symbols and formulae. Predicate calculus is *sound*: i.e., any theorem derived using the approach is true, and the calculus is also *complete*.

Scope of Quantifiers
The scope of the quantifier $(\forall x)$ in the well-formed formula $(\forall x)A$ is A. Similarly, the scope of the quantifier $(\exists x)$ in the well-formed formula $(\exists x)B$ is B. The variable

---

[2] This is stated more formally that if $H \cup \{P\} \vdash Q$ by a deduction containing no application of generalization to a variable that occurs free in P, then $H \vdash P \rightarrow Q$.

x that occurs within the scope of the quantifier is said to be a *bound variable*. If a variable is not within the scope of a quantifier, it is *free*.

Example 10.8 (*Scope of Quantifiers*)

(i) x is free in the well-formed formula $\forall y \, (x^2 + y > 5)$.
(ii) x is bound in the well-formed formula $\forall x \, (x^2 > 2)$.

A well-formed formula is *closed* if it has no free variables. The substitution of a term t for x in A can only take place only when no free variable in t will become bound by a quantifier in A through the substitution. Otherwise, the interpretation of A would be altered by the substitution.

A term t is free for x in A if no free occurrence of x occurs within the scope of a quantifier ($\forall y$) or ($\exists y$) where y is free in t. This means that the term t may be substituted for x without altering the interpretation of the well-formed formula A.

For example, suppose A is $\forall y \, (x^2 + y^2 > 2)$ and the term t is y, then t is not free for x in A as the substitution of t for x in A will cause the free variable y in t to become bound by the quantifier $\forall y$ in A, thereby altering the meaning of the formula to $\forall y \, (y^2 + y^2 > 2)$.

10.3.2 Interpretation and Valuation Functions

An *interpretation* gives meaning to a formula, and it consists of a *domain of discourse* and a *valuation function*. If the formula is a sentence (i.e., does not contain any free variables), then the given interpretation of the formula is either true or false. If a formula has free variables, then the truth or falsity of the formula depends on the values given to the free variables. A formula with free variables essentially describes a relation say, $R(x_1, x_2, \dots x_n)$ such that $R(x_1, x_2, \dots x_n)$ is true if $(x_1, x_2, \dots x_n)$ is in relation R. If the formula is true irrespective of the values given to the free variables, then the formula is true in the interpretation.

A *valuation* (meaning) *function gives meaning to the logical symbols and connectives*. Thus associated with each constant c is a constant c_Σ in some universe of values Σ; with each function symbol f of arity k, we have a function symbol f_Σ in Σ and $f_\Sigma : \Sigma^k \to \Sigma$; and for each predicate symbol P of arity k a relation $P_\Sigma \subseteq \Sigma^k$. *The valuation function, in effect, gives the semantics of the language of the predicate calculus L.*

The truth of a predicate P is then defined in terms of the meanings of the terms, the meanings of the functions, predicate symbols, and the normal meanings of the connectives.

Mendelson [3] provides a technical definition of truth in terms of *satisfaction* (with respect to an interpretation M). Intuitively a formula F is *satisfiable* if it is *true* (in the intuitive sense) for some assignment of the free variables in the formula F. If a formula F is satisfied for every possible assignment to the free variables in F, then it is *true* (in the technical sense) for the interpretation M. An analogous definition is provided for *false* in the interpretation M.

A formula is *valid* if it is true in every interpretation; however, as there may be an uncountable number of interpretations, it may not be possible to check this requirement in practice. M is said to be a model for a set of formulae if and only if every formula is true in M.

There is a distinction between proof theoretic and model theoretic approaches in predicate calculus. *Proof theoretic* is essentially syntactic, and there is a list of axioms with rules of inference. The theorems of the calculus are logically derived (i.e., $\vdash A$), and the logical truths are as a result of the syntax or form of the formulae, rather than the *meaning* of the formulae. *Model theoretical*, in contrast, is essentially semantic. The truth derives from the meaning of the symbols and connectives, rather than the logical structure of the formulae. This is written as $\vdash_M A$.

A calculus is *sound* if all of the logically valid theorems are true in the interpretation, i.e., proof theoretic $\Rightarrow$ model theoretic. A calculus is *complete* if all the truths in an interpretation are provable in the calculus, i.e., model theoretic $\Rightarrow$ proof theoretic. A calculus is *consistent* if there is no formula A such that $\vdash A$ and $\vdash \neg A$.

The predicate calculus is sound, complete, and consistent. *Predicate calculus is not decidable*: i.e., there is no algorithm to determine for any well-formed formula A whether A is a theorem of the formal system. The undecidability of the predicate calculus may be demonstrated by showing that if the predicate calculus is decidable, then the halting problem (of Turing machines) is solvable. The halting problem is discussed in Chap. 14.

10.3.3 Properties of Predicate Calculus

The following are properties of the predicate calculus.

(i) $(\forall x)\, P(x) \equiv (\forall y)\, P(y)$
(ii) $(\forall x)\, P(x) \equiv \neg\, (\exists x)\, \neg\, P(x)$
(iii) $(\exists x) P(x) \equiv \neg\, (\forall x)\, \neg\, P(x)$
(iv) $(\exists x) P(x) \equiv (\exists y) P(y)$
(v) $(\forall x)\, (\forall y)\, P(x,y) \equiv (\forall y)\, (\forall x)\, P(x,y)$
(vi) $(\exists x)(P(x) \vee Q(x)) \equiv (\exists x) P(x) \vee (\exists y) Q(y)$
(vii) $(\forall x)\, (P(x) \wedge Q(x)) \equiv (\forall x)\, P(x) \wedge (\forall y)\, Q(y)$

10.3.4 Applications of Predicate Calculus

The predicate calculus may be employed to formally state the system requirements of a proposed system. It may be used to conduct formal proof to verify the presence or absence of certain properties in a specification.

It may also be employed to define piecewise defined functions such as $f(x,y)$ where $f(x,y)$ is defined by

$$
\begin{aligned}
f(x, y) &= x^2 - y^2 \text{ where } x \leq 0 \wedge y < 0; \\
f(x, y) &= x^2 + y^2 \text{ where } x > 0 \wedge y < 0; \\
f(x, y) &= x + y \quad \text{ where } x \geq 0 \wedge y = 0; \\
f(x, y) &= x - y \quad \text{ where } x < 0 \wedge y = 0; \\
f(x, y) &= x + y \quad \text{ where } x \leq 0 \wedge y > 0; \\
f(x, y) &= x^2 + y^2 \text{ where } x > 0 \wedge y > 0
\end{aligned}
$$

The predicate calculus may be employed for program verification and to show that a code fragment satisfies its specification. The statement that a program F is correct with respect to its precondition P and postcondition Q is written as $P\{F\}Q$. The objective of program verification is to show that if the precondition is true before execution of the code fragment, then this implies that the postcondition is true after execution of the code fragment.

A program fragment a is *partially correct* for precondition P and postcondition Q if and only if whenever a is executed in any state in which P is satisfied and execution terminates, then the resulting state satisfies Q. Partial correctness is denoted by $P\{F\}Q$, and Hoare's axiomatic semantics is based on partial correctness. It requires proof that the postcondition is satisfied if the program terminates.

A program fragment a is *totally correct* for precondition P and postcondition Q, if and only if whenever a is executed in any state in which P is satisfied, then the execution terminates and the resulting state satisfies Q. It is denoted by $\{P\}F\{Q\}$, and Dijkstra's calculus of weakest preconditions is based on total correctness [2, 4]. It is required to prove that if the precondition is satisfied, then the program terminates and the postcondition is satisfied

10.3.5 Semantic Tableaux in Predicate Calculus

We discussed the use of semantic tableaux for determining the validity of arguments in propositional logic earlier in this chapter, and its approach is to negate the conclusion of an argument and to show that this results in inconsistency with the premises of the argument.

The use of semantic tableaux is similar with predicate logic, except that there are some additional rules to consider. As before, if all branches of a semantic tableau are closed, then the premises and the negation of the conclusion are mutually inconsistent. From this, we deduce that the conclusion must be true.

The rules of semantic tableaux for propositional logic were presented in Table 10.12, and the additional rules specific to predicate logic are detailed in Table 10.14.

Table 10.14 Extra rules of semantic tableaux (for predicate calculus)

Rule No	Definition	Description
1	$(\forall x)\,A(x)$ $A(t)$ where t is a term	Universal instantiation
2	$(\exists x)\,A(x)$ $A(t)$ where t is a term that has not been used in the derivation so far	Rule of existential instantiation. The term "t" is often a constant "a"
3	$\neg(\forall x)\,A(x)$ $(\exists x)\,\neg A(x)$	
4	$\neg(\exists x)\,A(x)$ $(\forall x)\neg A(x)$	

Example 10.9 (*Semantic Tableaux*) Show that the syllogism 'All Greeks are mortal; Socrates is a Greek; therefore Socrates is mortal' is a valid argument in predicate calculus.

Solution

We expressed this argument previously as $(\forall x)(G(x) \to M(x))$; $G(s)$; $M(s)$. Therefore, we negate the conclusion (i.e., $\neg M(s)$) and try to construct a closed tableau.

$$(\forall x)(G(x) \to M(x))$$
$$G(s)$$
$$\neg M(s).$$
$$G(s) \to M(s) \qquad\qquad \text{Universal Instantiation}$$
$$\wedge$$
$$\neg G(s) \quad M(s)$$
$$\text{-----} \qquad \text{--------}$$
$$closed$$

Therefore, as the tableau is closed we deduce that the negation of the conclusion is inconsistent with the premises, and that therefore the conclusion follows from the premises.

Example 10.10 (*Semantic Tableaux*) Determine whether the following argument is valid.

All lecturers are motivated.
Anyone who is motivated and clever will teach well.
Joanne is a clever lecturer.
Therefore, Joanne will teach well.

Solution

We encode the argument as follows:

$L(x)$ stands for 'x is a lecturer'.
$M(x)$ stands for 'x is motivated'.
$C(x)$ stands for 'x is clever'.
$W(x)$ stands for 'x will teach well'.

We therefore wish to show that
$(\forall x)(L(x) \to M(x)) \wedge (\forall x)((M(x) \wedge C(x)) \to W(x)) \wedge L(joanne) \wedge C(joanne) \models W(joanne)$

Therefore, we negate the conclusion (i.e., $\neg W(joanne)$) and try to construct a closed tableau.

1. $(\forall x)(L(x) \to M(x))$
2. $(\forall x)((M(x) \wedge C(x)) \to W(x))$
3. $L(joanne)$
4. $C(joanne)$
5. $\neg W(joanne)$
6. $L(joanne) \to M(joanne)$ Universal Instantiation (line 1)
7. $(M(joanne) \wedge C(joanne)) \to W(joanne)$ Universal Instantiation (line 2)

                  / \

8. $\neg L(joanne)$ $M(joanne)$ From line 6

   Closed

                  / \

9. $\neg (M(joanne) \wedge C(joanne))$ $W(joanne)$ From line 7

                     Closed

           / \

10. $\neg M(joanne)$ $\neg C(joanne)$
  ---------------   -------------
   Closed

Therefore, since the tableau is closed we deduce that the argument is valid.

10.4   Review Questions

1. Draw a truth table to show that $\neg (P \to Q) \equiv P \wedge \neg Q$.
2. Translate the sentence "Execution of program P begun with $x < 0$ will not terminate" into propositional form.

3. Prove the following theorems using the inference rules of natural deduction.
    (a) From b, infer $b \vee \neg c$.
    (b) From $b \Rightarrow (c \wedge d)$, b, infer d.
4. Explain the difference between the universal and the existential quantifier.
5. Express the following statements in the predicate calculus.
    (a) All natural numbers are greater than 10.
    (b) There is at least one natural number between 5 and 10.
    (c) There is a prime number between 100 and 200.
6. Which of the following predicates are true?
    (a) $\forall i \in \{10, \ldots, 50\}.\ i^2 < 2000 \wedge i < 100$
    (b) $\exists\, i \in \mathbb{N}.\ i > 5 \wedge i^2 = 25$
    (c) $\exists\, i \in \mathbb{N}.\ i^2 = 25$
7. Use semantic tableaux to show that $(A \rightarrow A) \vee (B \wedge \neg B)$ is true.
8. Determine if the following argument is valid.

If Pilar lives in Cork, she lives in Ireland. Pilar lives in Cork. Therefore, Pilar lives in Ireland.

10.5 Summary

Propositional logic is the study of propositions, and a proposition is a statement that is either true or false. A formula may contain several variables and a rich set of connectives is employed to combine propositions to build up the well-formed formulae of the calculus. This allows compound propositions to be formed, and the truth of these is determined from the truth-values of the constituent propositions and the rules associated with the logical connectives.

Propositional calculus is both complete and consistent with all true propositions deducible in the calculus, and there is no formula A such that both A and $\neg A$ are deducible in the calculus.

An argument in propositional logic consists of a sequence of formulae that are the premises of the argument and a formula that is the conclusion of the argument. One elementary way to see if the argument is valid is to produce a truth table to determine if the conclusion is true whenever all of the premises are true.

Predicates are statements involving variables, and these statements become propositions once the variables are assigned values. Predicate calculus allows expressions such as all members of the domain have a particular property or that there is at least one member that has a particular property.

Predicate calculus may be employed to specify the requirements of a proposed system and to give the definition of a piecewise defined function. Semantic tableaux may be used for determining the validity of arguments in propositional or predicate logic, and its approach is to negate the conclusion of an argument and to show that this results in inconsistency with the premises of the argument.

References

1. Kelly J (1997) The essence of logic. Prentice Hall
2. Gries D (1981) The science of programming. Springer. Berlin
3. Mendelson E (1987) Introduction to mathematical logic. Wadsworth and Cole/Brook, Advanced Books & Software
4. Dijkstra EW (1976) A disciple of programming. Prentice Hall

Advanced Topics in Logic

11

Key Topics

Fuzzy Logic

Intuitionist Logic

Temporal Logic

Undefined Values

Logic of Partial Functions

Logic and AI

11.1 Introduction

In this chapter we consider some advanced topics in logic including fuzzy logic, temporal logic, intuitionist logic, approaches that deal with undefined values, and logic and AI. Fuzzy logic is an extension of classical logic that acts as a mathematical model for vagueness, and it handles the concept of partial truth where truth-values lie between completely true and completely false. Temporal logic is concerned with the expression of properties that have time dependencies, and it allows temporal properties about the past, present, and future to be expressed.

Brouwer and others developed intuitionist logic which provided a controversial theory on the foundations of mathematics based on a rejection of the law of the excluded middle and an insistence on constructive existence. Martin Löf successfully applied intuitionist logic to type theory in the 1970s.

Partial functions arise naturally in computer science, and such functions may fail to be defined for one or more values in their domain. One approach to dealing with partial functions is to employ a precondition, which restricts the application

© The Author(s), under exclusive license to Springer Nature Switzerland AG 2023
G. O'Regan, *Mathematical Foundations of Software Engineering*,
Texts in Computer Science, https://doi.org/10.1007/978-3-031-26212-8_11

of the function to values where it is defined. We consider three approaches to deal with undefined values, including the logic of partial functions; Dijkstra's approach with his *cand* and *cor* operators; and Parnas's approach which preserves a classical two-valued logic.

We examine the contribution of logic to the AI field, with a short discussion of the work of John McCarthy and the Prolog logic programming language.

11.2   Fuzzy Logic

Fuzzy logic is a branch of *many-valued logic* that allows inferences to be made when dealing with vagueness, and it can handle problems with imprecise or incomplete data. It differs from classical two-valued propositional logic; in that it is based on degrees of truth, rather than on the standard binary truth-values of "true or false" (1 or 0) of propositional logic. That is, while statements made in propositional logic are either true or false (1 or 0), the truth-value of a statement made in fuzzy logic is a value between 0 and 1. Its value expresses the extent to which the statement is true, with a value of 1 expressing absolute truth and a value of 0 expressing absolute falsity.

Fuzzy logic uses *degrees of truth* as a mathematical model for vagueness, and this is useful since statements made in natural language are often vague and have a certain (rather than an absolute) degree of truth. It is an extension of classical logic to deal with the concept of partial truth, where the truth-value lies between completely true and completely false. Lofti Zadeh developed fuzzy logic at Berkley in the 1960s, and it has been successfully applied to expert systems and other areas of artificial intelligence.

For example, consider the statement "John is tall". If John is six feet, four inches, then we would say that this is a true statement (with a truth-value of 1) since John is well above average height. However, if John is five feet, nine inches tall (around average height) then this statement has a degree of truth, and this could be indicated by a fuzzy truth-value of 0.6. Finally, if John's height is four feet, ten inches then we would say that this is a false statement with truth-value 0. Similarly, the statement that today is sunny may be assigned a truth-value of 1 if there are no clouds, 0.8 if there are a small number of clouds, and 0 if it is raining all day.

Propositions in fuzzy logic may be combined together to form compound propositions. Suppose X and Y are propositions in fuzzy logic, then compound propositions may be formed from the conjunction, disjunction, and implication operators. The usual definition in fuzzy logic of the truth-values of the compound propositions formed from X and Y is given by:

$$\text{Truth}(\neg X) = 1 - \text{Truth}(X)$$
$$\text{Truth}(X \text{ and } Y) = \min(\text{Truth}(X), \text{Truth}(Y))$$
$$\text{Truth}(X \text{ or } Y) = \max(\text{Truth}(X), \text{Truth}(Y))$$
$$\text{Truth}(X \rightarrow Y) = \text{Truth}(\neg X \text{ or } Y))$$

Another way in which the operators may be defined is in terms of multiplication:

$$\text{Truth}(X \text{ and } Y) = \text{Truth}(X) * \text{Truth}(Y)$$

$$\text{Truth}(X \text{ or } Y) = 1 - (1 - \text{Truth}(X))^*(1 - \text{Truth}(Y))$$

$$\text{Truth}(X \rightarrow Y) = \max\{z | \text{Truth}(X) * z \leq \text{Truth}(Y)\} \text{ where } 0 \leq z \leq 1$$

Under these definitions, fuzzy logic is an extension of classical two-valued logic, which preserves the usual meaning of the logical connectives of propositional logic when the fuzzy values are just $\{0, 1\}$.

Fuzzy logic has been very useful in expert system and artificial intelligence applications. The first fuzzy logic controller was developed in England in the mid-1970s. Fuzzy logic has also been applied to the aerospace and automotive sectors and the medical, robotics, and transport sectors.

11.3 Temporal Logic

Temporal logic is concerned with the expression of properties that have time dependencies, and the various temporal logics can express facts about the past, present, and future. Temporal logic has been applied to specify temporal properties of natural language, artificial intelligence as well as the specification and verification of program and system behaviour. It provides a language to encode temporal properties in artificial intelligence applications, and it plays a useful role in the formal specification and verification of temporal properties (e.g., liveness and fairness) in safety critical systems.

The statements made in temporal logic can have a truth-value that varies over time. Another words, sometimes the statement is true and sometimes it is false, but it is never true or false at the same time. The two main types of temporal logics are *linear time logics* (reason about a single timeline) and *branching time logics* (reason about multiple timelines).

The roots of temporal logic lie in work done by Aristotle in the fourth century B.C., when he considered whether a truth-value should be given to a statement about a future event that may or may not occur. For example, what truth-value (if any) should be given to the statement that *"There will be a sea battle tomorrow"*? Aristotle argued against assigning a truth-value to such statements in the present time.

Newtonian mechanics assumes an absolute concept of time independent of space, and this viewpoint remained dominant until the development of the theory of relativity in the early twentieth century (when space–time became the dominant paradigm).

Arthur Prior began analysing and formalizing the truth-values of statements concerning future events in the 1950s, and he introduced tense logic (a temporal logic) in the early 1960s. Tense logic contains four modal operators (strong and weak) that express events in future or in the past:

- *P* (it has at some time been the case that)
- *F* (it will be at some time be the case that)
- *H* (it has always been the case that)
- *G* (it will always be the case that).

The *P* and *F* operators are known as weak tense operators, while the *H* and *G* operators are known as strong tense operators. The two pairs of operators are interdefinable via the equivalences:

$$P\phi \cong \neg H \neg \phi$$
$$H\phi, \cong \neg P \neg \phi$$
$$F\phi \cong \neg G \neg \phi$$
$$G\phi, \cong \neg F \neg \phi$$

The set of formulae in Prior's temporal logic may be defined recursively, and they include the connectives used in classical logic (e.g., $\neg, \wedge, \vee, \rightarrow, \leftrightarrow$). We can express a property ϕ that is always true as $A\phi \cong H\phi \wedge \phi \wedge G\phi$ and a property that is sometimes true as $E\phi \cong P\phi \vee \phi \vee F\phi$. Various extensions of Prior's tense logic have been proposed to enhance its expressiveness. These include the binary *since* temporal operator "*S*" and the binary *until* temporal operator "*U*". For example, the meaning of $\phi S \psi$ is that ϕ has been true since a time when ψ was true.

Temporal logics are applicable in the specification of computer systems, as a specification may require *safety, fairness,* and *liveness properties* to be expressed. For example, a fairness property may state that it will always be the case that a certain property will hold sometime in future. The specification of temporal properties often involves the use of special temporal operators.

The common temporal operators that may be used include an operator to express properties that will always be true; properties that will eventually be true; and a property that will be true in the next time instance. For example,

$\Box P$ *P* is always true
$\Diamond P$ *P* will be true sometime in future
$\bigcirc P$ *P* is true in the next time instant (*discrete time*)

Linear temporal logic (LTL) was introduced by Pnueli in the late 1970s and is useful in expressing safety and liveness properties. Branching time logics assume a non-deterministic branching future for time (with a deterministic, linear past). Computation tree logics (CTL and CTL*) were introduced in the early 1980s by Emerson and others.

It is also possible to express temporal operations directly in classical mathematics, and Parnas prefers this approach. He is critical of computer scientists for introducing unnecessary formalisms when classical mathematics has the ability to do this. For example, the value of a function f at a time instance prior to the current time t is defined as:

$$\text{Prior}(f, t) = \lim_{\varepsilon \to 0} f(t - \varepsilon)$$

For more detailed information on temporal logic the reader is referred to the excellent article in [1].

11.4 Intuitionist Logic

The school of intuitionist mathematics was founded by the Dutch mathematician, L. E. J. Brouwer, who was a famous topologist and well known for his fixpoint theorem in topology. Brouwer's constructive approach to mathematics proved to be highly controversial, as its acceptance as a foundation of mathematics would have led to the rejection of many accepted theorems in classical mathematics (including his own fixed-point theorem).

Brouwer was deeply interested in the foundations of mathematics and the problems arising from the paradoxes of set theory. He was determined to provide a secure foundation, and his view was that an existence theorem that demonstrates the proof of a mathematical object has no validity, unless the proof is constructive and accompanied by a procedure to construct the object. He therefore rejected indirect proof and the law of the excluded middle ($P \lor \neg P$) or equivalently ($\neg\neg P \to P$), and he insisted on an explicit construction of the mathematical object.

The problem with the law of the excluded middle (LEM) arises in dealing with properties of infinite sets. For finite sets, one can decide if all elements of the set possess a certain property P by testing each one. However, this procedure is no longer possible for infinite sets. We may know that a certain element of the infinite set does not possess the property, or it may be the actual method of construction of the set allows us to prove that every element has the property. However, the application of the law of the excluded middle is invalid for infinite sets, as we cannot conclude from the situation where not all elements of an infinite set possess a property P that there exists at least one element which does not have the property P. Another words, the law of the excluded middle may only be applied in cases where the conclusion can be reached in a finite number of steps.

Consequently, if the Brouwer view of the world were accepted then many of the classical theorems of mathematics (including his own well-known results in topology) could no longer be said to be true. His approach to the foundations of mathematics hardly made him popular with contemporary mathematicians (the differences were so fundamental that it was more like a civil war), and intuitionism never became mainstream in mathematics. It led to deep and bitter divisions between Hilbert[1] and Brouwer, with Hilbert accusing Brouwer (and

---

[1] David Hilbert was a famous German mathematician, and Hilbert's program is discussed in Chap. 14.

Weyl) of trying to overthrow everything that did not suit them in mathematics, and that intuitionism was treason to science. Hilbert argued that a suitable foundation for mathematics should aim to preserve most of mathematics. Brouwer described Hilbert's formalist program as a false theory that would produce nothing of mathematical value.

For Brouwer, "to exist" is synonymous with "constructive existence", and constructive mathematics is relevant to computer science, as a program may be viewed as the result obtained from a constructive proof of its specification. Brouwer developed one of the more unusual logics that have been invented (intuitionist logic), in which many of the results of classical mathematics were no longer true. Intuitionist logic may be considered the logical basis of constructive mathematics, and formal systems for intuitionist propositional and predicate logic were developed by Heyting and others [2].

Consider a hypothetical mathematical property $P(x)$ of which there is no known proof (i.e., it is unknown whether $P(x)$ is true or false for arbitrary x where x ranges over the natural numbers). Therefore, the statement $\forall x \ (P(x) \lor \neg P(x))$ cannot be asserted with the present state of knowledge, as neither $P(x)$ nor $\neg P(x)$ has been proved. That is, unproved statements in intuitionist logic are not given an intermediate truth-value, and they remain of an unknown truth-value until they have been either proved or disproved.

The intuitionist interpretation of the logical connectives is different from classical propositional logic. A sentence of the form $A \lor B$ asserts that either a proof of A or a proof of B has been constructed, and $A \lor B$ is not equivalent to $\neg \ (\neg A \land \neg B)$. Similarly, a proof of $A \land B$ is a pair whose first component is a proof of A and whose second component is a proof of B. The statement $\forall x \ \neg P(x)$ is not equivalent to $\exists x \ P(x)$ in intuitionist logic.

Intuitionist logic was applied to type theory by Martin Löf in the 1970s [3]. Intuitionist type theory is based on an analogy between propositions and types, where $A \land B$ is identified with $A \times B$, the Cartesian product of A and B. The elements in the set $A \times B$ are of the form (a, b) where $a \in A$ and $b \in B$. The expression $A \lor B$ is identified with $A + B$, the disjoint union of A and B. The elements in the set $A + B$ are got from tagging elements from A and B, and they are of the form inl(a) for $a \in A$ and inr(b) for $b \in B$. The left and right injections are denoted by inl and inr.

11.5 Undefined Values

Total functions $f: X \to Y$ are functions that are defined for every element in their domain, and total functions are widely used in mathematics. However, partial functions may be undefined for one or more elements in their domain, and one example is the function $y = 1/x$ which is undefined at $x = 0$.

Partial functions arise naturally in computer science, and such functions may fail to be defined for one or more values in their domain. One approach to dealing with partial functions is to employ a precondition, which restricts the application

∧	Q	T	F	⊥
P			P∧Q	
T		T	F	⊥
F		F	F	F
⊥		⊥	F	⊥

∨	Q	T	F	⊥
P			P∨Q	
T		T	T	T
F		T	F	⊥
⊥		T	⊥	⊥

Fig. 11.1   Conjunction and disjunction operators

of the function to where it is defined. This makes it possible to define a new set (a proper subset of the domain of the function) for which the function is total over the new set.

Undefined terms often arise[2] and need to be dealt with. Consider the example of the square root function $\sqrt{x}$ taken from [4]. The domain of this function is the positive real numbers, and the following expression is undefined:

$$((x > 0) \wedge (y = \sqrt{x})) \vee ((x \leq 0) \wedge (y = \sqrt{-x}))$$

The reason this is undefined is since the usual rules for evaluating such an expression involve evaluating each subexpression and then performing the Boolean operations. However, when $x < 0$ the subexpression $y = \sqrt{x}$ is undefined, whereas when $x > 0$ the subexpression $y = \sqrt{-x}$ is undefined. Clearly, it is desirable that such expressions be handled, and that for the example above, the expression would evaluate to true.

Classical two-valued logic does not handle this situation adequately, and there have been several proposals to deal with undefined values. Dijkstra's approach is to use the **cand** and **cor** operators in which the value of the left-hand operand determines whether the right-hand operand expression is evaluated or not. Jones logic of partial functions [5] uses a three-valued logic,[3] and Parnas's[4] approach is an extension to the predicate calculus to deal with partial functions that preserve the two-valued logic.

11.5.1 Logic of Partial Functions

Jones [5] has proposed the logic of partial functions (LPFs) as an approach to deal with terms that may be undefined. This is a three-valued logic, and a logical term may be true, false, or undefined (denoted ⊥). The truth tables for conjunction and disjunction are defined in Fig. 11.1.

The conjunction of P and Q is true when both P and Q are true; false if one of P or Q is false; and undefined otherwise. The operation is commutative. The

---

[2] It is best to avoid undefinedness by taking care with the definitions of terms and expressions.
[3] The above expression would evaluate to true under Jones three-valued logic of partial functions.
[4] The above expression evaluates to true for Parnas logic (a two-valued logic).

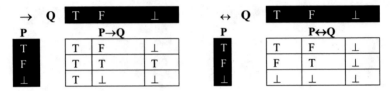

→ Q	T	F	⊥
P		P→Q	
T	T	F	⊥
F	T	T	T
⊥	T	⊥	⊥

↔ Q	T	F	⊥
P		P↔Q	
T	T	F	⊥
F	F	T	⊥
⊥	⊥	⊥	⊥

Fig. 11.2 Implication and equivalence operators

Fig. 11.3 Negation

A	¬A
T	F
F	T
⊥	⊥

disjunction of P and Q ($P \vee Q$) is true if one of P or Q is true; false if both P and Q are false; and undefined otherwise. The implication operation ($P \rightarrow Q$) is true when P is false or when Q is true; false when P is true and Q is false; and undefined otherwise. The equivalence operation ($P \leftrightarrow Q$) is true when both P and Q are true or false; it is false when P is true and Q is false (and vice versa); and it is undefined otherwise (Fig. 11.2).

The not operator ($\neg$) is a unary operator; such $\neg A$ is true when A is false, false when A is true, and undefined when A is undefined (Fig. 11.3).

The result of an operation may be known immediately after knowing the value of one of the operands (e.g., disjunction is true if P is true irrespective of the value of Q). The law of the excluded middle and several other well-known laws do not hold in the three-valued logic. Jones [5] argues that this is reasonable as one would not expect the following to be true:

$$(^1/_0 = 1) \vee (^1/_0 \neq 1)$$

11.5.2 Parnas Logic

Parnas's approach is based on classical two-valued logic with the philosophy that truth-values should be true or false only. His system is an extension to predicate calculus to deal with partial functions. The evaluation of a logical expression yields the value 'true' or 'false' irrespective of the assignment of values to the variables in the expression. This allows the expression: $((x > 0) \wedge (y = \sqrt{x})) \vee ((x \leq 0) \wedge (y = \sqrt{-x}))$ that is undefined in classical logic to yield the value true.[5]

The advantages of his approach are that no new symbols are introduced into the logic, and that the logical connectives retain their traditional meaning. This makes

---

[5] It seems strange to assign the value false to the primitive predicate calculus expression $y = 1/0$.

Table 11.1 Examples of Parnas evaluation of undefinedness

Expression	$x < 0$	$x \geq 0$
$y = \sqrt{x}$	False	True if $y = \sqrt{x}$, False otherwise
$y = {}^{1}/_{0}$	False	False
$y = x^2 + \sqrt{x}$	False	True if $y = x^2 + \sqrt{x}$, False otherwise

Table 11.2 Example of undefinedness in array

Expression	$i \in \{1...N\}$	$i \notin \{1...N\}$
$B[i] = x$	True if $B[i] = x$	False
$\exists i, B[i] = x$	True if $B[i] = x$ for some i, False otherwise	False

Fig. 11.4 Finding index in array

it easier for engineers and computer scientists to understand, as it is closer to their intuitive understanding of logic.

The meaning of predicate expressions is given by first defining the meaning of the primitive expressions. These are used as the building bocks for predicate expressions. The evaluation of a primitive expression $R_j(V)$ (where V is a comma that separated set of terms with some elements of V involving the application of partial functions) is false if the value of an argument of a function used in one of the terms of V is not in the domain of that function.[6] The following examples (Tables 11.1 and 11.2) should make this clearer.

These primitive expressions are used to build the predicate expressions, and the standard logical connectives are used to yield truth-values for the predicate expression. Parnas logic is defined in detail in [4].

The power of Parnas logic may be seen by considering a tabular expressions example. The table below specifies the behaviour of a program that searches the array B for the value x. It describes the properties of the values of j' and *present'*. There are two cases to consider (Fig. 11.4):

Clearly, from the example above the predicate expressions $\exists i, B[i] = x$, and $\neg(\exists i, B[i] = x)$ are defined. One disadvantage of the Parnas's approach is that some common relational operators (e.g., $>$, $\geq$, $\leq$, and $<$) are not primitive in the logic. However, these relational operators are then constructed from primitive operators. Further, the axiom of reflection does not hold in the logic.

---

[6] The approach avoids the undefined logical value ($\perp$) and preserves the two-valued logic.

Fig. 11.5 Edsger Dijkstra.
Courtesy of Brian Randell

11.5.3 Dijkstra and Undefinedness

The **cand** and **cor** operators were introduced by Dijkstra (Fig. 11.5) to deal with undefined values. These are non-commutative operators that allow the evaluation of predicates that contain undefined values.

Consider the following expression:

$$y = 0 \lor (x/y = 2)$$

Then this expression is undefined when $y = 0$ as x/y is undefined, since the logical disjunction operation is not defined when one of its operands is undefined. However, there is a case for giving meaning to such an expression when $y = 0$, since in that case the first operand of the logical or operation is true. Further, the logical *disjunction* operation is defined to be true if either of its operands is true. This motivates the introduction of the **cand** and **cor** operators. These operators are associative, and their truth tables are defined in Tables 11.3 and 11.4.

The order of the evaluation of the operands for the **cand** operation is to *evaluate the first operand*; if the first operand is true then the result of the operation is the

Table 11.3 *a cand b*

a	b	a cand b
T	T	T
T	F	F
T	U	U
F	T	F
F	F	F
F	U	F
U	T	U
U	F	U
U	U	U

Table 11.4 *a cor b*

a	*b*	*a cor b*
T	T	T
T	F	T
T	U	T
F	T	T
F	F	F
F	U	U
U	T	U
U	F	U
U	U	U

second operand; otherwise the result is false. The expression *a cand b* is equivalent to:

$$a \text{ cand } b \cong \textbf{if } a \textbf{ then } b \textbf{ else } F$$

The order of the evaluation of the operands for the *cor* operation is to evaluate the first operand. If the first operand is true then the result of the operation is true; otherwise the result of the operation is the second operand. The expression *a cor b* is equivalent to:

$$a \text{ cor } b \cong \textbf{if } a \textbf{ then } T \textbf{ else } b$$

11.6   Logic and AI

Artificial intelligence is a young field, and the term was coined at the Dartmouth conference in 1956. John McCarthy[7] has long advocated the use of logic in AI to formalize knowledge and to guide the design of mechanized reasoning systems (Fig. 11.6). Logic has been used as an analytic tool, as a knowledge representation formalism, and as a programming language.

McCarthy's goal was to formalize common-sense reasoning: i.e., the normal reasoning that is employed in problem solving and dealing with normal events in the real world. McCarthy [6] argues that it is reasonable for logic to play a key role in the formalization of common-sense knowledge, and this includes the formalization of basic facts about actions and their effects; facts about beliefs and desires; and facts about knowledge and how it is obtained. His approach allows common-sense problems to be solved by logical reasoning (Fig. 11.5).

---

[7] John McCarthy received the Turing Award in 1971 for his contributions to artificial intelligence. He also developed the programming language LISP.

Fig. 11.6   John McCarthy.
Courtesy of John McCarthy

Its formalization requires sufficient understanding of the common-sense world, and often the relevant facts to solve a particular problem are unknown. It may be that knowledge thought to be relevant is irrelevant and vice versa. A computer may have millions of facts stored in its memory, and the problem is how to determine which of these should be chosen from its memory to serve as premises in logical deduction.

McCarthy's influential 1959 paper discusses various common-sense problems such as getting home from the airport. Mathematical logic is the standard approach to express premises, and it includes rules of inferences that are used to deduce valid conclusions from a set of premises. Its rigorous deductive reasoning shows how new formulae may be logically deduced from a set or premises.

McCarthy's approach to programs with common sense has been criticized by Bar-Hillel and others on the grounds that common sense is fairly elusive and the difficulty that a machine would have in determining which facts are relevant to a particular deduction from its known set of facts. However, McCarthy's approach has showed how logical techniques can contribute to the solution of specific AI problems.

Artificial intelligence influenced the development of logic programming, and logic programming languages describe what is to be done, rather than how it should be done. These languages are concerned with the statement of the problem to be solved, rather than how the problem will be solved. These languages use mathematical logic as a tool in the statement of the problem definition.

Logic is a useful tool in developing a body of knowledge (or theory), and it allows rigorous mathematical deduction to derive further truths from the existing set of truths. The theory is built up from a small set of axioms or postulates, and rules of inference derive further truths logically.

Many problems are naturally expressed as a theory, and the statement of a problem to be solved is often equivalent to determining if a new hypothesis is consistent with an existing theory. Logic provides a rigorous way to determine this, as it includes a rigorous process for conducting proof.

Computation in logic programming is essentially logical deduction, and logic programming languages use first-order[8] predicate calculus. They employ theorem proving to derive a desired truth from an initial set of axioms. These proofs are constructive[9]; in that an actual object that satisfies the constraints is produced rather than a pure existence theorem. Logic programming specifies the objects, the relationships between them, and the constraints to be satisfied for the problem.

- The set of objects involved in the computation
- The relationships that hold between the objects
- The constraints of the particular problem.

The language interpreter decides how to satisfy the particular constraints. The first logic programming language was Planner developed by Carl Hewitt at MIT in 1969. It uses a procedural approach for knowledge representation rather than McCarthy's declarative approach.

The best-known logic programming language is Prolog, which was developed in the early 1970s by Alain Colmerauer and Robert Kowalski. It stands for programming in logic. It is a goal-oriented language that is based on predicate logic. Prolog became an ISO standard in 1995. The language attempts to solve a goal by tackling the subgoals that the goal consists of:

$$\text{goal} : - \quad \text{subgoal}_1, \dots, \text{subgoal}_n.$$

That is, in order to prove a particular goal it is sufficient to prove subgoal_1 through subgoal_n. Each line of a Prolog program consists of a rule or a fact, and the language specifies what exists rather than how. The following program fragment has one rule and two facts:

$$\text{grandmother}(G, S) : - \text{parent}(P, S), \text{mother}(G, P).$$
$$\text{mother}(\text{sarah, isaac}).$$
$$\text{parent}(\text{isaac, jacob}).$$

The first line in the program fragment is a rule that states that G is the grandmother of S if there is a parent P of S and G is the mother of P. The next two statements are facts stating that Isaac is a parent of Jacob, and that Sarah is the mother of Isaac. A particular goal clause is true if all of its subclauses are true:

$$\text{goalclause}(V_g) : - \text{clause}_1(V_1), \dots, \text{clause}_m(V_m)$$

---

[8] First-order logic allows quantification over objects but not functions or relations. Higher-order logics allow quantification of functions and relations.

[9] For example, the statement $\exists x$ such that $x = \sqrt{4}$ states that there is an x such that x is the square root of 4, and the constructive existence yields that the answer is that $x = 2$ or $x = -2$; i.e., constructive existence provides more the truth of the statement of existence, and an actual object satisfying the existence criteria is explicitly produced.

A Horn clause consists of a goal clause and a set of clauses that must be proven separately. Prolog finds solutions by *unification:* i.e., by binding a variable to a value. For an implication to succeed, all goal variables Vg on the left side of:- must find a solution by binding variables from the clauses which are activated on the right side. When all clauses are examined and all variables in Vg are bound, the goal succeeds. But if a variable cannot be bound for a given clause, then that clause fails. Following the failure, Prolog *backtracks*, and this involves going back to the left to previous clauses to continue trying to unify with alternative bindings. Backtracking gives Prolog the ability to find multiple solutions to a given query or goal.

Logic programming languages generally use a simple searching strategy to consider alternatives:

- If a goal succeeds and there are more goals to achieve, then remember any untried alternatives and go on to the next goal.
- If a goal is achieved and there are no more goals to achieve then stop with success.
- If a goal fails and there are alternative ways to solve it then try the next one.
- If a goal fails and there are no alternate ways to solve it, and there is a previous goal, then go back to the previous goal.
- If a goal fails and there are no alternate ways to solve it, and no previous goal, then stop with failure.

Constraint programming is a programming paradigm where relations between variables can be stated in the form of constraints. Constraints specify the properties of the solution and differ from the imperative programming languages in that they do not specify the sequence of steps to execute.

11.7 Review Questions

1. What is fuzzy logic?
2. What is intuitionist logic and how is it different from classical logic?
3. Discuss the problem of undefinedness and the advantages and disadvantages of three-valued logics. Describe the approaches of Parnas, Dijkstra, and Jones.
4. What is temporal logic?
5. Show how the temporal operators may be expressed in classical mathematics. Discuss the merits of temporal operators.
6. Discuss the applications of logic to AI.

11.8 Summary

We discussed some advanced topics in logic in this chapter, including fuzzy logic, temporal logic, intuitionist logic, undefined values, logic and AI, and theorem provers. Fuzzy logic is an extension of classical logic that acts as a mathematical model for vagueness, whereas temporal logic is concerned with the expression of properties that have time dependencies.

Intuitionism was a controversial school of mathematics that aimed to provide a solid foundation for mathematics. Its adherents rejected the law of the excluded middle and insisted that for an entity to exist that there must be a constructive proof of its existence.

Partial functions arise naturally in computer science, and such functions may fail to be defined for one or more values in their domain. There are a number of approaches to deal with undefined values, including the logic of partial functions; Dijkstra's approach with his cand and cor operators; and Parnas's approach which preserves a classical two-valued logic.

We discussed temporal logic and its applications to the specification of properties with time dependencies. We discussed the application of logic to the AI field, where logic has been used to formalize knowledge in AI systems.

References

1. Temporal logic. Stanford enclyopedia of philosophy. http://plato.stanford.edu/entries/logic-temporal/
2. Heyting A (1966) Intuitionist logic. An introduction. North-Holland Publishing
3. Löf PM. Intuitionist type theory. Notes by Giovanni Savin of lectures given
4. Parnas DL (1993) Predicate calculus for software engineering. IEEE Trans Softw Eng 19(9)
5. Jones C (1986) Systematic software development using VDM. Prentice Hall International
6. McCarthy J (1959) Programs with common sense. In: Proceedings of the Teddington conference on the mechanization of thought processes

Language Theory and Semantics

12

Key Topics

Alphabets

Grammars and Parse Trees

Axiomatic Semantics

Operational Semantics

Denotational Semantics

Lambda Calculus

Lattices and Partial Orders

Complete Partial Orders

Fixpoint Theory

12.1 Introduction

There are two key parts to any programming language, and these are its syntax and semantics. The syntax is the grammar of the language, and a program needs to be syntactically correct with respect to its grammar. The semantics of the language is deeper and determines the meaning of what has been written by the programmer.

The difference between syntax and semantics may be illustrated by an example in a natural language. A sentence may be syntactically correct but semantically meaningless, or a sentence may have semantic meaning but be syntactically incorrect. For example, consider the sentence:

© The Author(s), under exclusive license to Springer Nature Switzerland AG 2023
G. O'Regan, *Mathematical Foundations of Software Engineering*,
Texts in Computer Science, https://doi.org/10.1007/978-3-031-26212-8_12

I will go to Dublin yesterday.

Then this sentence is syntactically valid but semantically meaningless. Similarly, if a speaker utters the sentence "Me Dublin yesterday" we would deduce that the speaker had visited Dublin the previous day even though the sentence is syntactically incorrect.

The semantics of a programming language determines what a syntactically valid program will compute. A programming language is therefore given by:

Programming Language = Syntax + Semantics

Many programming languages have been developed since the birth of digital computing including Plankalkül which was developed by Zuse in the 1940s; Fortran developed by IBM in the 1950s; COBOL was developed by a committee in the late 1950s; ALGOL 60 and ALGOL 68 were developed by an international committee in the 1960s; Pascal was developed by Wirth in the early 1970s; Ada was developed for the US military in the late 1970s; the C language was developed by Richie and Thompson at Bell Labs in the early 1970s; C ++ was developed by Stroustrup at Bell Labs in the early 1980s; and Java developed by Gosling at Sun Microsystems in the mid-1990s. A short description of a selection of programming languages in use is in [1].

A programming language needs to have a well-defined syntax and semantics, and the compiler preserves the semantics of the language (rather than giving the semantics of a language). Compilers are programs that translate a program that is written in some programming language into another form. It involves syntax analysis and parsing to check the syntactic validity of the program; semantic analysis to determine what the program should do; optimization to improve the speed and performance; and code generation in some target language.

Alphabets are a fundamental building block in language theory, as words and language are generated from alphabets. They are discussed in the next section.

12.2   Alphabets and Words

An *alphabet* is a finite non-empty set A, and the elements of A are called letters. For example, consider the set A which consists of the letters a to z.

Words are finite strings of letters, and a set of words is generated from the alphabet. For example, the alphabet $A = \{a, b\}$ generates the following set of words[1]:

$$\{\varepsilon, a, b, aa, ab, bb, ba, aaa, bbb \ldots\}$$

---

[1] ε denotes the empty word.

Each word consists of an ordered list of one or more letters, and the set of words of length two consists of all ordered lists of two letters. It is given by

$$A^2 = \{aa, ab, bb, ba\}$$

Similarly, the set of words of length three is given by:

$$A^3 = \{aaa, aab, abb, aba, baa, bab, bbb, bba\}$$

The set of all words over the alphabet A is given by the positive closure A^+, and it is defined by:

$$A^+ = A \cup A^2 \cup A^3 \cup \ldots = \bigcup_{n=1}^{\infty} A^n$$

Given any two words $w_1 = a_1, a_2 \ldots a_k$ and $w_2 = b_1, b_2 \ldots b_r$ then the concatenation of w_1 and w_2 is given by:

$$w = w_1 w_2 = a_1 a_2 \ldots a_k b_1 b_2 \ldots b_r$$

The empty word is a word of length zero and is denoted by ε. Clearly, $\varepsilon w = w\varepsilon = w$ for all w, and so ε is the identity element under the concatenation operation. A^0 is used to denote the set containing the empty word $\{\varepsilon\}$, and the closure A^* ($= A^+ \cup \{\varepsilon\}$) denotes the infinite set of all words over A (including empty words). It is defined as:

$$A^* = \bigcup_{n=0}^{\infty} A^n$$

The mathematical structure $(A^*, \char`^, \varepsilon)$ forms a monoid,[2] where $\char`^$ is the concatenation operator for words and the identity element is ε. The length of a word w is denoted by $|w|$, and the length of the empty word is zero: i.e., $|\varepsilon| = 0$.

A subset L of A^* is termed a formal language over A. Given two languages L_1, L_2 then the concatenation (or product) of L_1 and L_2 is defined by:

$$L_1 L_2 = \{w | w = w_1 w_2 \text{ where } w_1 \in L_1 \text{ and } w_2 \in L_2\}$$

The positive closure of L and the closure of L may also be defined as:

$$L^+ = \bigcup_{n=1}^{\infty} L^n \quad L^* = \bigcup_{n=0}^{\infty} L^n$$

---

[2] Recall from Chap. 5 (see Sect. 5.8) that a monoid $(M, *, e)$ is a structure that is closed and associative under the binary operation "*", and it has an identity element "e".

12.3   Grammars

A formal grammar describes the syntax of a language, and we distinguish between *concrete* and *abstract syntaxes*. Concrete syntax describes the external appearance of programs as seen by the programmer, whereas abstract syntax aims to describe the essential structure of programs rather than its external form. In other words, abstract syntax aims to give the components of each language structure while leaving out the representation details (e.g., syntactic sugar). Backus Naur Form (BNF) notation is often used to specify the concrete syntax of a language. A grammar consists of

- A finite set of terminal symbols
- A finite set of non-terminal symbols
- A set of production rules
- A start symbol.

A formal grammar generates a formal language, which is set of finite length sequences of symbols created by applying the production rules of the grammar. The application of a production rule involves replacing symbols at the left-hand side of the rule with the symbols on the right-hand side of the rule. The formal language then consists of all words consisting of terminal symbols that are reached by a derivation (i.e., the application of production rules) starting from the start symbol of the grammar.

A construct that appears on the left-hand side of a production rule is termed a *non-terminal*, whereas a construct that only appears on the right-hand side of a production rule is termed a *terminal*. The set of non-terminals N is disjoint from the set of terminals A.

The theory of the syntax of programming languages is well established, and programming languages have a well-defined grammar that allows syntactically valid programs to be derived from the grammars.

Chomsky[3] (Fig. 12.1) is a famous linguist who classified a number of different types of grammar that occur. The Chomsky hierarchy (Table 12.1) consists of four levels including regular grammars; context-free grammars; context-sensitive grammars; and unrestricted grammars. The grammars are distinguished by the production rules, which determine the type of language that is generated.

Regular grammars are used to generate the words that may appear in a programming language. This includes the identifiers (e.g., names for variables, functions, and procedures); special symbols (e.g., addition, multiplication, etc.); and the reserved words of the language.

A rewriting system for context-free grammars is a finite relation between N and $(A \cup N)^*$: i.e., a subset of $N \times (A \cup N)^*$: A production rule $<N> \rightarrow w$ is one element

---

[3] Chomsky made important contributions to linguistics and the theory of grammars. He is more widely known today as a critic of US foreign policy.

Fig. 12.1 Noah Chomsky.
Public domain

Table 12.1 Chomsky hierarchy of grammars

Grammar type	Description
Type 0 grammar	Type 0 grammars include all formal grammars. They have production rules of the form $\alpha \rightarrow \beta$ where α and β are strings of terminals and non-terminals. They generate all languages that can be recognized by a turing machine (see Chap. 13)
Type 1 grammar (context sensitive)	These grammars generate the context-sensitive languages. They have production rules of the form $\alpha A\beta \rightarrow \alpha\gamma\beta$ where A is a non-terminal and α, β, and γ are strings of terminals and non-terminals. They generate all languages that can be recognized by a linear bounded automaton[4]
Type 2 grammar (context free)	These grammars generate the context-free languages. These are defined by rules of the form $A \rightarrow \gamma$ where A is a non-terminal and γ is a string of terminals and non-terminals. These languages are recognized by a pushdown automaton[5] and are used to define the syntax of most programming languages
Type 3 grammar (regular grammars)	These grammars generate the regular languages (or regular expressions). These are defined by rules of the form $A \rightarrow a$ or $A \rightarrow aB$ where A and B are non-terminals and a is a single terminal. A finite-state automaton recognizes these languages (see Chap. 13), and regular expressions are used to define the lexical structure of programming languages

of this relation and is an ordered pair ($<N>$, w) where w is a word consisting of zero or more terminal and non-terminal letters. This production rule means that $<N>$ may be replaced by w.

---

[4] A linear bounded automaton is a restricted form of a non-deterministic Turing machine in which a limited finite portion of the tape (a function of the length of the input) may be accessed.

[5] A pushdown automaton is a finite automaton that can make use of a stack containing data, and it is discussed in Chap. 13.

12.3.1 Backus Naur Form

Backus Naur Form[6] (BNF) provides an elegant means of specifying the syntax of programming languages. It was originally employed to define the grammar for the ALGOL 60 programming language [2], and a variant was used by Wirth to specify the syntax of the Pascal programming language. BNF is widely used to specify the syntax of programming languages.

BNF specifications essentially describe the external appearance of programs as seen by the programmer. The grammar of a context-free grammar may then be input into a parser (e.g., Yacc), and the parser is used to determine if a program is syntactically correct or not.

A BNF specification consists of a set of production rules with each production rule describing the form of a class of language elements such as expressions and statements. A production rule is of the form:

$$<symbol> ::= <expression\ with\ symbols>$$

where < symbol>is a *non-terminal*, and the expression consists of a sequence of terminal and non-terminal symbols. A construct that has alternate forms appears more than once, and this is expressed by sequences separated by the vertical bar "|" (which indicates a choice). In other words, there is more than one possible substitution for the symbol on the left-hand side of the rule. Symbols that never appear on the left-hand side of a production rule are called *terminals*.

The following example defines the syntax of various statements in a sample programming language:

$$<loop\ statement> ::= <while\ loop>|<for\ loop>$$
$$<while\ loop> ::= while(<condition>)<statement>$$
$$<for\ loop> ::= for\ (<expression>) < statement>$$
$$<statement> ::= <assignment\ statement>|<loop\ statement>$$
$$<assignment\ statement> ::= <variable> := <expression>$$

This is a partial definition of the syntax of various statements in the language. It includes various non-terminals such as < loop statement>,<while loop>. The terminals include "while", "for", ": = ", "(", and ")". The production rules for < condition>and<expression > are not included.

The grammar of a context-free language (e.g., LL(1), LL(k), LR(1), LR(k)) grammar expressed in BNF notation may be translated by a parser into a parse table. The parse table may then be employed to determine whether a particular program is valid with respect to its grammar.

---

[6] Backus Naur Form is named after John Backus and Peter Naur. It was created as part of the design of the ALGOL 60 programming language and is used to define the syntax rules of the language.

Example 12.1 (*Context-Free Grammar*) The example considered is that of paren-
thesis matching in which there are two terminal symbols and one non-terminal
symbol

$$S \rightarrow SS$$
$$S \rightarrow (S)$$
$$S \rightarrow ()$$

Then by starting with S and applying the rules we can construct:

$$S \rightarrow SS \rightarrow (S)S \rightarrow (())S \rightarrow (())()$$

Example 12.2 (*Context-Free Grammar*) The example considered is that of expressions in a programming language. The definition is ambiguous as there is more
than one derivation tree for some expressions (e.g., there are two parse trees for the
expression $5 \times 3 + 1$ discussed below).

$$<expr> ::= <numeral> | (<expr>)$$
$$| (< expr > < operator > < expr >)$$
$$< operator > ::= + | - | \times | /$$
$$< digit > ::= 0 | 1 | \ldots | 9$$
$$< numeral > ::= < digit > | < digit > < numeral >$$

Example 12.3 (*Regular Grammar*) The definition of an identifier in most programming languages is similar to:

$$<identifier> ::= <let> <letdig>$$
$$<letdig> ::= <let> | <dig> | \varepsilon$$
$$<letdig> ::= <let> <letdig> | <dig> <letdig>$$
$$<let> ::= a | b | c | \ldots | z$$
$$<dig> ::= 0 | 1 | \ldots . | 9$$

12.3.2 Parse Trees and Derivations

Let A and N be the terminal and non-terminal alphabet of a rewriting system and
let $<X> \rightarrow w$ be a production. Let x be a word in $(A \cup N)^*$ with $x = u<X>v$ for
some words $u, v \in (A \cup N)^*$. Then x is said to directly yield uwv, and this is written
as $x \Rightarrow uwv$.

This single substitution ($\Rightarrow$) can be extended by a finite number of productions
($\Rightarrow^*$), and this gives the set of words that can be obtained from a given word. This

derivation is achieved by applying several production rules (one production rule is applied at a time) in the grammar.

That is, given $x, y \in (A \cup N)^*$ then x yields y (or y is a derivation of x) if $x = y$, or there exists a sequence of words $w_1, w_2, \ldots w_n \in (A \cup N)^*$ such that $x = w_1$, $y = w_n$ and $w_i \Rightarrow w_{i+1}$ for $1 \leq i \leq n - 1$. This is written as $x \Rightarrow^* y$.

The expression grammar presented in Example 12.2 is ambiguous, and this means that an expression such as $5 \times 3 + 1$ has more than one interpretation. (Figs. 12.2 and 12.3). It is not clear from the grammar whether multiplication is performed first and then addition, or whether addition is performed first and then multiplication.

The interpretation of the parse tree in Fig. 12.2 is that multiplication is performed first and then addition (this is the normal interpretation of such expressions in programming languages as multiplication is a higher precedence operator than addition).

The interpretation of the second parse tree is that addition is performed first and then multiplication (Fig. 12.3). It may seem a little strange that one expression has two parse trees and it shows that the grammar is ambiguous. This means that there is a choice for the compiler in evaluating the expression, and the compiler needs

Fig. 12.2 Parse tree $5 \times 3 + 1$

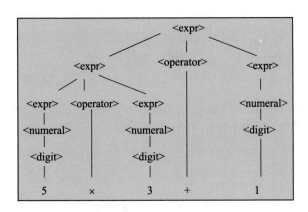

Fig. 12.3 Parse tree $5 \times 3 + 1$

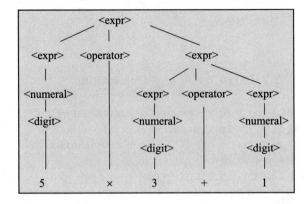

to assign the right meaning to the expression. For the expression grammar one solution would be for the language designer to alter the definition of the grammar to remove the ambiguity.

12.4 Programming Language Semantics

The formal semantics of a programming language is concerned with defining the actual meaning of a language. Language semantics is deeper than syntax, and the theory of the syntax of programming languages is well established. A programmer writes a program according to the rules of the language. The compiler first checks the program for syntactic correctness: i.e., it determines whether the program as written is valid according to the rules of the grammar of the language. If the program is syntactically correct, then the compiler determines the meaning of what has been written and generates the corresponding machine code.[7]

The compiler must preserve the semantics of the language: i.e., the semantics is not defined by the compiler, but rather the function of the compiler is to preserve the semantics of the language. Therefore, there is a need to have an unambiguous definition of the meaning of the language independently of the compiler, and the meaning is then preserved by the compiler.

A program's syntax[8] gives no information as to the meaning of the program, and therefore there is a need to supplement the syntactic description of the language with a formal unambiguous definition of its semantics.

We mentioned that it is possible to utter syntactically correct but semantically meaningless sentences in a natural language. Similarly, it is possible to write syntactically correct programs that behave in quite a different way from the intention of the programmer.

The formal semantics of a language is given by a mathematical model that describes the possible computations described by the language. There are three main approaches to programming language semantics, namely axiomatic semantics, operational semantics, and denotational semantics (Table 12.2).

There are several applications of programming language semantics including language design, program verification, compiler writing, and language standardization. The three main approaches to semantics are described in more detail below.

---

[7] It is possible that what the programmer has written is not be what the programmer had intended.

[8] There are attribute (or affix) grammars that extend the syntactic description of the language with supplementary elements covering the semantics. The process of adding semantics to the syntactic description is termed decoration.

Table 12.2 Programming language semantics

Approach	Description
Axiomatic semantics	This involves giving meaning to phrases of the language using logical axioms
	It employs *pre-* and *postcondition assertion*s to specify what happens when the statement executes. The relationship between the initial assertion and the final assertion essentially gives the semantics of the code
Operational semantics	This approach describes how a valid program is interpreted as sequences of computational steps. These sequences then define the meaning of the program
	An abstract machine (SECD machine) may be defined to give meaning to phrases, and this is done by describing the transitions they induce on states of the machine
Denotational semantics	This approach provides meaning to programs in terms of mathematical objects such as integers, tuples, and functions
	Each phrase in the language is translated into a mathematical object that is the *denotation* of the phrase

12.4.1 Axiomatic Semantics

Axiomatic semantics gives meaning to phrases of the language by describing the logical axioms that apply to them. It was developed by C.A.R. Hoare[9] in a famous paper "*An axiomatic basis for computer programming*" [3]. His axiomatic theory consists of *syntactic elements*, *axioms*, and *rules of inference*.

The well-formed formulae that are of interest in axiomatic semantics are pre- and postassertion formulae of the form $P\{a\}Q$, where a is an instruction in the language and P and Q are assertions: i.e., properties of the program objects that may be true or false.

An *assertion* is essentially a predicate that may be true in some states and false in other states. For example, the assertion $(x - y > 5)$ is true in the state in which the values of x and y are 7 and 1, respectively, and false in the state where x and y have values 4 and 2.

The pre- and postcondition assertions are employed to specify what happens when the statement executes. The relationship between the initial assertion and the final assertion gives the semantics of the code statement. The *pre-* and *postcondition* assertions are of the form:

$$P\{a\}Q$$

The precondition P is a predicate (input assertion), and the postcondition Q is a predicate (output assertion). The braces separate the assertions from the program

---

[9] Hoare was influenced by earlier work by Floyd on assigning meanings to programs using flowcharts [8].

fragment. The well-formed formula $P\{a\}Q$ is itself a predicate that is either true or false.

This notation expresses the *partial correctness*[10] of a with respect to P and Q, and its meaning is that if statement a is executed in a state in which the predicate P is true and execution terminates, then it will result in a state in which assertion Q is satisfied.

The axiomatic semantics approach is described in more detail in [4], and the axiomatic semantics of a selection of statements is presented below.

- **Skip**

  The skip statement does nothing, and whatever condition is true on entry to the command is true on exit from the command. Its meaning is given by:

  $$P\{\text{skip}\}P$$

- **Assignment**

  The meaning of the assignment statement is given by the axiom:

  $$P_e^x\{x := e\}P$$

  The meaning of the assignment statement is that P will be true after execution of the assignment statement if and only if the predicate $P^x{}_e$ with the value of x replaced by e in P is true before execution (since x will contain the value of e after execution).

  The notation $P^x{}_e$ denotes the expression obtained by substituting e for all free occurrences of x in P.

- **Compound**

  The meaning of the conditional command is:

  $$\frac{P\{S_1\}Q,\ Q\{S_2\}R}{P\{S_1;\ S_2\}R}$$

  The compound statement involves the execution of S_1 followed by the execution of S_2. The meaning of the compound statement is that R will be true after the execution of the compound statement $S_1;\ S_2$ provided that P is true, if it is established that Q will be true after the execution of S_1 provided that P is true, and that R is true after the execution of S_2 provided Q is true.

  There needs to be at least one rule associated with every construct in the language in order to give its axiomatic semantics. The semantics of other programming language statements such as the "while" statement and the "if" statement is described in [4].

---

[10] Total correctness is expressed using $\{P\}a\{Q\}$, and program fragment a is totally correct for precondition P and postcondition Q if and only if whenever a is executed in any state in which P is satisfied then execution terminates, and the resulting state satisfies Q.

12.4.2 Operational Semantics

The operational semantics definition is similar to that of an interpreter, where the semantics of the programming language is expressed using a mechanism that makes it possible to determine the effect of any program written in the language. The meaning of a program is given by the evaluation history that an interpreter produces when it interprets the program. The interpreter may be close to an executable programming language, or it may be a mathematical language.

The operational semantics for a programming language describes how a valid program is interpreted as sequences of computational steps. The evaluation history defines the meaning of the program, and this is a sequence of internal interpreter configurations.

John McCarthy did early work on operational semantics in the late 1950s with his work on the semantics of LISP in terms of the lambda calculus. The use of lambda calculus allows the meaning of a program to be expressed using a mathematical interpreter, which gives precision through the use of mathematics.

The meaning of a program may be given in terms of a hypothetical or virtual machine that performs the set of actions that corresponds to the program. An abstract machine (SECD machine[11]) may be defined to give meaning to phrases in the language, and this is done by describing the transitions that they induce on states of the machine.

Operational semantics gives an intuitive description of the programming language being studied, and its descriptions are close to real programs. It can play a useful role as a testing tool during the design of new languages, as it is relatively easy to design an interpreter to execute the description of example programs. This allows the effects of new languages or new language features to be simulated and studied through actual execution of the semantic descriptions prior to writing a compiler for the language. Another words, operational semantics can play a role in rapid prototyping during language design and to get early feedback on the suitability of the language.

One disadvantage of the operational approach is that the meaning of the language is understood in terms of execution: i.e., in terms of interpreter configurations, rather than in an explicit *machine independent specification*. An operational description is just one way to execute programs. Another disadvantage is that the interpreters for non-trivial languages often tend to be large and complex. A more detailed account of operational semantics is in [5, 6].

---

[11] The stack, environment, code, and dump (SECD) virtual stack-based machine was originally designed by Peter Landin (a British computer scientist) to evaluate lambda calculus expressions, and it has since been used as a target for several compilers. Landin was influenced by McCarthy's LISP.

Fig. 12.4 Denotational
semantics

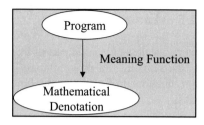

12.4.3 Denotational Semantics

Denotational semantics expresses the semantics of a programming language by a translation schema that associates a meaning (denotation) with each program in the language [6]. It maps a program directly to its meaning, and it was originally called mathematical semantics as it provides meaning to programs in terms of mathematical values such as integers, tuples, and functions. That is, the meaning of a program is a mathematical object, and an interpreter is not employed. Instead, a valuation function is employed to map a program directly to its meaning, and the denotational description of a programming language is given by a set of *meaning functions M* associated with the constructs of the language (Fig. 12.4).

Each meaning function is of the form $M_T: T \rightarrow D_T$ where T is some construct in the language and D_T is some semantic domain. Many of the meaning functions will be *"higher order"*: i.e., functions that yield functions as results. The signature of the meaning function is from syntactic domains (i.e., T) to semantic domains (i.e., D_T). A valuation map $V_T: T \rightarrow \mathbf{B}$ may be employed to check the static semantics prior to giving a meaning of the language construct.[12]

A denotational definition is more abstract than an operational definition. It does not specify the computational steps, and its exclusive focus is on the programs to the exclusion of the state and other data elements. The state is less visible in denotational specifications.

It was developed by Christopher Strachey and Dana Scott at the Programming Research Group at Oxford, England, in the mid-1960s, and their approach to semantics is known as the Scott–Strachey approach [7]. It provides a mathematical foundation for the semantics of programming languages.

Dana Scott's contributions included the formulation of domain theory, which allows programs containing recursive functions and loops to be given a precise semantics. Each phrase in the language is translated into a mathematical object that is the *denotation* of the phrase. Denotational semantics has been applied to language design and implementation.

---

[12] This is similar to what a compiler does in that if errors are found during the compilation phase, the compiler halts and displays the errors and does not continue with code generation.

12.5   Lambda Calculus

Functions are an essential part of mathematics, and they play a key role in specifying the semantics of programming language constructs. We discussed partial and total functions in Chap. 3, and a function was defined as a special type of relation, and simple finite functions may be defined as an explicit set of pairs: e.g.,

$$f \underline{\Delta} \{(a, 1), (b, 2), (c, 3)\}$$

However, for more complex functions there is a need to define the function more abstractly, rather than listing all of its member pairs. This may be done in a similar manner to set comprehension, where a set is defined in terms of a characteristic property of its members.

Functions may be defined (by comprehension) through a powerful abstract notation known as lambda calculus. This notation was introduced by Alonzo Church in the 1930s to study computability (discussed in Chap. 14), and lambda calculus provides an abstract framework for describing mathematical functions and their evaluation. It may be used to study function definition, function application, parameter passing, and recursion.

Any computable function can be expressed and evaluated using lambda calculus or Turing machines, as these are equivalent formalisms. Lambda calculus uses a small set of transformation rules, and these include:

- Alpha-conversion rule (α-conversion)[13]
- Beta-reduction rule (β-reduction)[14]
- Eta-conversion (η-conversion).[15]

Every expression in the λ-calculus stands for a function with a single argument. The argument of the function is itself a function with a single argument, and so on. The definition of a function is anonymous in the calculus. For example, the function that adds one to its argument is usually defined as $f(x) = x + 1$. However, in λ-calculus the function is defined as:

$$succ \underline{\Delta} \lambda x \cdot x + 1$$

The name of the formal argument x is irrelevant, and an equivalent definition of the function is $\lambda z. z + 1$. The evaluation of a function f with respect to an argument (e.g., 3) is usually expressed by $f(3)$. In λ-calculus this would be written as $(\lambda x. x + 1)\ 3$, and this evaluates to $3 + 1 = 4$. Function application is *left associative*: i.e., $f\ x\ y = (f\ x)\ y$. A function of two variables is expressed in lambda

---

[13] This essentially expresses that the names of bound variables are unimportant.

[14] This essentially expresses the idea of function application.

[15] This essentially expresses the idea that two functions are equal if and only if they give the same results for all arguments.

calculus as a function of one argument, which returns a function of one argument. This is known as *currying*: e.g., the function $f(x, y) = x + y$ is written as $\lambda\ x.\ \lambda\ y.\ x + y$. This is often abbreviated to $\lambda\ xy.\ x + y$.

λ-calculus is a simple mathematical system, and its syntax is defined as follows:

$$
\begin{array}{ll}
< \exp > ::= < \text{identifier}> & | \\
\check{}< \text{identifier}>.< \exp > & |\ \text{- - abstraction} \\
< \exp >< \exp > & |\ \text{- - application} \\
(< \exp>) &
\end{array}
$$

λ-calculus's four lines of syntax plus *conversion* rules are sufficient to define Booleans, integers, data structures, and computations on them. It inspired LISP and modern functional programming languages. The original calculus was untyped, but typed lambda calculi were later introduced. The typed lambda calculus allows the sets to which the function arguments apply to be specified. For example, the definition of the *plus* function is given as:

$$\text{plus}\ \underline{\Delta}\lambda a, b : \mathbb{N} \cdot a + b$$

The lambda calculus makes it possible to express properties of the function without reference to members of the base sets on which the function operates. It allows functional operations such as function composition to be applied, and one key benefit is that the calculus provides powerful support for higher-order functions. This is important in the expression of the denotational semantics of the constructs of programming languages.

12.6   Lattices and Order

This section considers some the mathematical structures used in the definition of the semantic domains used in denotational semantics. These mathematical structures may also be employed to give a secure foundation for recursion (discussed in Chap. 6), and it is essential that the conditions in which recursion may be used safely be understood.

It is natural to ask when presented with a recursive definition whether it means anything at all, and in some cases the answer is negative. Recursive definitions are a powerful and elegant way of giving the denotational semantics of language constructs. The mathematical structures considered in this section include partial orders, total orders, lattices, complete lattices, and complete partial orders.

12.6.1 Partially Ordered Sets

A *partial order* $\leq$ on a set P is a binary relation such that for all $x,\ y,\ z \in P$ the following properties hold (Fig. 12.5):

Fig. 12.5 Pictorial
representation of a partial
order

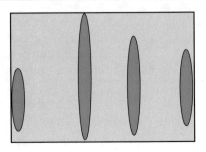

(i) $x \leq x$ (reflexivity)
(ii) $x \leq y$ and $y \leq x \Rightarrow x = y$ (anti-isymmetry)
(iii) $x \leq y$ and $y \leq z \Rightarrow x \leq z$ (transitivity)

A set P with an order relation $\leq$ is said to be a *partially ordered* set.

Example 12.4 Consider the power set $\mathbb{P}X$, which consists of all the subsets of the
set X with the ordering defined by set inclusion. That is, $A \leq B$ if and only if $A \subseteq B$
then $\subseteq$ is a partial order on $\mathbb{P}X$.

A partially ordered set is a *totally ordered* set (*also called chain*) if for all $x, y \in P$
then either $x \leq y$ or $y \leq x$. That is, any two elements of P are directly comparable.
A partially ordered set P is an *anti-chain* if for any x, y in P then $x \leq y$ only if
$x = y$. That is, the only elements in P that are comparable to a particular element
are the element itself.

Maps between Ordered Sets
Let P and Q be partially ordered sets then a map ϕ from P to Q may preserve the
order in P and Q. We distinguish among order preserving, order embedding, and
order isomorphism. These terms are defined as follows:

Order Preserving (or *Monotonic Increasing Function*)

A mapping $\phi\colon P \to Q$ is said to be order preserving if

$$x \leq y \Rightarrow \phi(x) \leq \phi(y)$$

Order Embedding
A mapping $\phi\colon P \to Q$ is said to be an order embedding if
   $x \leq y$ in P if and only if $\phi(x) \leq \phi(y)$ in Q.

Order Isomorphism
The mapping $\phi\colon P \to Q$ is an order isomorphism if and only if it is an order
embedding mapping onto Q.

Dual of a Partially Ordered Set

The dual of a partially ordered set P (denoted P^{∂}) is a new partially ordered set formed from P where $x \leq y$ holds in P^{∂} if and only if $y \leq x$ holds in P (i.e., P^{∂} is obtained by reversing the order on P).

For each statement about P there is a corresponding statement about P^{∂}. Given any statement Φ about a partially ordered set, then the dual statement Φ^{∂} is obtained by replacing each occurrence of $\leq$ by $\geq$ and vice versa.

Duality Principle

Given that statement Φ is true of a partially ordered set P, then the statement Φ^{∂} is true of P^{∂}.

Maximal and Minimum Elements

Let P be a partially ordered set and let $Q \subseteq P$ then

(i) $a \in Q$ is a *maximal* element of Q if $a \leq x \in Q \Rightarrow a = x$.

(ii) $a \in Q$ is the *greatest (or maximum)* element of Q if $a \geq x$ for every $x \in Q$, and in that case we write $a = \max Q$.

A *minimal* element of Q and the *least (or minimum)* are defined dually by reversing the order. The greatest element (if it exists) is called the top element and is denoted by $\top$. The least element (if it exists) is called the bottom element and is denoted by $\bot$.

Example 12.5 Let X be a set and consider $\mathbb{P}X$ the set of all subsets of X with the ordering defined by set inclusion. The top element $\top$ is given by X, and the bottom element $\bot$ is given by $\emptyset$.

A finite totally ordered set always has top and bottom elements, but an infinite chain need not have.

12.6.2 Lattices

Let P be a partially ordered set and let $S \subseteq P$. An element $x \in P$ is an upper bound of S if $s \leq x$ for all $s \in S$. A lower bound is defined similarly.

The set of all upper bounds for S is denoted by S^u, and the set of all lower bounds for S is denoted by S^l.

$$S^u = \{x \in P \mid (\forall s \in S) \, s \leq x\}$$
$$S^l = \{x \in P \mid (\forall s \in S) \, s \geq x\}$$

If S^u has a least element x then x is called the *least upper bound* of S. Similarly, if S^l has a greatest element x then x is called the *greatest lower bound* of S.

Another words, x is the least upper bound of S if.

(i) x is an upper bound of S.

(ii) $x \leq y$ for all upper bounds y of S.

The least upper bound of S is also called the *supremum* of S denoted (sup S), and the greatest lower bound is also called the infimum of S and is denoted by inf S.

Join and Meet Operations

The *join* of x and y (denoted by $x \vee y$) is given by sup$\{x, y\}$ when it exists. The *meet* of x and y (denoted by $x \wedge y$) is given by inf$\{x, y\}$ when it exists.

The supremum of S is denoted by $\bigvee S$, and the infimum of S is denoted by $\bigwedge S$.

Definition Let P be a non-empty partially ordered set then

(i) If $x \vee y$ and $x \wedge y$ exist for all $x, y \in P$ then P is called a *lattice*.

(ii) If $\bigvee S$ and $\bigwedge S$ exist for all $S \subseteq P$ then P is called a *complete lattice*.

Every non-empty finite subset of a lattice has a meet and a join (inductive argument can be used), and every finite lattice is a complete lattice. Further, any complete lattice is bounded: i.e., it has top and bottom elements (Fig. 12.6).

Example 12.6 Let X be a set, and consider $\mathbb{P}X$ the set of all subsets of X with the ordering defined by set inclusion. Then $\mathbb{P}X$ is a complete lattice in which

$$\vee \{A_i | i \in I\} = \cup A_i$$
$$\wedge \{A_i | i \in I\} = \cap A_i$$

Consider the set of natural numbers $\mathbb{N}$ and consider the usual ordering of $<$. Then $\mathbb{N}$ is a lattice with the join and meet operations defined as:

$$x \vee y = \max(x, y)$$

Fig. 12.6 Pictorial representation of a complete lattice

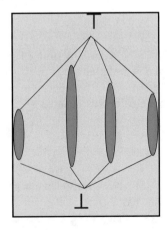

$$x \wedge y = \min(x, y)$$

Another possible definition of the meet and join operations are in terms of the greatest common multiple and lease common divisor.

$$x \vee y = \mathrm{lcm}(x, y)$$
$$x \wedge y = \mathrm{gcd}(x, y)$$

12.6.3 Complete Partial Orders

Let S be a non-empty subset of a partially ordered set P. Then

(i) S is said to be a *directed set* if for every finite subset F of S there exists $z \in S$ such that $z \in F^u$.
(ii) S is said to be *consistent* if for every finite subset F of S there exists $z \in P$ such that $z \in F^u$.

A partially ordered set P is a *complete partial order* (CPO) if:

(i) P has a bottom element $\bot$.
(ii) $\bigvee D$ exists for each directed subset D of P.

The simplest example of a directed set is a chain, and we note that any complete lattice is a complete partial order, and that any finite lattice is a complete lattice.

12.6.4 Recursion

Recursive definitions arise frequently in programs and offer an elegant way to define routines and data types. A recursive routine contains a direct or indirect call to itself, and a recursive data type contains a direct or indirect reference to specimens of the same type. Recursion needs to be used with care, as there is always a danger that the recursive definition may be circular (i.e., defines nothing). It is therefore important to investigate when a recursive definition may be used safely and to give a mathematical definition of recursion.

The control flow in a recursive routine must contain at least one non-recursive branch since if all possible branches included a recursive form the routine could never terminate. The value of at least one argument in the recursive call is different from the initial value of the formal argument as otherwise the recursive call would result in the same sequence of events and therefore would never terminate.

The mathematical meaning of recursion is defined in terms of *fixed-point theory*, which is concerned with determining solutions to equations of the form $x = \tau(x)$, where the function τ is of the form $\tau \colon X \to X$.

A recursive definition may be interpreted as a fixpoint equation of the form $f = \Phi(f)$; i.e., the fixpoint of a high-level functional Φ that takes a function as an argument. For example, consider the functional Φ defined as follows:

$$\Phi \triangleq \lambda f \, \lambda n \cdot \text{if } n = 0 \text{ then } 1 \text{ else } n * f(n-1)$$

Then a fixpoint of Φ is a function f such that $f = \Phi(f)$ or another words

$$f = \lambda n \cdot \text{if } n = 0 \text{ then } 1 \text{ else } n * f(n-1)$$

Clearly, the factorial function is a fixpoint of Φ, and it is the only total function that is a fixpoint. The solution of the equation $f = \Phi(f)$ (where Φ has a fixpoint) is determined as the limit f of the sequence of functions $f_0, f_1, f_2, \ldots$, where the f_i is defined inductively as:

$$f_0 \triangleq \emptyset \quad \text{(the empty partial function)}$$
$$f_i \triangleq \Phi(f_{i-1})$$

Each f_i may be viewed as a successive approximation to the true solution f of the fixpoint equation, with each f_i bringing a little more information on the solution than its predecessor f_{i-1}.

The function f_i is defined for one more value than f_{i-1} and gives the same result for any value for which they are both defined. The definition of the factorial function is thus built up as follows:

$$f_0 \triangleq \emptyset \qquad \text{(the empty partial function)}$$
$$f_1 \triangleq \{0 \rightarrow 1\}$$
$$f_2 \triangleq \{0 \rightarrow 1, 1 \rightarrow 1\}$$
$$f_3 \triangleq \{0 \rightarrow 1, 1 \rightarrow 1, 2 \rightarrow 2\}$$
$$f_4 \triangleq \{0 \rightarrow 1, 1 \rightarrow 1, 2 \rightarrow 2, 3 \rightarrow 6\}$$

For every i, the domain of f_i is the interval $1, 2, \ldots i-1$ and $f_i(n) = n!$ for any n in this interval. Another words f_i is the factorial function restricted to the interval $1, 2, \ldots i-1$. The sequence of f_i may be viewed as successive approximations of the true solution of the fixpoint equation (which is the factorial function), with each f_i bringing defined for one more value that is its predecessor f_{i-1} and defining the same result for any value for which they are both defined.

The candidate fixpoint f_∞ is the limit of the sequence of functions f_i, and is the union of all the elements in the sequence. It may be written as follows:

$$f_\infty \triangleq \emptyset \cup \Phi(\emptyset) \cup \Phi(\Phi(\emptyset)) \cup \ldots = \cup_{i:N} f_i$$

where the sequence f_i is defined inductively as

$$f_0 \triangleq \emptyset \qquad \text{(the empty partial function)}$$

$$f_{i+1} \underline{\Delta} f_i \cup \Phi(f_i)$$

This forms a subset chain where each element is a subset of the next, and it follows by induction that:

$$f_{i+1} = \cup_{j:0...i} \Phi(f_i)$$

A general technique for solving fixpoint equations of the form $h = \tau(h)$ for some functional τ is to start with the least defined function $\emptyset$ and iterate with τ. The union of all the functions obtained as successive sequence elements is the fixpoint.

The condition in which f_∞ is a fixpoint of Φ is the requirement for $\Phi(f_\infty) = f_\infty$. This is equivalent to:

$$\Phi(\cup_{i:\mathbb{N}} f_i) = \cup_{i:\mathbb{N}} f_i$$
$$\Phi(\cup_{i:\mathbb{N}} f_i) = \cup_{i:\mathbb{N}} \Phi(f_i)$$

A sufficient point for Φ to have a fixpoint is that the property $\Phi(\cup_{i:\mathbb{N}} f_i) = \cup_{i:\mathbb{N}} \Phi(f_i)$ holds for any subset chain f_i.

A more detailed account on the mathematics of recursion is in Chap. 8 of [6].

12.7 Review Questions

1. Explain the difference between syntax and semantics.
2. Describe the Chomsky hierarchy of grammars and give examples of each type.
3. Show that a grammar may be ambiguous leading to two difference parse trees. What problems does this create and how should it be dealt with?
4. Describe axiomatic semantics, operation semantics, and denotational semantics and explain the differences between them.
5. Explain partial orders, lattices, and complete partial orders. Give examples of each.
6. Show how the meaning of recursion is defined with fixpoint theory.

12.8 Summary

This chapter considered the syntax and semantics of programming languages. The syntax of the language is concerned with the production of grammatically correct programs in the language, whereas the semantics of the language is deeper and is concerned with the meaning of what has been written by the programmer.

The semantics of programming languages may be given by axiomatic, operational, and denotational semantics. Axiomatic semantics is concerned with defining properties of the language in terms of axioms; operational semantics is concerned with defining the meaning of the language in terms of an interpreter; and denotational semantics is concerned with defining the meaning of the phrases in a language by the denotation or mathematical meaning of the phrase.

Compilers are programs that translate a program that is written in some programming language into another form. It involves syntax analysis and parsing to check the syntactic validity of the program; semantic analysis to determine what the program should do; optimization to improve the speed and performance of the compiler; and code generation in some target language.

Various mathematical structures including partial orders, total orders, lattices, and complete partial orders were considered. These are useful in the definition of the denotational semantics of a language and in giving a mathematical interpretation of recursion.

References

1. O'Regan G (2016) Introduction to the history of computing. Springer
2. Naur P (ed) (1960) Report on the algorithmic language, ALGOL 60. Commun ACM 3(5):299–314
3. Hoare CAR (1969) An axiomatic basis for computer programming. Commun ACM 12(10):576–585
4. O'Regan G (2006) Mathematical approaches to software quality. Springer
5. Plotkin G (1981) A structural approach to operational semantics. Technical Report DAIM FN-19. Computer Science Department. Aarhus University, Denmark
6. Meyer B (1990) Introduction to the theory of programming languages. Prentice Hall
7. Stoy J (1977) Denotational semantics. The Scott-Strachey approach to programming language theory. MIT Press
8. Floyd R (1967) Assigning meanings to programs. In: Proceedings of symposia in applied mathematics, vol 19, pp 19–32

Automata Theory

13

Key Topics

Finite-State Automata

State Transition Table

Deterministic FSA

Non-deterministic FSA

Pushdown Automata

Turing Machine

13.1 Introduction

Automata theory is the branch of computer science that is concerned with the study of abstract machines and automata. These include finite-state machines, pushdown automata, and Turing machines. Finite-state machines are abstract machines that may be in one of a finite number of states. These machines are in only one state at a time (current state), and the input symbol causes a transition from the current state to the next state. Finite-state machines have limited computational power due to memory and state constraints, but they have been applied to a number of fields including communication protocols, neurological systems, and linguistics.

Pushdown automata have greater computational power than finite-state machines, and they contain extra memory in the form of a stack from which symbols may be pushed or popped. The state transition is determined from the current state of the machine, the input symbol, and the element on the top of the stack. The action may be to change the state and/or push/pop an element from the stack.

G. O'Regan, *Mathematical Foundations of Software Engineering*,
Texts in Computer Science, https://doi.org/10.1007/978-3-031-26212-8_13

The Turing machine is the most powerful model for computation, and this theoretical machine is equivalent to an actual computer in the sense that it can compute exactly the same set of functions. The memory of the Turing machine is a tape that consists of a potentially infinite number of one-dimensional cells. It provides a mathematical abstraction of computer execution and storage, as well as providing a mathematical definition of an algorithm. However, Turing machines are not suitable for programming, and therefore they do not provide a good basis for studying programming and programming languages.

13.2   Finite-State Machines

Warren McCulloch and Walter Pitts (two neurophysiologists) published early work on finite-state automata in 1943. They were interested in modelling the thought process for humans and machines. Moore and Mealy developed this work further in the mid-1950s, and their finite-state machines are referred to as the "*Mealy machine*" and the "*Moore machine*". The Mealy machine determines its outputs from the current state and the input, whereas the output of Moore's machine is based upon the current state alone.

Definition 13.1 (*Finite-State Machine*) A finite-state machine (FSM) is an abstract mathematical machine that consists of a finite number of states. It includes a start state q_0 in which the machine is in initially; a finite set of states Q; an input alphabet Σ; a state transition function δ; and a set of final accepting states F (where $F \subseteq, Q$).

The state transition function δ takes the current state and an input symbol and returns the next state. That is, the transition function is of the form:

$$\delta : Q \times \Sigma \rightarrow Q$$

The transition function provides rules that define the action of the machine for each input symbol, and its definition may be extended to provide output as well as a transition of the state. State diagrams are used to represent finite-state machines, and each state accepts a finite number of inputs. A finite-state machine (Fig. 13.1) may be deterministic or non-deterministic, and a *deterministic machine* changes to exactly (or at most)[1] one state for each input transition, whereas a *non-deterministic machine* may have a choice of states to move to for a particular input symbol.

Finite-state automata can compute only very primitive functions, and so they are not adequate as a model for computing. There are more powerful automata such as the Turing machine that is essentially a finite automaton with a potentially

---

[1] The transition function may be undefined for a particular input symbol and state.

Fig. 13.1 Finite-state machine with output

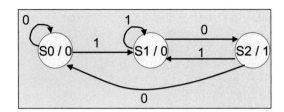

Fig. 13.2 Deterministic FSM

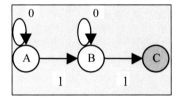

infinite storage (memory). Anything that is computable is computable by a Turing machine.

A finite-state machine can model a system that has a finite number of states and a finite number of inputs/events that trigger transitions between states. The behaviour of the system at a point in time is determined from its current state and input, with behaviour defined for the possible input to that state. The system starts in the initial state.

A finite-state machine (also known as finite-state automata) is a quintuple (Σ, Q, δ, q_0, F). The alphabet of the FSM is given by Σ; the set of states is given by Q; the transition function is defined by $\delta: Q \times \Sigma \rightarrow Q$; the initial state is given by q_0; and the set of accepting states is given by F (where F is a subset of Q). A string is given by a sequence of alphabet symbols: i.e., $s \in \Sigma^*$, and the transition function δ can be extended to $\delta^*: Q \times \Sigma^* \rightarrow Q$.

A string $s \in \Sigma^*$ is accepted by the finite-state machine if $\delta^*(q_0, s) = q_f$ where $q_f \in F$, and the set of all strings accepted by a finite-state machine is the language generated by the machine. A finite-state machine is termed *deterministic* (Fig. 13.2) if the transition function δ is a function,[2] and otherwise (where it is a relation) it is said to be *non-deterministic*. A non-deterministic automaton is one for which the next state is not uniquely determined from the present state and input symbol, and the transition may be to a set of states rather than to a single state.

For the example above the input alphabet is given by $\Sigma = \{0,1\}$; the set of states by $\{A, B, C\}$; the start state by A; the accepting states by $\{C\}$; and the transition function is given by the state transition table below (Table 13.1). The language accepted by the automata is the set of all binary strings that end with a one that contain exactly two ones.

---

[2] It may be a total or a partial function (as discussed in Chap. 3).

Table 13.1 State transition table

State	0	1
A	A	B
B	B	C
C	–	–

Fig. 13.3 Non-deterministic finite-state machine

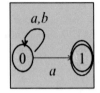

A *non-deterministic* automaton (NFA) or non-deterministic finite-state machine is a finite-state machine where from each state of the machine and any given input, the machine may go to several possible next states. However, a non-deterministic automaton (Fig. 13.3) is equivalent to a deterministic automaton, in that they both recognize the same formal language (i.e., regular languages as defined in Chomsky's classification in Chap. 12). For any non-deterministic automaton, it is possible to construct the equivalent deterministic automaton using power set construction.

NFAs were introduced by Scott and Rabin in 1959, and a NFA is defined formally as a 5-tuple $(Q, \Sigma, \delta, q_0, F)$ as in the definition of a deterministic automaton, and the only difference is in the transition function δ.

$$\delta : Q \times \Sigma \to \mathbb{P}Q$$

The non-deterministic finite-state machine $M_1 = (Q, \Sigma, \delta, q_0, F)$ may be converted to the equivalent deterministic machine $M_2 = (Q', \Sigma, \delta', q_0', F')$ where:

$$Q' = \mathbb{P}Q \text{ (the set of all subsets of Q)}$$
$$q_0' = \{q_0\}$$
$$F' = \{q \in Q' \text{ and } q \cap F \neq \emptyset\}$$
$$\delta'(q, \sigma) = \cup_{p \in q} \delta(p, \sigma) \text{ for each state } q \in Q' \text{ and } \sigma \in \Sigma.$$

The set of strings (or language) accepted by an automaton M is denoted $L(M)$. That is, $L(M) = \{s: \mid \delta^*(q_0, s) = q_f \text{ for some } q_f \in F\}$. A language is termed regular if it is accepted by some finite-state machine. Regular sets are closed under union, intersection, concatenation, complement, and transitive closure. That is, for regular sets A, B $\subseteq \Sigma^*$ then:

- A $\cup$ B and A $\cap$ B are regular.
- $\Sigma^* \setminus$ A (i.e., A^c) is regular.

- AB and A* is regular.

The proof of these properties is demonstrated by constructing finite-state machines to accept these languages. The proof for $A \cap B$ is to construct a machine $M_{A \cap B}$ that mimics the execution of M_A and M_B and is in a final state if and only if both M_A and M_B are in a final state. Finite-state machines are useful in designing systems that process sequences of data.

13.3 Pushdown Automata

A pushdown automaton (PDA) is essentially a finite-state machine with a stack, and its three components (Fig. 13.4) are an input tape; a control unit; and a potentially infinite stack. The stack head scans the top symbol of the stack, and two operations (push or pop) may be performed on the stack. The *push* operation adds a new symbol to the top of the stack, whereas the *pop* operation reads and removes an element from the top of the stack.

A pushdown automaton may remember a potentially infinite amount of information, whereas a finite-state automaton remembers only a finite amount of information. A PDA also differs from a FSM in that it may use the top of the stack to decide on which transition to take, and it may manipulate the stack as part of performing a transition. The input and current state determine the transition of a finite-state machine, and the FSM has no stack to work with.

A pushdown automaton is defined formally as a 7-tuple (Σ, Q, Γ, δ, q_0, Z, F). The set Σ is a finite set which is called the input alphabet; the set Q is a finite set of states; Γ is the set of stack symbols; δ, is the transition function which maps

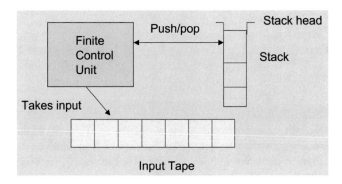

Fig. 13.4 Components of pushdown automata

Fig. 13.5 Transition in
pushdown automata

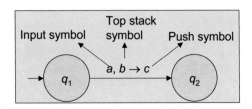

$Q \times \{\Sigma \cup \{\varepsilon\}\}^3 \times \Gamma$ into finite subsets of $Q \times \Gamma^{*4}$; q_0 is the initial state; Z is the
initial stack top symbol on the stack (i.e., $Z \in \Gamma$); and F is the set of accepting
states (i.e., $F \subseteq Q$).

Fig. 13.5 shows a transition from state q_1 to q_2, which is labelled as a, $b \to c$.
This means that if the input symbol a occurs in state q_1, and the symbol on the
top of the stack is b, then b is popped from the stack and c is pushed onto the
stack. The new state is then q_2.

In general, a pushdown automaton has several transitions for a given input
symbol, and so pushdown automata are mainly *non-deterministic*. If a pushdown
automaton has at most one transition for the same combination of state, input symbol, and top of stack symbol it is said to be a *deterministic PDA* (DPDA). The set
of strings (or language) accepted by a pushdown automaton M is denoted $L(M)$.

The class of languages accepted by pushdown automata is the context-free languages, and every context-free grammar can be transformed into an equivalent
non-deterministic pushdown automaton. We discussed the Chomsky classification
of grammars in Chap. 12.

Example (Pushdown Automata)
Construct a non-deterministic pushdown automaton which recognizes the language
$\{0^n 1^n \mid n \geq 0\}$.

Solution
We construct a pushdown automaton $M = (\Sigma, Q, \Gamma, \delta, q_0, Z, F)$ where $\Sigma = \{0, 1\}$;
$Q = \{q_0, q_1, q_f\}$; $\Gamma = \{A, Z\}$; q_0 is the start state; the start stack symbol is Z; and
the set of accepting states is given by $\{q_f\}$:. The transition function (relation) δ is
defined by:

The transition function (Fig. 13.6) essentially says that whenever the value 0
occurs in state q_0 then A is pushed onto the stack. Parts (3) and (4) of the transition
function essentially state that the automaton may move from state q_0 to state q_1 at
any moment. Part (5) states when the input symbol is 1 in state q_1 then one symbol A
is popped from the stack. Finally, part (6) states the automaton may move from state

---

[3] The use of $\{\Sigma \cup \{\varepsilon\}\}$ is to formalize that the PDA can either read a letter from the input or proceed
leaving the input untouched.
[4] This could also be written as $\delta : Q \times \{\Sigma \cup \{\varepsilon\}\} \times \Gamma \to \mathbb{P}(Q \times \Gamma^*)$. It may also be described as
a transition relation.

Fig. 13.6 Transition function for pushdown automata M

1. $(q_0, 0, Z) \rightarrow (q_0, AZ)$
2. $(q_0, 0, A) \rightarrow (q_0, AA)$
3. $(q_0, \varepsilon, Z) \rightarrow (q_1, Z)$
4. $(q_0, \varepsilon, A) \rightarrow (q_1, A)$
5. $(q_1, 1, A) \rightarrow (q_1, \varepsilon)$
6. $(q_1, \varepsilon, Z) \rightarrow (q_f, Z)$

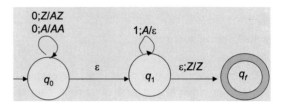

q_1 to the accepting state q_f only when the stack consists of the single stack symbol Z.

For example, it is easy to see that the string 0011 is accepted by the automaton, and the sequence of transitions is given by:

$$(q_0, 0011, Z) \vdash (q_0, 011, AZ) \vdash (q_0, 11, AAZ)$$
$$\vdash (q_1, 11, AAZ) \vdash (q_1, 1, AZ) \vdash (q_1, \varepsilon, Z) \vdash (q_f, Z).$$

13.4 Turing Machines

Turing introduced the theoretical Turing machine (TM) in 1936, and this abstract mathematical machine consists of a head and a potentially infinite tape that is divided into frames (Fig. 13.7). Each frame may be either blank or printed with a symbol from a finite alphabet of symbols. The input tape may initially be blank or have a finite number of frames containing symbols. At any step, the head can read the contents of a frame; the head may erase a symbol on the tape, leave it unchanged, or replace it with another symbol. It may then move one position to the right, one position to the left, or not at all. If the frame is blank, the head can either leave the frame blank or print one of the symbols.

Turing believed that a human with finite equipment and with an unlimited supply of paper to write on could do every calculation. The unlimited supply of paper is formalized in the Turing machine by a paper tape marked off in squares, and the tape is potentially infinite in both directions. The tape may be used for intermediate calculations as well as input and output. The finite number of configurations of the Turing machine was intended to represent the finite states of mind of a human calculator.

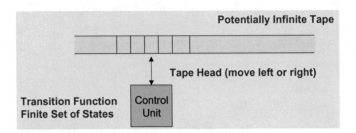

Fig. 13.7 Turing machine

The transition function determines for each state and the tape symbol what the next state to move to and what should be written on the tape, and where to move the tape head. The Turing machine is defined formally as follows:

Definition 13.2 (*Turing Machine*) A Turing machine $M = (Q, \Gamma, b, \Sigma, \delta, q_0, F)$ is a 7-tuple is defined as follows in [1]:

- Q is a finite set of *states.*
- Γ is a finite set of the *tape alphabet/symbols.*
- $b \in \Gamma$ is the *blank symbol* (This is the only symbol that is allowed to occur infinitely often on the tape during each step of the computation).
- Σ is the set of input symbols and is a subset of Γ (i.e., $\Gamma = \Sigma \cup \{b\}$).
- $\delta: Q \times \Gamma \to Q \times \Gamma \times \{L, R\}^5$ is the transition function. This is a partial function where L is left shift and R is right shift.
- $q_0 \in Q$ is the initial state.
- $F \subseteq Q$ is the set of final or accepting states.

The Turing machine is a simple machine that is equivalent to an actual physical computer in the sense that it can compute exactly the same set of functions. It is much easier to analyse and prove things about than a real computer, but it is not suitable for programming and does not provide a good basis for studying programming and programming languages.

Fig. 13.8 illustrates the behaviour when the machine is in state q_1 and the symbol under the tape head is a, where b is written to the tape and the tape head moves to the left and the state changes to q_2.

A Turing machine is essentially a finite-state machine (FSM) with an unbounded tape. The tape is potentially infinite and unbounded, whereas real computers have a large but finite store. The machine may read from and write to the tape. The FSM is essentially the control unit of the machine, and the tape is

---

[5] We may also allow no movement of the tape head to be represented by adding the symbol "N" to the set.

Fig. 13.8 Transition on turing machine

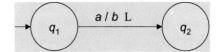

essentially the store. However, the store in a real computer may be extended with backing tapes and discs and in a sense may be regarded as unbounded. However, the maximum amount of tape that may be read or written within n steps is n.

A Turing machine has an associated set of rules that defines its behaviour. Its actions are defined by the transition function. It may be programmed to solve any problem for which there is an algorithm. However, if the problem is unsolvable then the machine will either stop or compute forever. The solvability of a problem may not be determined beforehand. There is, of course, some answer (i.e., either the machine halts or it computes forever). The applications of the Turing machine to computability and decidability are discussed in Chap. 14.

Turing also introduced the concept of a Universal Turing Machine, and this machine is able to simulate any other Turing machine. For more detailed information on automata theory see [1].

13.5 Review Questions

1. What is a finite-state machine?
2. Explain the difference between the deterministic and non-deterministic finite-state machines.
3. Show how to convert the non-deterministic finite-state automaton in Fig. 7.3 to a deterministic automaton.
4. What is a pushdown automaton?
5. What is a Turing machine?
6. Explain what is meant by the language accepted by an automaton.
7. Give an example of a language accepted by a pushdown automaton but not by a finite-state machine.
8. Describe the applications of the Turing machine to computability and decidability.

13.6 Summary

Automata theory is concerned with the study of abstract machines and automata. These include finite-state machines, pushdown automata, and Turing machines. Finite-state machines are abstract machines that may be in one of a finite number of states. These machines are in only one state at a time (current state), and the state transition function determines the new state from the current state and the input symbol. Finite-state machines have limited computational power due to memory and state constraints, but they have been applied to a number of fields including communication protocols and linguistics.

Pushdown automata have greater computational power than finite-state machines, and they contain extra memory in the form of a stack from which symbols may be pushed or popped. The state transition is determined from the current state of the machine, the input symbol, and the element on the top of the stack. The action may be to change the state and/or push/pop an element from the stack.

The Turing machine is the most powerful model for computation, and it is equivalent to an actual computer in the sense that it can compute exactly the same set of functions. The Turing machine provides a mathematical abstraction of computer execution and storage, as well as providing a mathematical definition of an algorithm.

Reference

1. Hopcroft JE, Ullman, JD (1979) Introduction to automata theory, languages and computation. Addison-Wesley, Boston

Computability and Decidability

14

Key Topics

Computability

Completeness

Decidability

Formalism

Logicism

14.1 Introduction

It is impossible for a human or machine to write out all of the members of an infinite countable set, such as the set of natural numbers $\mathbb{N}$. However, humans can do something quite useful in the case of certain enumerable infinite sets: they can give explicit instructions (that may be followed by a machine or another human) to produce the nth member of the set for an arbitrary finite n. The problem remains that for all but a finite number of values of n it will be physically impossible for any human or machine to actually carry out the computation, due to the limitations on the time available for computation, the speed at which the individual steps in the computation may be carried out, and due to finite materials.

The intuitive meaning of computability is in terms of an algorithm (or effective procedure) that specifies a set of instructions to be followed to complete the task. Another words, a function f is *computable* if there exists an algorithm that produces the value of f correctly for each possible argument of f. The computation of f for a particular argument x just involves following the instructions in the algorithm, and it produces the result $f(x)$ in a finite number of steps if x is in

© The Author(s), under exclusive license to Springer Nature Switzerland AG 2023
G. O'Regan, *Mathematical Foundations of Software Engineering*,
Texts in Computer Science, https://doi.org/10.1007/978-3-031-26212-8_14

the domain of f. If x is not in the domain of f then the algorithm may produce an answer saying so or it might run forever never halting. A computer program implements an algorithm.

The concept of computability may be made precise in several equivalent ways such as Church's *lambda calculus, recursive function theory,* or by the theoretical *Turing machines.*[1] These are all equivalent, and perhaps the most well known is the Turing machine (discussed in Chap. 13). This is a mathematical machine with a potentially infinite tape divided into frames (or cells) in which very basic operations can be carried out. The set of functions that are computable are those that are computable by a Turing machine.

Decidability is an important topic in contemporary mathematics. Church and Turing independently showed in 1936 that mathematics is not decidable. In other words, there is no mechanical procedure (i.e., algorithm) to determine whether an arbitrary mathematical proposition is true or false, and so the only way to determine the truth or falsity of a statement is try to solve the problem. That is, it is impossible to of prove or disprove certain statements within a formal system, and there is no a general method to solve all instances of a specific problem.

14.2 Logicism and Formalism

Gottlob Frege (Fig. 9.2) was a 19th-century German mathematician and logician who invented a formal system which is the basis of modern predicate logic. It included axioms, definitions, universal and existential quantification, and formalization of proof. His objective was to show that mathematics was reducible to logic (logicism) but his project failed, and one of the axioms that he had added to his system led to inconsistency.

The inconsistency was pointed out by Bertrand Russell, and it is known as *Russell's paradox.*[2] Russell later introduced his theory of types to deal with the paradox, and he jointly published *Principia Mathematica* with Alfred North White- head as an attempt to derive the truths of arithmetic from a set of logical axioms and rules of inference.

The sentences of Frege's logical system denote the truth-values of true or false. The sentences may include expressions such as equality $(x = y)$, and this returns true if x is the same as y and false otherwise. Similarly, a more complex expression such as $f(x,y,z) = w$ is true if $f(x,y,z)$ is identical with w and false otherwise. Frege represented statements such as "5 is a prime" by "$P(5)$" where $P()$ is termed a concept. The statement $P(x)$ returns true if x is prime and false otherwise. His approach was to represent a predicate as a function of one variable which returns a Boolean value of true or false.

---

[1] The Church–Turing thesis states that anything that is computable is computable by a Turing machine.

[2] Russell's paradox considers the question as to whether the set of all sets that contain themselves as members is a set. In either case there is a contradiction.

Fig. 14.1 David Hilbert

Formalism was proposed by Hilbert (Fig. 14.1) as a foundation for mathematics in the early twentieth century. The motivation for the program was to provide a secure foundations for mathematics and to resolve the contradictions in the formalization of set theory identified by Russell's paradox. The presence of a contradiction in a theory means the collapse of the whole theory, and so it was seen as essential that there be a proof of the consistency of the formal system. The methods of proof in mathematics are formalized with axioms and rules of inference.

Formalism is a formal system that contains meaningless symbols together with rules for manipulating them. The individual formulae are certain finite sequences of symbols obeying the syntactic rules of the formal language. A formal system consists of:

- A formal language
- A set of axioms
- Rules of inference.

The expressions in a formal system are terms, and a term may be simple or complex. A simple term may be an object such as a number, and a complex term may be an arithmetic expression such as $4^3 + 1$. A complex term is formed via functions, and the expression above uses two functions, namely the cube function with argument 4 and the plus function with two arguments.

A formal system is generally intended to represent some aspect of the real world. A rule of inference relates a set of formulae $(P_1, P_2, \ldots P_k)$ called the premises to the consequence formula P called the conclusion. For each rule of inference there is a finite procedure for determining whether a given formula Q is an immediate consequence of the rule from the given formulae $(P_1, P_2, \ldots P_k)$. A *proof* in a formal system consists of a finite sequence of formulae, where each formula is either an axiom or derived from one or more preceding formulae in the sequence by one of the rules of inference.

Hilbert's program was concerned with the formalization of mathematics (i.e., the axiomatization of mathematics) together with a proof that the axiomatization is consistent (i.e., there is no formula A such that both A and $\neg A$ are deducible in the calculus). Its specific objectives were to:

- Provide a formalism of mathematics.
- Show that the formalization of mathematics is *complete*: i.e., all mathematical truths can be proved in the formal system.
- Provide a proof that the formal system is *consistent* (i.e., that no contradictions may be derived).
- Show that mathematics is *decidable*: i.e., there is an algorithm to determine the truth of falsity of any mathematical statement.

The formalist movement in mathematics led to the formalization of large parts of mathematics, where theorems could be proved using just a few mechanical rules. The two most comprehensive formal systems developed were *Principia Mathematica* by Russell and Whitehead and the axiomatization of set theory by Zermelo-Fraenkel (subsequently developed further by von Neumann).

Principia Mathematica is a comprehensive three-volume work on the logical foundations of mathematics written by Bertrand Russel and Alfred Whitehead between 1910 and 1913. Its goal was to show that all of the concepts of mathematics can be expressed in logic, and that all of the theorems of mathematics can be proved using only the logical axioms and rules of inference of logic. It covered set theory, ordinal numbers, and real numbers, and it showed that in principle that large parts of mathematics could be developed using *logicism*.

It avoided the problems with contradictions that arose with Frege's system by introducing the theory of types in the system. The theory of types meant that one could no longer speak of the set of all sets, as a set of elements is of a different type from that of each of its elements, and so Russell's paradox was avoided. It remained an open question at the time as to whether the *Principia* was consistent and complete. That is, is it possible to derive all the truths of arithmetic in the system and is it possible to derive a contradiction from the Principia's axioms? However, it was clear from the three-volume work that the development of mathematics using the approach of the Principia was extremely lengthy and time-consuming.

14.3 Decidability

The question remained whether these axioms and rules of inference are sufficient to decide any mathematical question that can be expressed in these systems. Hilbert believed that every mathematical problem could be solved, and that the truth or falsity of any mathematical proposition could be determined in a finite number of steps. He outlined twenty-three key problems in 1900 that needed to be solved by mathematicians in the twentieth century.

He believed that the formalism of mathematics would allow a mechanical procedure (or algorithm) to determine whether a particular statement was true or false.

The problem of the decidability of mathematics is known as the decision problem (*Entscheidungsproblem*).

The question of the decidability of mathematics had been considered by Leibnitz in the seventeenth century. He had constructed a mechanical calculating machine and wondered if a machine could be built that could determine whether particular mathematical statements are true or false.

Definition 14.1 (*Decidability*) Mathematics is decidable if the truth or falsity of any mathematical proposition may be determined by an algorithm.

Church and Turing independently showed this to be impossible in 1936. Church developed the lambda calculus in the 1930s as a tool to study computability,[3] and he showed that anything that is computable is computable by the lambda calculus. Turing showed that decidability was related to the halting problem for Turing machines, and that therefore if first-order logic were decidable then the halting problem for Turing machines could be solved. However, he had already proved that there was no general algorithm to determine whether a given Turing machine halts. Therefore, first-order logic is undecidable.

The question as to whether a given Turing machine halts or not can be formulated as a first-order statement. If a general decision procedure exists for first-order logic, then the statement of whether a given Turing machine halts or not is within the scope of the decision algorithm. However, Turing had already proved that the halting problem for Turing machines is not computable: i.e., it is not possible algorithmically to decide whether or not any given Turing machine will halt or not. Therefore, since there is no general algorithm that can decide whether any given Turing machine halts, there is no general decision procedure for first-order logic. The only way to determine whether a statement is true or false is to try to solve it. However, if one tries but does not succeed this does not prove that an answer does not exist.

There are first-order theories that are decidable. However, first-order logic that includes Peano's axioms of arithmetic (or any formal system that includes addition and multiplication) cannot be decided by an algorithm. That is, there is no algorithm to determine whether an arbitrary mathematical proposition is true or false. Propositional logic is decidable as there is a procedure (e.g., using a truth table) to determine whether an arbitrary formula is valid[4] in the calculus.

Gödel (Fig. 14.2) proved that first-order predicate calculus is *complete*; i.e., all truths in the predicate calculus can be proved in the language of the calculus.

---

[3] The Church–Turing thesis states that anytime that is computable is computable by lambda calculus or equivalently by a Turing machine.

[4] A well-formed formula is valid if it follows from the axioms of first-order logic. A formula is valid if and only if it is true in every interpretation of the formula in the model.

Fig. 14.2 Kurt Gödel

Definition 14.2 (*Completeness*) A formal system is complete if all the truths in the system can be derived from the axioms and rules of inference.

Gödel's *first incompleteness theorem* showed that first-order arithmetic is incomplete; i.e., there are truths in first-order arithmetic that cannot be proved in the language of the axiomatization of first-order arithmetic. Gödel's *second incompleteness theorem* showed that any formal system extending basic arithmetic cannot prove its own consistency within the formal system.

Definition 14.3 (*Consistency*) A formal system is consistent if there is no formula A such that A and ¬A are provable in the system (i.e., there are no contradictions in the system).

14.4   Computability

Alonzo Church (Fig. 14.3) developed the lambda calculus in the mid-1930s, as part of his work into the foundations of mathematics. Turing published a key paper on computability in 1936, which introduced the theoretical machine known as the Turing machine. This machine is computationally equivalent to the lambda calculus and is capable of performing any conceivable mathematical problem that has an algorithm.

Definition 14.4 (*Algorithm*) An algorithm (or effective procedure) is a finite set of unambiguous instructions to perform a specific task.

A function is *computable* if there is an effective procedure or algorithm to compute f for each value of its domain. The algorithm is finite in length and sufficiently detailed so that a person can execute the instructions in the algorithm. The execution of the algorithm will halt in a finite number of steps to produce the value of $f(x)$ for all x in the domain of f. However, if x is not in the domain of

Fig. 14.3 Alonzo Church

f then the algorithm may produce an answer saying so, or it may get stuck, or it may run forever never halting.

The *Church–Turing thesis* states that *any computable function may be computed by a Turing machine.* There is overwhelming evidence in support in support of this thesis, including the fact that alternative formalizations of computability in terms of lambda calculus, recursive function theory, and Post systems have all been shown to be equivalent to Turing machines.

A Turing machine consists of a head and a potentially infinite tape that is divided into cells. Each cell on the tape may be either blank or printed with a symbol from a finite alphabet of symbols. The input tape may initially be blank or have a finite number of cells containing symbols.

At any step, the head can read the contents of a frame. The head may erase a symbol on the tape, leave it unchanged, or replace it with another symbol. It may then move one position to the right, one position to the left, or not at all. If the frame is blank, the head can either leave the frame blank or print one of the symbols.

Turing believed that a human with finite equipment and with an unlimited supply of paper could do every calculation. The unlimited supply of paper is formalized in the Turing machine by a tape marked off in cells.

The Turing machine is a simple theoretical machine, but it is equivalent to an actual physical computer in the sense that they both compute exactly the same set of functions. A Turing machine is easier to analyse and prove things about than a real computer. The formal definition of a Turing machine as *a* 7-tuple $M = (Q, \Gamma, b, \Sigma, \delta, q_0, F)$ is given in Chap. 13.

A Turing machine is essentially a finite state machine (FSM) with an unbounded tape. The machine may read from and write to the tape, and the tape provides memory and acts as the store. The finite state machine is essentially the control unit of the machine, whereas the tape is a potentially infinite and unbounded store. A real computer has a large but finite store, whereas the store in a Turing machine is potentially infinite. However, the store in a real computer may be extended with backing tapes and discs and in a sense may be regarded as unbounded. The maximum amount of tape that may be read or written within n steps is n.

A Turing machine has an associated set of rules that defines its behaviour. These rules are defined by the transition function that specify the actions that a machine will perform with respect to a particular input. The behaviour will depend on the current state of the machine and the contents of the tape.

A Turing machine may be programmed to solve any problem for which there is an algorithm. However, if the problem is unsolvable then the machine will either stop in a non-accepting state or compute forever. The solvability of a problem may not be determined beforehand, but there is, of course, some answer (i.e., either the machine either halts or computes forever).

Turing showed that there was no solution to the decision problem (*Entscheidungsproblem*) posed by Hilbert. Hilbert believed that the truth or falsity of a mathematical problem may always be determined by a mechanical procedure, and he believed that first-order logic is decidable: i.e., there is a decision procedure to determine if an arbitrary formula is a theorem of the logical system.

Turing also introduced the concept of a Universal Turing Machine, and this machine is able to simulate any other Turing machine. Turing's results on computability were proved independently of Church's lambda calculus equivalent results in computability. Turing's studied at Princeton University in 1937 and 1938 and was awarded a PhD from the university in 1938. His research supervisor was Alonzo Church.[5]

Question 14.1 (Halting Problem)
Given an arbitrary program is there an algorithm to decide whether the program will finish running or will it continue running forever? Another words, given a program and an input will the program eventually halt and produce an output or will it run forever?

Note (Halting Problem)
The halting problem was one of the first problems that was shown to be undecidable: i.e., there is no general decision procedure or algorithm that may be applied to an arbitrary program and input to decide whether the program halts or not when run with that input.

---

[5] Alonzo Church was a famous American mathematician and logician who developed the lambda calculus. He also showed that Peano arithmetic and first-order logic were undecidable. Lambda calculus is equivalent to Turing machines and whatever may be computed is computable by Lambda calculus or a Turing machine.

Proof We assume that there is an algorithm (i.e., a computable function $H(i, j)$) that takes any program i (program i refers to the ith program in the enumeration of all the programs) and arbitrary input j to the program such that:

$$H(i, j) = \begin{cases} 1 & \text{If program } i \text{ halts on input } j. \\ 0 & \text{otherwise} \end{cases}$$

We then employ a diagonalization argument[6] to show that every computable total function f with two arguments differs from the desired function H. First, we construct a partial function g from any computable function f with two arguments such that g is computable by some program e.

$$g(i) = \begin{cases} 0 & \text{if } f(i, i) = 0 \\ \text{undefined} & \text{otherwise} \end{cases}$$

There is a program e that computes g and this program is one of the programs in which the halting problem is defined. One of the following two cases must hold:

$$g(e) = f(e, e) = 0 \tag{14.1}$$

In this case $H(e, e) = 1$ because e halts on input e.

$$g(e) \text{ is undefined and } f(e, e) \neq 0. \tag{14.2}$$

In this case $H(e, e) = 0$ because the program e does not halt on input e.

In either cases, f is not the same function as H. Further, since f was an arbitrary total computable function all such functions must differ from H. Hence, the function H is not computable, and there is no such algorithm to determine whether an arbitrary Turing machine halts for an input x. Therefore, the halting problem is not decidable.

14.5   Computational Complexity

An algorithm is of little practical use if it takes millions of years to compute particular instances. There is a need to consider the efficiency of the algorithm due to practical considerations. Chapter 20 discusses cryptography and the RSA algorithm, and the security of the RSA encryption algorithm is due to the fact that there is no known efficient algorithm to determine the prime factors of a large number.

---

[6] This is similar to Cantor's diagonalization argument that shows that the real numbers are uncountable. This argument assumes that it is possible to enumerate all real numbers between 0 and 1, and it then constructs a number whose nth decimal differs from the nth decimal position in the nth number in the enumeration. If this holds for all n then the newly defined number is not among the enumerated numbers.

There are often slow and fast algorithms for the same problem, and a measure of complexity is the number of steps in a computation. An algorithm is of *time complexity* $f(n)$ if for all n and all inputs of length n the execution of the algorithm takes at most $f(n)$ steps.

An algorithm is said to be *polynomially bounded* if there is a polynomial $p(n)$ such that for all n and all inputs of length n the execution of the algorithm takes at most $p(n)$ steps. The notation P is used for all problems that can be solved in polynomial time.

A problem is said to be *computationally intractable* if it may not be solved in polynomial time—that is, there is no known algorithm to solve the problem in polynomial time.

A problem L is said to be in the set *NP* (non-deterministic polynomial time problems) if any given solution to L can be verified quickly in polynomial time. A problem is *NP complete* if it is in the set NP of non-deterministic polynomial time problems, and it is also in the class of *NP hard* problems. A key characteristic to NP complete problems is that there is no known fast solution to them, and the time required to solve the problem using known algorithms increases quickly as the size of the problem grows. Often, the time required to solve the problem is in billions or trillions of years. Although any given solution can be verified quickly there is no known efficient way to find a solution.

14.6   Review Questions

1. Explain computability and decidability.
2. What were the goals of logicism and formalism and how successful were these movement in mathematics?
3. What is a formal system?
4. Explain the difference among consistency, completeness, and decidability.
5. Describe a Turing machine and explain its significance in computability.
6. Describe the halting problem and show that it is undecidable.
7. Discuss the complexity of an algorithm and explain terms such as "poly-nomial bounded", "computationally intractable", and "NP complete".

14.7   Summary

This chapter provided an introduction to computability and decidability. The intuitive meaning of computability is that in terms of an algorithm (or effective procedure) that specifies a set of instructions to be followed to solve the problem. Another words, a function f is computable if there exists an algorithm that produces the value of f correctly for each possible argument of f. The computation of f for a particular argument x just involves following the instructions in the algorithm, and it produces the result $f(x)$ in a finite number of steps if x is in the domain of f.

The concept of computability may be made precise in several equivalent ways such as Church's lambda calculus, recursive function theory, or by the theoretical Turing machines. The Turing machine is a mathematical machine with a potentially infinite tape divided into frames (or cells) in which very basic operations can be carried out. The set of functions that are computable are those that are computable by a Turing machine.

A formal system contains meaningless symbols together with rules for manipulating them and is generally intended to represent some aspect of the real world. The individual formulae are certain finite sequences of symbols obeying the syntactic rules of the formal language. A formal system consists of a formal language, a set of axioms, and rules of inference.

Church and Turing independently showed in 1936 that mathematics is not decidable. In other words it is not possible to determine the truth or falsity of any mathematical proposition by an algorithm.

Turing had already proved that the halting problem for Turing machines is not computable: i.e., it is not possible algorithmically to decide whether a given Turing machine will halt or not. He then applied this result to first-order logic to show that it is undecidable. That is, the only way to determine whether a statement is true or false is to try to solve it.

The complexity of an algorithm was discussed, and it was noted that an algorithm is of little practical use if it takes millions of years to compute the solution. There is a need to consider the efficiency of the algorithm due to practical considerations. The class of polynomial time bound problems and non-deterministic polynomial time problems were considered, and it was noted that the security of various cryptographic algorithms is due to the fact that there are no time-efficient algorithms to determine the prime factors of large integers.

The reader is referred to [1] for a more detailed account of decidability and computability.

Reference

1. Rozenberg G, Salomaa A (1994) Cornerstones of undecideability. Prentice Hall

Software Reliability and Dependability

15

Key Topics

Software reliability

Software reliability model

System availability

Dependability

Computer Security

Safety critical systems

Cleanroom

15.1 Introduction

This chapter gives an introduction to the important area of software reliability and dependability, and it discusses important topics in software engineering such as software reliability; software availability; software reliability models; the Cleanroom methodology; dependability and its various dimensions; security engineering; and safety critical systems.

Software reliability is the probability that the program works without failure for a period of time, and it is usually expressed as the mean time to failure. It is different from hardware reliability, which is characterized by components that physically wear out over time, whereas software is intangible and software failures are due to design and implementation errors. In other words, software is either correct or incorrect when it is designed and developed, and it does not physically deteriorate with time.

© The Author(s), under exclusive license to Springer Nature Switzerland AG 2023
G. O'Regan, *Mathematical Foundations of Software Engineering*,
Texts in Computer Science, https://doi.org/10.1007/978-3-031-26212-8_15

Harlan Mills and others at IBM developed the Cleanroom approach to software development, and the process is described in [1]. It involves the application of statistical techniques to calculate a software reliability measure based on the expected usage of the software.[1] This involves executing tests chosen from the population of all possible uses of the software in accordance with the probability of its expected use. Statistical usage testing has been shown to be more effective in finding defects that lead to failure than coverage testing.

Models are simplifications of the reality, and a good model allows accurate predictions of future behaviour to be made. A model is judged effective if there is good empirical evidence to support it, and a good software reliability model will have good theoretical foundations and realistic assumptions. The extent to which the software reliability model can be trusted depends on the accuracy of its predictions, and empirical data will need to be gathered to judge its accuracy. A good software reliability model will give good predictions of the reliability of the software.

It is essential that software that is widely used is dependable, which means that the software is available whenever required, and that it operates safely and reliably without any adverse side effects (e.g., the software problems with the Therac-25 radiography machine led to several patients receiving massive overdoses in radiation in the mid-1980s leading to serious injury and death of several patients as discussed in [2]).

Today, billions of computers are connected to the Internet, and this has led to a growth in attacks on computers. It is essential that computer security is carefully considered, and developers need to be aware of the threats facing a system and techniques to eliminate them. The developers need to be able to develop secure systems that are able to deal with and recover from external attacks.

15.2   Software Reliability

The design and development of high-quality software has become increasingly important for society. The hardware field has been very successful in developing sound reliability models, which allow useful predictions of how long a hardware component (or product) will function to be provided. This has led to a growing interest in the software field in the development of a sound software reliability model. Such a model would provide a sound mechanism to predict the reliability of the software prior to its deployment at the customer site, as well as confidence that the software is fit for purpose and safe to use.

Definition 15.1 (*Software reliability*) It is the probability that the program works without failure for a specified length of time, and it is a statement of the future

---

[1] The expected usage of the software (or operational profile) is a quantitative characterization (usually based on probability) of how the system will be used.

behaviour of the software. It is generally expressed in terms of the *mean time to failure* (MTTF) or the *mean time between failure* (MTBF).

Statistical sampling techniques are often employed to predict the reliability of hardware, as it is not feasible to test all items in a production environment. The quality of the sample is then used to make inferences on the quality of the entire population, and this approach is effective in manufacturing environments where variations in the manufacturing process often lead to defects in the physical products.

There are similarities and differences between hardware and software reliability. A hardware failure generally arises due to a component wearing out due to its age, and often a replacement component is required. Many hardware components are expected to last for a certain period of time, and the variation in the failure rate of a hardware component is often due to variations in the manufacturing process and to the operating environment of the component. Good hardware reliability predictors have been developed, and each hardware component has an expected mean time to failure. The reliability of a product may then be determined from the reliability of the individual components.

Software is an intellectual undertaking involving a team of designers and programmers. It does not physically wear out as such, and software failures manifest themselves from particular user inputs. Each copy of the software code is identical, and the software code is either correct or incorrect. That is, software failures are due to design and implementation errors, rather than to the software physically wearing out over time. The software community has not yet developed a sound software reliability predictor model.

The software population to be sampled consists of all possible execution paths of the software, and since this is potentially infinite it is generally not possible to perform exhaustive testing. The way in which the software is used (i.e., the inputs entered by the users) will impact upon its perceived reliability. Let I_f represent the fault set of inputs (i.e., $i_f \in I_f$ and only if the input of i_f by the user leads to failure). The randomness of the time to software failure is due to the unpredictability in the selection of an input $i_f \in I_f$. It may be that the elements in I_f are inputs that are rarely used, and therefore the software will be perceived as reliable.

Statistical usage testing may be used to make predictions on the future performance and reliability of the software. This requires an understanding of the expected usage profile of the system, as well as the population of all possible usages of the software. The sampling is done in accordance with the expected usage profile, and a software reliability measure is calculated.

15.2.1 Software Reliability and Defects

The release of an unreliable software product may result in damage to property or injury (including loss of life) to a third party. Consequently, companies need to be confident that their software products are fit for use prior to their release. The

Table 15.1 Adam's 1984 study of software failures of IBM products

Rare						Frequent		
	1	2	3	4	5	6	7	8
MTTF (years)	5000	1580	500	158	50	15.8	5	1.58
Avg % fixes	33.4	28.2	18.7	10.6	5.2	2.5	1.0	0.4
Prob failure	0.008	0.021	0.044	0.079	0.123	0.187	0.237	0.300

project team needs to conduct extensive inspections and testing of the software, as well as considering all associated risks prior to its release.

Objective product quality criteria may be set (e.g., 100% of tests performed and passed) that must be satisfied prior to the release of the product. This provides a degree of confidence that the software has the desired quality, and is fit for purpose. However, these results are historical in the sense that they are a statement of past and present quality. The question is whether the past behaviour and performance provide a sound indication of future behaviour.

Software reliability models are an attempt to predict the future reliability of the software and to assist in deciding on whether the software is ready for release. A defect does not always result in a failure, as it may occur on a rarely used execution path. Studies indicate that many observed failures arise from a small proportion of the existing defects.

Adam's 1984 case study of defects in IBM software [3] indicates that over 33% of the defects led to an observed failure with mean time to failure greater than 5000 years, whereas less than 2% of defects led to an observed failure with a mean time to failure of less than five years. This suggests that a small proportion of defects often lead to almost all of the observed failures (Table 15.1).

The analysis shows that 61.6% of all fixes (Group 1 and 2) were for failures that will be observed less than once in 1580 years of expected use, and that these constitute only 2.9% of the failures observed by typical users. On the other hand, groups 7 and 8 constitute 53.7% of the failures observed by typical users and only 1.4% of fixes.

This case study indicates that *coverage testing* is not cost effective in increasing MTTF. *Usage testing*, in contrast, would allocate 53.7% of the test effort to fixes that will occur 53.7% of the time for a typical user. Harlan Mills has argued [4] that the data in the table shows that usage testing is 21 times more effective than coverage testing.

There is a need to be careful with *reliability growth models*, as there is no tangible growth in reliability unless the corrected defects are likely to manifest themselves as a failure.[2] Many existing software reliability growth models assume that all remaining defects in the software have an equal probability of failure and

---

[2] We are assuming that the defect has been corrected perfectly with no new defects introduced by the changes made.

Table 15.2 New and old version of software

Similarities and differences between new/old version
• The new version of the software is identical to the previous version except that the identified defects have been corrected
• The new version of the software is identical to the previous version, except that the identified defects have been corrected, but the developers have introduced some new defects
• No assumptions can be made about the behaviour of the new version of the software until further data is obtained

that the correction of a defect leads to an increase in software reliability. These assumptions are questionable.

The defect count and defect density may be poor predictors of operational reliability, and an emphasis on removing a large number of defects from the software may not be sufficient to achieve high reliability.

The correction of defects in the software leads to a newer version of the software, and many software reliability models assume reliability growth: i.e., the new version is more reliable than the older version as several identified defects have been corrected. However, in some sectors such as the safety critical field the view is that the new version of a program is a new entity and that no inferences may be drawn until further investigation has been done. There are a number of ways to interpret the relationship between the new version of the software and the older version (Table 15.2).

The safety critical industry (e.g., the nuclear power industry) takes the conservative viewpoint that any change to a program creates a new program. The new program is therefore required to demonstrate its reliability, and so extensive testing needs to be performed.

15.2.2 Cleanroom Methodology

Harlan Mills and others at IBM developed the Cleanroom methodology as a way to develop high-quality software [4]. Cleanroom helps to ensure that the software is released only when it has achieved the desired quality level, and the probability of zero defects is very high.

The way in which the software is used will impact on its perceived quality and reliability. Failures will manifest themselves on certain input sequences, and as the input sequences will vary among users, the result will be different perceptions of the reliability of the software among the users. The knowledge of how the software will be used allows the software testing to focus on verifying the correctness of common everyday tasks carried out by users.

Therefore, it is important to determine the operational profile of the users to enable effective software testing to be performed. This may be difficult to determine and could change over time, as users may potentially change their behaviour as their needs evolve. The determination of the operational profile involves identifying the common operations to be performed, and the probability of each operation being performed.

Cleanroom employs *statistical usage testing* rather than coverage testing, and this involves executing tests chosen from the population of all possible uses of the software in accordance with the probability of its expected use. The software reliability measure is calculated by statistical techniques based on the expected usage of the software, and Cleanroom provides a certified mean time to failure of the software.

Coverage testing involves designing tests that cover every path through the program, and this type of testing is as likely to find a rare execution failure as well as a frequent execution failure. However, it is essential to find failures that occur on frequently used parts of the system.

The advantage of usage testing (that matches the actual execution profile of the software) is that it has a better chance of finding execution failures on frequently used parts of the system. This helps to maximize the expected mean time to failure of the software.

The Cleanroom software development process and calculation of the software reliability measure are described in [1], and the Cleanroom development process enables engineers to deliver high-quality software on time and on budget. Some of the benefits of the use of Cleanroom on projects at IBM are described in [4] and summarized in Table 15.3.

Table 15.3 Cleanroom results in IBM

Project	Results
Flight control project (1987) 33KLOC	Completed ahead of schedule Error-fix effort reduced by factor of five 2.5 errors KLOC before any execution
Commercial product (1988)	Deployment failures of 0.1/KLOC Certification testing failures 3.4/KLOC Productivity 740 LOC/month
Satellite control (1989) 80 KLOC (Partial cleanroom)	50% improvement in quality Certification testing failures of 3.3/KLOC Productivity 780 LOC/month 80% improvement in productivity
Research project (1990) 12 KLOC	Certified to 0.9978 with 989 test cases

Table 15.4 Characteristics of good software reliability model

Characteristics of good software reliability model
Good theoretical foundation
Realistic assumptions
Good empirical support
As simple as possible (Ockham's Razor)
Trustworthy and accurate

15.2.3 Software Reliability Models

Models are simplifications of the reality, and a good model allows accurate predictions of future behaviour to be made. It is important to determine the adequacy of the model, and this is done by model exploration, and determining the extent to which it explains the actual manifested behaviour, as well as the accuracy of its predictions.

A model is judged effective if it has accurate predictions and has good empirical evidence to support it, and more accurate models are sought to replace inadequate models. Models are often modified (or replaced) over time, as further facts and observations lead to aberrations that cannot be explained with the current model. A good software reliability model will have the following characteristics (Table 15.4).

The underlying mathematics used in the calculation of software reliability (i.e., probability and statistics) is discussed in Chaps. 22 and 23. There are several existing software reliability predictor models employed (Table 15.5) with varying degrees of success. Some of these models just compute defect counts rather than estimating software reliability in terms of mean time to failure. They may be categorized into:

- *Size and Complexity Metrics*
  These are used to predict the number of defects that a system will reveal in operation or testing.
- *Operational Usage Profile*
  These predict failure rates based on the expected operational usage profile of the system. The number of failures encountered is determined, and the software reliability is predicted (e.g., Cleanroom and its prediction of the MTTF).
- *Quality of the Development Process*
  These predict failure rates based on the process maturity of the software development process in the organization (e.g., CMMI maturity).

The extent to which the software reliability model can be trusted depends on the accuracy of its predictions, and empirical data will need to be gathered to make a judgement. It may be acceptable to have a little inaccuracy during the early stages of prediction, provided the predictions of operational reliability are close to the

Table 15.5 Software reliability models

Model	Description	Comments
Jelinski/Moranda model	The failure rate is a Poisson process[a] and is proportional to the current defect content of program. The initial defect count is N; the initial failure rate is $N\varphi$; it decreases to $(N-1)\varphi$ after the first fault is detected and eliminated, and so on. The constant φ is termed the proportionality constant	Assumes defects are corrected perfectly and no new defects are introduced. Assumes each fault contributes the same amount to failure rate
Littlewood/Verrall model	Successive execution time between failures is independent exponentially distributed random variables.[b] Software failures are the result of the particular inputs and faults introduced from the correction of defects	Does not assume perfect correction of defects
Seeding and tagging	This is analogous to estimating the fish population of a lake (Mills). A known number of defects are inserted into a software program, and the proportion of these identified during testing determined. Another approach (Hyman) is to regard the defects found by one tester as tagged, and then to determine the proportion of tagged defects found by a 2nd independent tester	Estimate of the total number of defects in the software but not a not s/w reliability predictor. Assumes all faults equally likely to be found and introduced faults representative of existing
Generalized poisson model	The number of failures observed in ith time interval τ_i has a Poisson distribution with mean $\phi(N-M_{i-1})\,\tau_i^{\alpha}$ where N is the initial number of faults; M_{i-1} is the total number of faults removed up to the end of the $(i-1)$th time interval; and ϕ is the proportionality constant	Assumes faults are removed perfectly at end of time interval

[a]The Poisson process is a widely used counting process, and especially in counting the occurrence of certain events that appear to happen at a certain rate but at random. A Poisson random variable is of the form $P\{X = i\} = e^{-\lambda}\,\lambda^i / i!$.

[b]The exponential distribution is used to model the time between the occurrence of events in an interval of time. The density function is given by $f(x) = \lambda e^{-\lambda x}$.

observations. A model that gives overly optimistic results is termed 'optimistic', whereas a model that gives overly pessimistic results is termed 'pessimistic'.

The assumptions in the reliability model need to be examined to determine whether they are realistic. Several software reliability models have questionable assumptions such as:

- All defects are corrected perfectly.
- Defects are independent of one another.
- Failure rate decreases as defects are corrected.
- Each fault contributes the same amount to the failure rate.

15.3 Dependability

Software is ubiquitous and is important to all sections of society, and so it is essential that widely used software is dependable (or trustworthy). In other words, the software should be available whenever required, as well as operating properly, safely and reliably, without any adverse side effects or security concerns. It is essential that the software used in systems in the safety critical and security critical fields is dependable, as the consequence of failure (e.g., the failure of a nuclear power plant) could be massive damage leading to loss of life or endangering the lives of the public.

Dependability engineering is concerned with techniques to improve the dependability of systems, and it involves the use of a rigorous design and development process to minimize the number of defects in the software. A dependable system is generally designed for fault tolerance, where the system can deal with (and recover from) faults that occur during software execution. Such a system needs to be secure and able to protect itself from accidental or deliberate external attacks. Table 15.6 lists several dimensions of dependability.

Modern software systems are subject to attack by malicious software such as viruses that change the behaviour of the software, or corrupt data causing the system to become unreliable. Other malicious attacks include a denial-of-service attack that negatively impacts the system's availability.

Table 15.6 Dimensions of dependability

Dimension	Description
Availability	System is available for use at any time
Reliability	The system operates correctly and is trustworthy
Safety	The system does not injure people or damage the environment
Security	The system prevents unauthorized intrusions

The design and development of dependable software needs to include protection measures that protect against external attacks that could compromise the availability and security of the system. Further, a dependable system needs to include recovery mechanisms to enable normal service to be restored as quickly as possible following an attack.

Dependability engineering is concerned with techniques to improve the dependability of systems, and in designing dependable systems. A dependable system will generally be developed using an explicitly defined repeatable process, and it may employ *redundancy* (spare capacity) and *diversity* (different types) to achieve reliability.

There is a trade-off between dependability and the performance of the system, as dependable systems often need to carry out extra checks to monitor themselves and to check for erroneous states, and to recover from faults before failure occurs. This inevitably leads to increased costs in the design and development of dependable systems.

Software availability is the percentage of the time that the software system is running, and is a measure of the uptime/downtime of the software during a particular time period. The downtime refers to a period of time when the software is unavailable for use (including planned and unplanned outages), and many companies aim to develop software that is available for use 99.999% of the time in the year (i.e., a downtime of less than five minutes per annum). This goal is known as *five nines*, and it is a common goal in the telecommunications sector.

Safety critical systems are systems where it is essential that the system is safe for the public and that people or the environment are not harmed in the event of system failure. These include aircraft control systems and process control systems for chemical and nuclear power plants. The failure of a safety critical system could in some situations lead to loss of life or serious economic damage.

Formal methods are discussed in Chap. 16, and they provide a precise way of specifying the requirements of the proposed system, and demonstrating (using mathematics) that key properties are satisfied in the formal specification. Further, they may be used to show that the implemented program satisfies its specification. The use of formal methods generally leads to increased confidence in the correctness of safety critical and security critical systems.

The security of the system refers to its ability to protect itself from accidental or deliberate external attacks, which are common today since most computers are networked and connected to the Internet. There are various security threats in any networked system including threats to the confidentiality and integrity of the system and its data, and threats to the availability of the system.

Therefore, controls are required to enhance security and to ensure that attacks are unsuccessful. Encryption is one way to reduce system vulnerability, as encrypted data is unreadable to the attacker. There may be controls that detect and repel attacks, and these controls are used to monitor the system and to take action to shut down parts of the system or restrict access in the event of an attack. There may be controls that limit exposure (e.g., insurance policies and automated backup strategies) that allow recovery from the problems introduced.

It is important to have a reasonable level of security as otherwise all of the other dimensions of dependability (reliability, availability, and safety) are compromised. Security loopholes may be introduced in the development of the system, and so care needs to be taken to prevent hackers from exploiting security vulnerabilities.

Risk analysis plays a key role in the specification of security and dependability requirements, and this involves identifying risks that can result in serious incidents. This leads to the generation of specific security requirements as part of the system requirements to ensure that these risks do not materialize, or if they do materialize then serious incidents will not materialize.

15.4   Computer Security

The introduction of the World Wide Web in the early 1990s transformed the world of computing, and it led inexorably to more and more computers being connected to the Internet. This has subsequently led to an explosive growth in attacks on computers and systems, as hackers and malicious software seek to exploit known security vulnerabilities. It is therefore essential to develop secure systems that can deal with and recover from such external attacks.

Hackers will often attempt to steal confidential data and to disrupt the services being offered by a system. Security engineering is concerned with the development of systems that can prevent such malicious attacks and recover from them. It has become an important part of software and system engineering, and software developers need to be aware of the threats facing a system and develop solutions to eliminate them.

Hackers may probe parts of the system for weaknesses, and system vulnerabilities may lead to attackers gaining unauthorized access to the system. There is a need to conduct a risk assessment of the security threats facing a system early in the software development process, and this will lead to several security requirements for the system.

The system needs to be designed for security, as it is difficult to add security after it has been implemented. Security loopholes may be introduced in the development of the system, and so care needs to be taken to prevent these as well as prevent hackers from exploiting security vulnerabilities. There may be controls that detect and repel attacks, and these monitor the system and take appropriate action to restrict access in the event of an attack.

The choice of architecture and how the system is organized are fundamental to the security of the system, and different types of systems will require different technical solutions to provide an acceptable level of security to its users. There following guidelines for designing secure systems are described in [5]:

- Security decisions should be based on the security policy.
- A security critical system should fail securely.
- A secure system should be designed for recoverability.
- A balance is needed between security and usability.

- A single point of failure should be avoided.
- A log of user actions should be maintained.
- Redundancy and diversity should be employed.
- Organization information in system into compartments.

It is important to have a reasonable level of security, as otherwise all of the other dimensions of dependability are compromised.

15.5   System Availability

System availability is the percentage of time that the software system is running without downtime, and robust systems will generally aim to achieve 5-nines availability (i.e., 99.999% availability). This is equivalent to approximately five minutes of downtime (including planned / unplanned outages) per year. The availability of a system is measured by its performance when a subsystem fails, and its ability to resume service in a state close to the original state. A fault-tolerant system continues to operate correctly (possibly at a reduced level) after some part of the system fails, and it aims to achieve 100% availability.

System availability and software reliability are related, with availability measuring the percentage of time that the system is operational, and reliability measuring the probability of failure-free operation over a period of time. The consequence of a system failure may be to freeze or crash the system, and system availability is measured by how long it takes to recover and restart after a failure. A system may be unreliable and yet have good availability metrics (fast restart after failure), or it may be highly reliable with poor availability metrics (taking a long time to recover after a failure).

Software that satisfies strict availability constraints is usually reliable. The downtime generally includes the time needed for activities such as rebooting a machine, upgrading to a new version of software, planned and unplanned outages. It is theoretically possible for software to be highly unreliable but yet to have good availability metrics or for software that is highly reliable to have poor availability metrics. Consequently, care is required before drawing conclusions between software reliability and software availability metrics.

15.6   Safety Critical Systems

A safety critical system is a system whose failure could result in significant economic damage or loss of life. There are many examples of safety critical systems including aircraft flight control systems and missile systems. It is therefore essential to employ rigorous processes in their design and development, and testing alone is usually insufficient to verifying the correctness of a safety critical system.

The safety critical industry takes the view that any change to safety critical software creates a new program. The new program is therefore required to demonstrate that it is reliable and safe to the public, and so extensive testing needs to be performed. Other techniques such as formal verification and model checking may be employed to provide an extra level of assurance in the correctness of the safety critical system.

Safety critical systems need to be dependable and available for use whenever required. Safety critical software must operate correctly and reliably without any adverse side effects. The consequence of failure (e.g., the failure of a weapons system) could be massive damage, leading to loss of life or endangering the lives of the public.

Safety critical systems are generally designed for fault tolerance, where the system can deal with (and recover from) faults that occur during execution. Fault tolerance is achieved by anticipating exceptional events and designing the system to handle them. A fault-tolerant system is designed to fail safely, and programs are designed to continue working (possibly at a reduced level of performance) rather than crashing after the occurrence of an error or exception. Many fault-tolerant systems mirror all operations, where each operation is performed on two or more duplicate systems, and so if one fails then the other system can take over.

The development of a safety critical system needs to be rigorous, and subject to strict quality assurance to ensure that the system is safe to use and that the public will not be in danger. This involves rigorous design and development processes to minimize the number of defects in the software, as well as comprehensive testing to verify its correctness. Formal methods are often employed in the development of safety critical systems (Chap. 16).

15.7 Review Questions

1. Explain the difference between software reliability and system availability.
2. What is software dependability?
3. Explain the significance of Adam's 1984 study of failures at IBM.
4. Describe the Cleanroom methodology.
5. Describe the characteristics of a good software reliability model.
6. Explain the relevance of security engineering.
7. What is a safety critical system?

15.8 Summary

This chapter gave an introduction to some important topics in software engineering including software reliability and the Cleanroom methodology; dependability; availability; security; and safety critical systems.

Software reliability is the probability that the program works without failure for a period of time, and it is usually expressed as the mean time to failure. Cleanroom involves the application of statistical techniques to calculate software reliability, and it is based on the expected usage of the software.

It is essential that software used in the safety and security critical fields is dependable, with the software available when required, as well as operating safely and reliably without any adverse side effects. Many of these systems are fault tolerant and are designed to deal with (and recover) from faults that occur during execution.

Such a system needs to be secure and able to protect itself from external attacks and needs to include recovery mechanisms to enable normal service to be restored as quickly as possible. In other words, it is essential that if the system fails then it fails safely.

Today, billions of computers are connected to the Internet, and this has led to a growth in attacks on computers. It is essential that developers are aware of the threats facing a system and are familiar with techniques to eliminate them.

References

1. O'Regan G (2006) Mathematical approaches to software quality. Springer
2. O'Regan G (2022) Concise guide to software engineering, 2nd edn. Springer
3. Adams E (1984) Optimizing preventive service of software products. IBM Res J 28(1):2–14
4. Cobb RH, Mills HD (1990) Engineering software under statistical quality control. IEEE Software
5. Sommerville I (2011) Software engineering, 9th edn. Pearson

Overview of Formal Methods

16

> **Key Topics**
>
> Formal Specification
>
> Vienna Development Method
>
> Z Specification Language
>
> B Method
>
> Model-oriented approach
>
> Axiomatic approach
>
> Process Calculus
>
> Refinement
>
> Finite State Machines
>
> Model Checking
>
> Usability of Formal Methods

16.1 Introduction

The term "*formal methods*" refer to various mathematical techniques used for the formal specification and development of software. They consist of a formal specification language and employ a collection of tools to support the syntax checking of the specification, as well as the proof of properties of the specification. They allow questions to be asked about what the system does independently of the implementation.

© The Author(s), under exclusive license to Springer Nature Switzerland AG 2023
G. O'Regan, *Mathematical Foundations of Software Engineering*,
Texts in Computer Science, https://doi.org/10.1007/978-3-031-26212-8_16

The use of mathematical notation avoids speculation about the meaning of phrases in an imprecisely worded natural language description of a system. Natural language is inherently ambiguous, whereas mathematics employs a precise rigorous notation. Spivey [1] defines formal specification as:

Definition 16.1 (*Formal Specification*) Formal specification is the use of mathematical notation to describe in a precise way the properties that an information system must have, without unduly constraining the way in which these properties are achieved.

The formal specification thus becomes the key reference point for the different parties involved in the construction of the system. It may be used as the reference point for the requirements, program implementation, testing, and program documentation. It thus promotes a common understanding for all those concerned with the system. The term "*formal methods*" is used to describe a formal specification language and a method for the design and implementation of a computer system. Formal methods may be employed at a number of levels:

- Formal specification only (program developed informally)
- Formal specification, refinement, and verification (some proofs)
- Formal specification, refinement, and verification (with extensive theorem proving).

The specification is written in a mathematical language, and the implementation may be derived from the specification via stepwise refinement.[1] The refinement step makes the specification more concrete and closer to the actual implementation. There is an associated proof obligation to demonstrate that the refinement is valid and that the concrete state preserves the properties of the abstract state. Thus, assuming that the original specification is correct and the proof of correctness of each refinement step is valid, then there is a very high degree of confidence in the correctness of the implemented software.

Stepwise refinement is illustrated as follows: the initial specification S is the initial model M_0; it is then refined into the more concrete model M_1, and M_1 is then refined into M_2, and so on until the eventual implementation $M_n = E$ is produced.

$$S = M_0 \sqsubseteq M_1 \sqsubseteq M_2 \sqsubseteq M_3 \sqsubseteq \ldots \sqsubseteq M_n = E$$

---

[1] It is questionable whether stepwise refinement is cost effective in mainstream software engineering, as it involves rewriting a specification *ad nauseum*. It is time-consuming to proceed in refinement steps with significant time also required to prove that the refinement step is valid. It is more relevant to the safety critical field. Others in the formal methods field may disagree with this position.

Requirements are the foundation of the system to be built, and irrespective of the best design and development practices, the product will be incorrect if the requirements are incorrect. The objective of requirements validation is to ensure that the requirements reflect what is actually required by the customer (in order to build the right system). Formal methods may be employed to model the requirements, and the model exploration yields further desirable or undesirable properties.

Formal methods provide the facility to prove that certain properties are true of the specification, and this is valuable, especially in safety critical and security critical applications. The properties are a logical consequence of the mathematical definition, and the requirements may be amended where appropriate. Thus, formal methods may be employed in a sense to debug the requirements during requirements validation.

The use of formal methods generally leads to more robust software and increased confidence in its correctness. Formal methods may be employed at different levels (e.g., just for specification with the program developed informally). The challenges involved in the deployment of formal methods in an organization include the education of staff in formal specification, as the use of these mathematical techniques may be a culture shock to many staff.

Formal methods have been applied to a diverse range of applications, including the safety and security critical fields to develop dependable software. The applications include the railway sector, microprocessor verification, the specification of standards, and the specification and verification of programs. Parnas and others have criticized formal methods (Table 16.1)

However, formal methods are potentially quite useful and reasonably easy to use. The use of a formal method such as Z or VDM forces the software engineer to be precise and helps to avoid ambiguities present in natural language. Clearly, a formal specification should be subject to peer review to provide confidence in its correctness. New formalisms need to be intuitive to be usable by practitioners, and an advantage of the use of classical mathematics is that it is familiar to students.

16.2   Why Should We Use Formal Methods?

There is a strong motivation to use best practice in software engineering in order to produce software adhering to high-quality standards. Quality problems with software may cause minor irritations or major damage to a customer's business including loss of life. Formal methods are a leading-edge technology that may be of benefit to companies in reducing the occurrence of defects in software products. Brown [2] argues that for the safety critical field that:

Comment 16.1 (Missile Safety)
Missile systems must be presumed dangerous until shown to be safe and that the absence of evidence for the existence of dangerous errors does not amount to evidence for the absence of danger.

Table 16.1 Criticisms of formal methods

No.	Criticism
1	Often the formal specification is as difficult to read as the program[a]
2	Many formal specifications are wrong[b]
3	Formal methods are strong on syntax but provide little assistance in deciding on what technical information should be recorded using the syntax[c]
4	Formal specifications provide a model of the proposed system. However, a precise unambiguous mathematical statement of the requirements is what is needed[d]
5	Stepwise refinement is unrealistic[e]. It is like, for example, deriving a bridge from the description of a river and the expected traffic on the bridge. There is always a need for the creative step in design
6	Many unnecessary mathematical formalisms have been developed rather than using the available classical mathematics[f]

[a] Of course, others might reply by saying that some of Parnas's tables are not exactly intuitive and that the notation he employs in some of his tables is quite unfriendly. The usability of all of the mathematical approaches needs to be enhanced if they are to be taken seriously by industrialists

[b] Obviously, the formal specification must be analysed using mathematical reasoning and tools to provide confidence in its correctness. The validation of a formal specification can be carried out using mathematical proof of key properties of the specification; software inspections; or specification animation

[c] Approaches such as VDM include a method for software development as well as the specification language

[d] Models are extremely valuable as they allow simplification of the reality. A mathematical study of the model demonstrates whether it is a suitable representation of the system. Models allow properties of the proposed requirements to be studied prior to implementation

[e] Stepwise refinement involves rewriting a specification with each refinement step producing a more concrete specification (that includes code and formal specification) until eventually the detailed code is produced. It is difficult and time-consuming but tool support may make refinement easier

[f] Approaches such as VDM or Z are useful in that they add greater rigour to the software development process. They are reasonably easy to learn, and there have been some good results obtained by their use. Classical mathematics is familiar to students, and therefore it is desirable that new formalisms are introduced only where absolutely necessary

This suggests that companies in the safety critical field will need to demonstrate that every reasonable practice was taken to prevent the occurrence of defects. One such practice is the use of formal methods, and its exclusion may need to be justified in some domains. It is quite possible that a software company may be sued for software which injures a third party, and this suggests that companies will need a rigorous quality assurance system to prevent the occurrence of defects.

There is some evidence to suggest that the use of formal methods provides savings in the cost of the project. For example, a 9% cost saving is attributed to the use of formal methods during the CICS project; the T800 project attributes a 12-month reduction in testing time to the use of formal methods. These are discussed in more detail in chapter one of [3].

The use of formal methods is mandatory in certain circumstances. The Ministry of Defence (MOD) in the United Kingdom issued two safety critical standards in the early 1990s related to the use of formal methods in the software development lifecycle.

The first is Defence Standard 00-55, *"The Procurement of safety critical software in defense equipment"* [4] which makes it mandatory to employ formal methods in the development of safety critical software in the UK. The standard mandates the use of formal proof that the most crucial programs correctly implement their specifications.

The other is Def. Stan 00-56 *"Hazard analysis and safety classification of the computer and programmable electronic system elements of defense equipment"* [5]. The objective of this standard is to provide guidance to identify which systems or parts of systems being developed are safety critical and thereby require the use of formal methods. This proposed system is subject to an initial hazard analysis to determine whether there are safety critical parts.

The reaction to these defence standards 00-55 and 00-56 was quite hostile initially, as most suppliers were unlikely to meet the technical and organizational requirements of the standard [6]. The U.K. Defence Standards 0055 and 0056 were later revised to be less prescriptive on the use of formal methods.

16.3   Industrial Applications of Formal Methods

Formal methods have been employed in several domains such as the transport sector, the nuclear sector, the space sector, the defence sector, the semiconductor sector, the financial sector, and the telecoms sector. The extent of the application of formal methods has varied from formal specification only, to specification with inspections, to proofs, to refinement, to test generation, and to model checking. Formal methods are applicable to the regulated sector, and it has been applied to real-time applications in the nuclear industry, the aerospace industry, the security technology area, and the railroad domain. These sectors are subject to stringent regulatory controls to ensure that safety and security are properly addressed.

Several organizations have piloted formal methods with varying degrees of success. IBM developed the VDM specification language at its laboratory in Vienna, and it piloted the Z and B formal specification languages on the CICS (Customer Information Control System) project at its plant in Hursley, England.

The mathematical techniques developed by Parnas (i.e., his requirements model and tabular expressions) were employed to specify the requirements of the A-7 aircraft (as part of a research project for the US Navy).[2] Tabular expressions were also employed for the software inspection of the automated shutdown software of

---

[2] However, the resulting software was never actually deployed on the A-7 aircraft.

the Darlington Nuclear power plant in Canada.[3] These are two successful uses of mathematical techniques in software engineering.

There are examples of the use of formal methods in the railway domain, with GEC Alstom and RATP using B for the formal specification and verification of the computerized signalling system on the Paris Metro. Several examples dealing with the modelling and verification of a railroad gate controller and railway signalling are described in [3]. Clearly, it is essential to verify safety critical properties such as *"when the train goes through the level crossing then the gate is closed"*.

PVS is a mechanized environment for formal specification and verification, and it was developed at SRI in California. It includes a specification language integrated with support tools and an interactive theorem prover. The specification language is based on higher-order logic, and the theorem prover is guided by the user in conducting proof. It has been applied to the verification of hardware and software, and PVS has been used for the formal specification and partial verification of the micro-code of the AAMP5 microprocessor.

A selection of applications of formal methods to industry is presented in [8].

16.4   Industrial Tools for Formal Methods

Formal methods have been criticized for the limited availability of tools to support the software engineer in writing the formal specification and in conducting proof. Many of the early tools were criticized as not being of industrial strength. However, in recent years more advanced tools have become available to support the software engineer's work in formal specification and formal proof, and this is likely to continue in the coming years.

The tools include syntax checkers that determine whether the specification is syntactically correct; specialized editors which ensure that the written specification is syntactically correct; tools to support refinement; automated code generators that generate a high-level language corresponding to the specification; theorem provers to demonstrate the correctness of refinement steps and to identify and resolve proof obligations, as well proving the presence or absence of key properties; and specification animation tools where the execution of the specification can be simulated.

The *B*-Toolkit[4] from *B*-Core is an integrated set of tools that supports the *B*-Method. It provides functionality for syntax and type checking, specification animation, proof obligation generator, an auto-prover, a proof assistant, and code generation. This, in theory, allows the complete formal development from the initial specification to the final implementation, with every proof obligation justified,

---

[3] This was an impressive use of mathematical techniques, and it has been acknowledged that formal methods must play an important role in future developments at Darlington. However, given the time and cost involved in the software inspection of the shutdown software some managers have less enthusiasm in shifting from hardware to software controllers [7].

[4] The source code for the *B*-Toolkit is now available.

leading to a provably correct program. There is also the Atelier *B* tool to support formal specification and development in *B*.

The IFAD Toolbox[5] is a support tool for the VDM-SL specification language, and it provides support for syntax and type checking, an interpreter and debugger to execute and debug the specification, and a code generator to convert from VDM-SL to C++. The Overture Integrated Development Environment (IDE) is an open-source tool for formal modelling and analysis of VDM-SL specifications.

There are various tools for model checking including Spin, Bandera, SMV, and UppAal. These tools perform a systematic check on property *P* in all states and are applicable if the system generates a finite behavioural model. Spin is an open-source tool, and it checks finite-state systems with properties specified by linear temporal logic. It generates a counterexample trace if determines that a property is violated.

There are tools to support theorem provers (see Chap. 19) such as the Boyer-Moore Theorem prover (NQTHM) which was developed at the University of Texas in the late 1970s. It is far more automated than many other interactive theorem provers, but it requires detailed human guidance (with suggested lemmas) for difficult proofs. The user therefore needs to understand the proof being sought and the internals of the theorem prover. Many mathematical theorems have been proved including Gödel's incompleteness theorem.

The HOL system was developed at the University of Cambridge, and it is an environment for interactive theorem proving in a higher-order logic. It requires skilled human guidance and has been used for the verification of microprocessor design. It is a widely used theorem prover.

16.5 Approaches to Formal Methods

There are two key approaches to formal methods: namely the *model-oriented approach* of VDM or Z, and the *algebraic* or *axiomatic approach* of the process calculi such as the calculus communicating systems (CCS) or communicating sequential processes (CSP).

16.5.1 Model-Oriented Approach

The model-oriented approach to specification is based on mathematical models, where a model is a simplification or abstraction of the real world that contains only the essential details. For example, the model of an aircraft will not include the colour of the aircraft, and the objective may be to model the aerodynamics

---

[5] The IFAD Toolbox has been renamed to VDMTools as IFAD sold the VDM Tools to CSK in Japan. The CSK VDM tools are available for worldwide use.

of the aircraft. There are many models employed in the physical world, such as meteorological models that allow weather forecasts to be made.

The importance of models is that they serve to explain the behaviour of a particular entity and may also be used to predict future behaviour. Different models may vary in their ability to explain aspects of the entity under study. One model may be good at explaining some aspects of the behaviour, whereas another model might be good at explaining other aspects. The *adequacy* of a model is a key concept in modelling and reflects the effectiveness of the model in representing the underlying behaviour, and in its ability to predict future behaviour. Model exploration consists of asking questions, and determining whether the model is able to give an effective answer to the particular question. A good model is chosen as a representation of the real world and is referred to whenever there are questions in relation to the aspect of the real world.

It is fundamental to explore the model to determine its adequacy, and to determine the extent to which it explains the underlying physical behaviour, and allows accurate predictions of future behaviour to be made. There may be more than one possible model of a particular entity, for example, the Ptolemaic model and the Copernican model are different models of the solar system. This leads to the question as to which is the best or most appropriate model to use, and on the criteria to use to determine the most suitable model. The ability of the model to explain the behaviour, its simplicity, and its elegance will be part of the criteria. The principle of *"Ockham's Razor"* (law of parsimony) is often used in modelling, and it suggests that the simplest model with the least number of assumptions required should be selected.

The adequacy of the model will determine its acceptability as a representation of the physical world. Models that are ineffective will be replaced with models that offer a better explanation of the manifested physical behaviour. There are many examples in science of the replacement of one theory by a newer one. For example, the Copernican model of the universe replaced the older Ptolemaic model, and Newtonian physics was replaced by Einstein's theories of relativity. The structure of the revolutions that take place in science is described in [9].

Modelling can play a key role in computer science, as computer systems tend to be highly complex, whereas a model allows simplification or an abstraction of the underlying complexity, and it enables a richer understanding of the underlying reality to be gained. The model-oriented approach to software development involves defining an abstract model of the proposed software system, and the model is then explored to determine its suitability as a representation of the system. This takes the form of model interrogation, i.e., asking questions, and determining the extent to which the model can answer the questions. The modelling in formal methods is typically performed via elementary discrete mathematics, including set theory, sequences, functions, and relations.

Various models have been applied to assist with the complexities in software development. These include the Capability Maturity Model (CMM), which is employed as a framework to enhance the capability of the organization in software development; UML, which has various graphical diagrams that are employed to model the requirements and design; and mathematical models that are employed for formal specification.

VDM and *Z* are model-oriented approaches to formal methods. VDM arose from work done at the IBM laboratory in Vienna in formalizing the semantics for the PL/1 compiler in the early 1970s, and it was later applied to the specification of software systems. The origin of the *Z* specification language is in work done at Oxford University in the early 1980s.

16.5.2 Axiomatic Approach

The axiomatic approach focuses on the properties that the proposed system is to satisfy, and there is no intention to produce an abstract model of the system. The required properties and behaviour of the system are stated in mathematical notation. The difference between the axiomatic specification and a model-based approach may be seen in the example of a stack.

The stack includes operators for pushing an element onto the stack and popping an element from the stack. The properties of *pop* and *push* are explicitly defined in the axiomatic approach. The model-oriented approach constructs an explicit model of the stack, and the operations are defined in terms of the effect that they have on the model. The axiomatic specification of the *pop* operation on a stack is given by properties, for example, $pop(push(s, x)) = s$.

Comment 16.2 (Axiomatic Approach)
The property-oriented approach has the advantage that the implementer is not constrained to a particular choice of implementation, and the only constraint is that the implementation must satisfy the stipulated properties.

The emphasis is on specifying the required properties of the system, and implementation issues are avoided. The properties are typically stated using mathematical logic or higher-order logics. Mechanized theorem-proving techniques may be employed to prove results.

One potential problem with the axiomatic approach is that the properties specified may not be satisfied in any implementation. Thus, whenever a "formal axiomatic theory" is developed a corresponding "model" of the theory must be identified, in order to ensure that the properties may be realized in practice. That is, when proposing a system that is to satisfy some set of properties, there is a need to prove that there is at least one system that will satisfy the set of properties.

16.6   Proof and Formal Methods

The nature of theorem proving is discussed in Chap. 19. A mathematical proof typically includes natural language and mathematical symbols, and often many of the tedious details of the proof are omitted. The proof may employ a "*divide and conquer*" technique; i.e., breaking the conjecture down into sub-goals and then attempting to prove each of the sub-goals.

Many proofs in formal methods are concerned with crosschecking the details of the specification, checking the validity of the refinement steps, or checking that certain properties are satisfied by the specification. There are often many tedious lemmas to be proved, and theorem provers[6] play a key role in dealing with them. Machine proof is explicit, and reliance on some brilliant insight is avoided. Proofs by hand are notorious for containing errors or jumps in reasoning, while machine proofs are explicit but are often extremely lengthy and unreadable. The infamous machine proof of the correctness of the VIPER microprocessor[7] consisted of several million formulae [6].

A formal mathematical proof consists of a sequence of formulae, where each element is either an axiom or derived from a previous element in the series by applying a fixed set of mechanical rules.

The application of formal methods in an industrial environment requires the use of machine-assisted proof, since thousands of proof obligations arise from a formal specification, and mechanized theorem provers are essential in resolving these efficiently. Automated theorem proving is difficult, as often mathematicians prove a theorem with an initial intuitive feeling that the theorem is true. Human intervention to provide guidance or intuition improves the effectiveness of the theorem prover.

The proof of various properties about a program increases confidence in its correctness. However, an absolute proof of correctness[8] is unlikely except for the most trivial of programs. A program may consist of legacy software that is assumed to work; a compiler that is assumed to work correctly creates it. Theorem provers are programs that are assumed to function correctly. The best that formal methods can claim is increased confidence in correctness of the software, rather than an absolute proof of correctness.

16.7 Debate on Formal Methods in Software Engineering

The debate concerning the level of use of formal methods in software engineering is still ongoing. Many practitioners are against the use of mathematics and avoid its use. They argue that in the current competitive industrial environment where time to market is a key driver that the use of such formal techniques would

---

[6] Most existing theorem provers are difficult to use and are for specialist use only. There is a need to improve the usability of theorem provers.

[7] This verification was controversial with RSRE and Charter overselling VIPER as a chip design that conforms to its formal specification.

[8] This position is controversial with others arguing that if correctness is defined mathematically then the mathematical definition (i.e., formal specification) is a theorem, and the task is to prove that the program satisfies the theorem. They argue that the proofs for nontrivial programs exist and that the reason why there are not many examples of such proofs is due to a lack of mathematical specifications.

seriously impact the market opportunity. Industrialists often need to balance conflicting needs such as quality, cost, and delivering on time. They argue that the commercial necessities require methodologies and techniques that allow them to achieve their business goals effectively.

The other camp argues that the use of mathematics is essential in the delivery of high-quality and reliable software and that if a company does not place sufficient emphasis on quality it will pay the price in terms of poor quality and loss of reputation.

It is unrealistic to expect companies to deploy formal methods unless they have clear evidence that it will support them in delivering commercial products to the marketplace ahead of their competition, at the right price and with the right quality. Formal methods need to prove that it can do this if it wishes to be taken seriously in mainstream software engineering.

16.8   The Vienna Development Method

VDM was developed by a research team at the IBM research laboratory in Vienna in the early 1970s. This group[9] was specifying the semantics of the PL/1 programming language using an operational semantic approach. That is, the semantics of the language were defined in terms of a hypothetical machine which interprets the programs of that language [10, 11]. Later work led to the Vienna Development Method (VDM) with its specification language, Meta IV. This was used to give the denotational semantics of programming languages; i.e., a mathematical object (set, function, etc.) is associated with each phrase of the language. The mathematical object is termed the *denotation* of the phrase.

VDM is a *model-oriented approach* and this means that an explicit model of the state of an abstract machine is given, and operations are defined in terms of the state. Operations may act on the system state, taking inputs, and producing outputs as well as a new system state. Operations are defined in a precondition and postcondition style. Each operation has an associated proof obligation to ensure that if the precondition is true, then the operation preserves the system invariant. The initial state itself is, of course, required to satisfy the system invariant.

VDM uses keywords to distinguish different parts of the specification, e.g., preconditions, postconditions, as introduced by the keywords *pre* and *post,* respectively. In keeping with the philosophy that formal methods specify *what* a system does as distinct from *how*, VDM employs postconditions to stipulate the effect of the operation on the state. The previous state is then distinguished by employing *hooked variables*, e.g., $v^{\leftarrow}$ and the postcondition specifies the new state which is defined by a logical predicate relating the pre-state to the poststate.

---

[9] The IBM research laboratory was set up by Dr. Heinz Zamenek, and its members included Peter Lucas, Cliff Jones, Dines Bjørner, and others.

VDM is more than its specification language VDM-SL and is, in fact, a software development method, with rules to verify the steps of development. The rules enable the executable specification, i.e., the detailed code, to be obtained from the initial specification via refinement steps. Thus, we have a sequence $S = S_0, S_1, ..., S_n = E$ of specifications, where S is the initial specification, and E is the final (executable) specification.

Retrieval functions enable a return from a more concrete specification to the more abstract specification. The initial specification consists of an initial state, a system state, and a set of operations. The system state is a particular domain, where a domain is built out of primitive domains such as the set of natural numbers and integers or constructed from primitive domains using domain constructors such as Cartesian product and disjoint union. A domain-invariant predicate may further constrain the domain, and a *type* in VDM reflects a domain obtained in this way. Thus, a type in VDM is more specific than the signature of the type and thus represents values in the domain defined by the signature, which satisfy the domain invariant. In view of this approach to types, it is clear that VDM types may not be "statically type checked".

VDM specifications are structured into modules, with a module containing the module name, parameters, types, operations, etc. Partial functions occur frequently in computer science as many functions, may be undefined, or fail to terminate for some arguments in their domain. VDM addresses partial functions by employing non-standard logical operators, namely the logic of partial functions (LPFs), which was discussed in Chap. 11.

VDM has been used in industrial projects, and its tool support includes the IFAD Toolbox.[10] VDM is described in more detail in [12]. There are several variants of VDM, including VDM++, the object-oriented extension of VDM, and the Irish school of the VDM, which is discussed in the next section.

16.9   VDM♣, the Irish School of VDM

The Irish School of VDM is a variant of standard VDM and is characterized by its constructive approach, classical mathematical style, and its terse notation [13]. This method aims to combine the *what* and *how* of formal methods in that its terse specification style stipulates in concise form *what* the system should do; furthermore, the fact that its specifications are constructive (or functional) means that the *how* is included with the *what*.

However, it is important to qualify this by stating that the how as presented by VDM♣ is not directly executable, as several of its mathematical data types have no corresponding structure in high-level programming languages or functional languages. Thus, a conversion or reification of the specification into a functional

---

[10] The VDM Tools are now available from the CSK Group in Japan.

or higher-level language must take place to ensure a successful execution. Further, the fact that a specification is constructive is no guarantee that it is a good implementation strategy, if the construction itself is naive.

The Irish school follows a similar development methodology as in standard VDM, and it is a model-oriented approach. The initial specification is presented, with the initial state and operations defined. The operations are presented with preconditions; however, no postcondition is necessary as the operation is "functionally" (i.e., explicitly) constructed.

There are proof obligations to demonstrate that the operations preserve the invariant. That is, if the precondition for the operation is true, and the operation is performed, then the system invariant remains true after the operation. The philosophy is to exhibit existence *constructively* rather than providing a theoretical proof of existence that demonstrates the existence of a solution without presenting an algorithm to construct the solution.

The school avoids the existential quantifier of predicate calculus, and reliance on logic in proof is kept to a minimum, with emphasis instead placed on equational reasoning. Structures with nice algebraic properties are sought, and one nice algebraic structure employed is the monoid, which has closure, associative, and a unit element. The concept of isomorphism is powerful, reflecting that two structures are essentially identical, and thus we may choose to work with either, depending on which is more convenient for the task in hand.

The school has been influenced by the work of Polya and Lakatos. The former [14] advocated a style of problem-solving characterized by first considering an easier sub-problem and considering several examples. This generally leads to a clearer insight into solving the main problem. Lakatos's approach to mathematical discovery [15] is characterized by heuristic methods. A primitive conjecture is proposed and if global counterexamples to the statement of the conjecture are discovered, then the corresponding *hidden lemma* for which this global counterexample is a local counterexample is identified and added to the statement of the primitive conjecture. The process repeats, until no more global counterexamples are found. A sceptical view of absolute truth or certainty is inherent in this.

Partial functions are the norm in VDM♣, and as in standard VDM, the problem is that functions may be undefined or fail to terminate for several of the arguments in their domain. The logic of partial functions (LPFs) is avoided, and instead care is taken with recursive definitions to ensure termination is achieved for each argument. Academic and industrial projects have been conducted using VDM, but tool support is limited. The Irish School of VDM is discussed in more detail in [ORg17b].

16.10 The Z Specification Language

Z is a formal specification language founded on Zermelo set theory, and it was developed by Abrial at Oxford University in the early 1980s. It is used for the formal specification of software and is a model-oriented approach. An explicit model of the state of an abstract machine is given, and the operations are defined in terms of the effect on the state. It includes a mathematical notation that is similar to VDM and the visually striking schema calculus. The latter consists essentially of boxes (or schemas), and these are used to describe operations and states. The schema calculus enables schemas to be used as building blocks and combined with other schemas. The Z specification language was published as an ISO standard (ISO/IEC 13568:2002) in 2002.

The schema calculus is a powerful means of decomposing a specification into smaller pieces or schemas. This helps to make Z specification highly readable, as each individual schema is small in size and self-contained. Exception handling is done by defining schemas for the exception cases, and these are then combined with the original operation schema. Mathematical data types are used to model the data in a system, and these data types obey mathematical laws. These laws enable simplification of expressions and are useful with proofs.

Operations are defined in a precondition/postcondition style. However, the precondition is implicitly defined within the operation; i.e., it is not separated out as in standard VDM. Each operation has an associated proof obligation to ensure that if the precondition is true, then the operation preserves the system invariant. The initial state itself is, of course, required to satisfy the system invariant. Postconditions employ a logical predicate which relates the pre-state to the poststate, and the poststate of a variable v is given by priming, e.g., v'. Various conventions are employed, e.g., $v?$ indicates that v is an input variable and $v!$ indicates that v is an output variable. The symbol Ξ Op operation indicates that this operation does not affect the state, whereas Δ Op indicates that this operation affects the state.

Many data types employed in Z have no counterpart in standard programming languages. It is therefore important to identify and describe the concrete data structures that will ultimately represent the abstract mathematical structures. The operations on the abstract data structures may need to be refined to yield operations on the concrete data structure that yield equivalent results. For simple systems, direct refinement (i.e., one step from abstract specification to implementation) may be possible; in more complex systems, deferred refinement is employed, where a sequence of increasingly concrete specifications is produced to eventually yield the executable specification.

Z has been successfully applied in industry, and one of its well-known successes is the CICS project at IBM Hursley in England. Z is described in more detail in Chap. 17.

16.11 The *B*-Method

The *B-Technologies* [16] consist of three components: a method for software development, namely the *B*-Method; a supporting set of tools, namely the *B*-Toolkit; and a generic program for symbol manipulation, namely the *B*-Tool (from which the *B*-Toolkit is derived). The *B*-Method is a model-oriented approach and is closely related to the Z specification language. Abrial developed the B specification language, and every construct in the language has a set theoretic counterpart, and the method is founded on Zermelo set theory. Each operation has an explicit precondition.

A key role of the *abstract machine* in the B-Method is to provide encapsulation of variables representing the state of the machine and operations that manipulate the state. Machines may refer to other machines, and a machine may be introduced as a refinement of another machine. The abstract machines are specification machines, refinement machines, or implementable machines. The B-Method adopts a layered approach to design where the design is gradually made more concrete by a sequence of design layers. Each design layer is a refinement that involves a more detailed implementation in terms of the abstract machines of the previous layer. The design refinement ends when the final layer is implemented purely in terms of library machines. Any refinement of a machine by another has associated proof obligations, and proof is required to verify the validity of the refinement step.

Specification animation of the Abstract Machine Notation (AMN) specification is possible with the B-Toolkit, and this enables typical usage scenarios to be explored for requirements validation. This is, in effect, an early form of testing, and it may be used to demonstrate the presence or absence of desirable or undesirable behaviour. Verification takes the form of a proof to demonstrate that the invariant is preserved when the operation is executed within its precondition, and this is performed on the AMN specification with the B-Toolkit.

The B-Toolkit provides several tools that support the B-Method, and these include syntax and type checking; specification animation, proof obligation generator, auto-prover, proof assistor, and code generation. Thus, in theory, a complete formal development from initial specification to final implementation may be achieved, with every proof obligation justified, leading to a provably correct program.

The B-Method and toolkit have been successfully applied in industrial applications, including the CICS project at IBM Hursley in the United Kingdom [17]. The automated support provided has been cited as a major benefit of the application of the B-Method and the B-Toolkit.

16.12 Predicate Transformers and Weakest Preconditions

The precondition of a program S is a predicate, i.e., a statement that may be true or false, and it is usually required to prove that if the precondition Q is true then execution of S is guaranteed to terminate in a finite amount of time in a state satisfying R. This is written as $\{Q\}\ S\ \{R\}$.

The weakest precondition of a command S with respect to a postcondition R [18] represents the set of all states such that if execution begins in any one of these states, then execution will terminate in a finite amount of time in a state with R true. These set of states may be represented by a predicate Q', so that $wp(S,R) = wp_S(R) = Q'$, and so wp_S is a predicate transformer: i.e., it may be regarded as a function on predicates. The weakest precondition is the precondition that places the fewest constraints on the state than all of the other preconditions of (S,R). That is, all of the other preconditions are stronger than the weakest precondition.

The notation $Q\{S\}R$ is used to denote partial correctness and indicates that if execution of S commences in any state satisfying Q, and if execution terminates, then the final state will satisfy R. Often, a predicate Q which is stronger than the weakest precondition $wp(S, R)$ is employed, especially where the calculation of the weakest precondition is non-trivial. Thus, a stronger predicate Q such that $Q \Rightarrow wp(S,R)$ is often employed.

There are many properties associated with the weakest preconditions, and these may be used to simplify expressions involving weakest preconditions, and in determining the weakest preconditions of various program commands such as assignments and iterations. Weakest preconditions may be used in developing a proof of correctness of a program in parallel with its development [19].

An imperative program may be regarded as a predicate transformer. This is since a predicate P characterizes the set of states in which the predicate P is true, and an imperative program may be regarded as a binary relation on states, which leads to the Hoare triple $P\{F\}Q$. That is, the program F acts as a predicate transformer with the predicate P regarded as an input assertion, i.e., a Boolean expression that must be true before the program F is executed, and the predicate Q is the output assertion, which is true if the program F terminates (where F commenced in a state satisfying P).

16.13 The Process Calculi

The objectives of the process calculi [20] are to provide mathematical models which provide insight into the diverse issues involved in the specification, design, and implementation of computer systems which continuously act and interact with their environment. These systems may be decomposed into sub-systems that interact with each other and their environment.

The basic building block is the *process*, which is a mathematical abstraction of the interactions between a system and its environment. A process that lasts indefinitely may be specified recursively. Processes may be assembled into systems; they may execute concurrently or communicate with each other. Process communication may be synchronized, and this takes the form of one process outputting a message simultaneously to another process inputting a message. Resources may be shared among several processes. Process calculi such as CSP [20] and CCS [21] have been developed and they enrich the understanding of communication and concurrency, and they obey several mathematical laws.

The expression $(a \,?\, P)$ in CSP describes a process which first engages in event a, and then behaves as process P. A recursive definition is written as $(\mu X)\bullet F(X)$ and an example of a simple chocolate vending machine is:

$$\text{VMS} = \mu X : \{\text{coin, choc}\} \bullet (\text{coin} \,?\, (\text{choc} \,?\, X))$$

The simple vending machine has an alphabet of two symbols, namely *coin* and *choc*. The behaviour of the machine is that a coin is entered into the machine, and then a chocolate is selected and provided, and the machine is ready for further use. CSP processes use channels to communicate values with their environment, and input on channel c is denoted by $(c?.x\ P_x)$. This describes a process that accepts any value x on channel c and then behaves as process P_x. In contrast, $(c!e\ P)$ defines a process which outputs the expression e on channel c and then behaves as process P.

The π-calculus is a process calculus based on names. Communication between processes takes place between known channels, and the name of a channel may be passed over a channel. There is no distinction between channel names and data values in the π-calculus. The output of a value v on channel a is given by $\bar{a}v$; i.e., output is a negative prefix. Input on a channel a is given by $a(x)$ and is a positive prefix. Private links or restrictions are denoted by $(x)P$.

16.14 Finite-State Machines

Warren McCulloch and Walter Pitts published early work on finite-state automata in 1943. Moore and Mealy developed this work further, and these machines are referred to as the "*Moore machine*" and the "*Mealy machine*".

A finite-state machine (FSM) is an abstract mathematical machine that consists of a finite number of states. It includes a start state q_0 in which the machine is in initially; a finite set of states Q; an input alphabet Σ; a state transition function δ; and a set of final accepting states F (where $F \subseteq Q$).

The state transition function takes the current state and an input and returns the next state. It provides rules that define the action of the machine for each input, and it may be extended to provide output as well as a state transition. State diagrams are used to represent finite-state machines, and each state accepts a finite number of inputs.

A *deterministic machine* changes to exactly one state for each input transition, whereas a *non-deterministic machine* may have a choice of states to move to for a particular input.

Finite-state automata compute very primitive functions and are not an adequate model for computing. There are more powerful automata such as the *Turing machine* that is essentially a finite automaton with an infinite storage (memory). Anything that is computable is computable by a Turing machine. The Turing machine provides a mathematical abstraction of computer execution and storage, as well as providing a mathematical definition of an algorithm. Automata theory was discussed in Chap. 13.

Table 16.2 Parnas's contributions to software engineering

Area	Contribution
Tabular expressions	These are mathematical tables for specifying requirements and enable complex predicate logic expressions to be represented in a simpler form
Mathematical documentation	He advocates the use of precise mathematical documentation for requirements and design
Requirements specification	He advocates the use of mathematical relations to specify the requirements precisely
Software design	He developed *information hiding* that is used in object-oriented design[a] and allows software to be designed for change
Software inspections	His approach requires the reviewers to take an active part in the inspection. They are provided with a list of questions by the author and their analysis involves the production of mathematical table to justify the answers
Predicate logic	He developed an extension of the predicate calculus to deal with partial functions, and it preserves the classical two-valued logic when dealing with undefined values

[a] It is surprising that many in the object-oriented world seem unaware that information hiding goes back to the early 1970s and many have never heard of Parnas.

16.15 The Parnas Way

Parnas has been influential in the computing field, and his ideas on the specification, design, implementation, maintenance, and documentation of computer software remain important. He advocates a solid engineering approach and argues that the role of the engineer is to apply scientific principles and mathematics to design and develop products. He argues that computer scientists need to be educated as engineers to ensure that they have the appropriate background to build software correctly.

His tabular expressions were used for the specification of the requirements of the A-7 aircraft for the US Navy, and his mathematical inspections were used to verify the correctness of the shutdown software at the Darlington Nuclear power plant in Canada. His contributions to software engineering include (Table 16.2).

16.16 Model Checking

Model checking is an automated technique such that given a finite-state model of a system and a formal property, (expressed in temporal logic) then it systematically checks whether the property is true or false in a given state in the model. It is an effective technique to identify potential design errors, and it increases the confidence in the correctness of the system design. Model checking is a highly effective verification technology and is widely used in the hardware and software fields. It has been employed in the verification of microprocessors; in security protocols; in the transportation sector (trains); and in the verification of software in the space sector.

Model checking is a formal verification technique based on graph algorithms and formal logic. It allows the desired behaviour (specification) of a system to be verified, and its approach is to employ a suitable model of the system and to carry out a systematic and exhaustive inspection of all states of the model to verify that the desired properties are satisfied. These properties are generally safety properties such as the absence of deadlock, request-response properties, and invariants. Its systematic search determines whether a given system model truly satisfies a particular property or not. Model checking is discussed in more detail in Chap. 18.

16.17 Usability of Formal Methods

There are practical difficulties associated with the industrial use of formal methods. It seems to be assumed that programmers and customers are willing to become familiar with the mathematics used in formal methods, but this is true in only some domains.[11] It is usually possible to get a developer to learn a formal method, as a programmer has some experience of mathematics and logic. However, it is more difficult to get a customer to learn a formal method, and this makes it more difficult to perform a rigorous validation of the formal specification.

This often means that often a formal specification of the requirements and an informal definition of the requirements using a natural language are maintained. It is essential that both of these are consistent and that there is a rigorous validation of the formal specification. Otherwise, if the programmer proves the correctness of the code with respect to the formal specification, and the formal specification is incorrect, then the formal development of the software will be incorrect. There are several techniques to validate a formal specification including:

- Proof that the formal specification satisfies key properties
- Software inspections to compare formal specification and informal set of requirements
- Specification animation to validate the formal specification.

Formal methods are perceived as being difficult to use, and of providing limited value in mainstream software engineering. Programmers receive education in mathematics as part of their studies, but many never use formal methods again once they take an industrial position. Some of the reasons for this are:

---

[11] The domain in which the software is being used will influence the willingness or otherwise of the customers to become familiar with the mathematics required. There appears to be little interest in mainstream software engineering, and their perception is that formal methods are unusable. However, there is a greater interest in the mathematical approach in the safety critical field.

- The notation is not intuitive.
- It is difficult to write a formal specification.
- Validation of a formal specification is difficult.
- Refinement and proof are difficult.
- Limited tool support.

It is important to investigate ways by which formal methods can be made more usable to software engineers and to design more usable notations and better tools to support the process. Practical training and coaching to employees can help. Some of the characteristics of a usable formal method are:

- A formal method should be intuitive.
- It should have tool support.
- A formal method should be teachable.
- It should be able to adapt to change.
- The technology transfer path should be defined.
- A formal method should be cost-effective.

16.18 Review Questions

1. What are formal methods and describe their potential benefits? How essential is tool support?
2. What is stepwise refinement and how realistic is it in mainstream software engineering?
3. Discuss Parnas's criticisms of formal methods and discuss whether his views are valid.
4. Discuss the industrial applications of formal methods and which areas have benefited most from their use? What problems have arisen?
5. Describe a technology transfer path for the deployment of formal methods in an organization.
6. Explain the difference between the model-oriented approach and the axiomatic approach.
7. Discuss the nature of proof in formal methods and tools to support proof.
8. Discuss the Vienna Development Method and explain the difference between standard VDM and VDM♣.
9. Discuss Z and B. Describe the tools in the B-Toolkit.
10. Discuss process calculi such as CSP, CCS, or π–calculus.

16.19 Summary

Formal methods provide a mathematical approach to the development of high-quality software. They consist of a formal specification language; a methodology for formal software development; and a set of tools to support the syntax checking of the specification, as well as the proof of properties of the specification.

The model-oriented approach includes formal methods such as VDM, Z, and B, and the axiomatic approach includes the process calculi such as CSP, CCS, and the π calculus. VDM was developed at the IBM lab in Vienna and has been used in academia and industry.

Formal methods allow questions to be asked and answered about what the system does independently of the implementation. They offer a way to debug the requirements and to show that certain desirable properties are true of the specification, whereas certain undesirable properties are absent.

The use of formal methods generally leads to increased confidence in its correctness. There are challenges involved in the deployment of formal methods, as mathematical techniques may be a culture shock to staff. The usability of existing formal methods was considered, and reasons for their perceived difficulty were considered. The characteristics of a usable formal method were explored.

There are various tools to support formal methods including syntax checkers; specialized editors; tools to support refinement; automated code generators to generate a high-level language corresponding to the specification; theorem provers; and specification animation tools for simulation of the specification.

References

1. Spivey JM (1992) The Z notation. A reference manual. Prentice Hall International Series in Computer Science
2. Brown (1990) Rational for the development of the U.K. defence standards for safety critical software. Compass Conference
3. Hinchey M, Bowen J (eds) (1995) Applications of formal methods. Prentice Hall International Series in Computer Science
4. Ministry of Defence (1991) 00-55 (Part 1)/Issue 1 The procurement of safety critical software in defence equipment. Part 1: requirements. Interim Defence Standard. UK
5. Ministry of Defence (1991) 00-55 (Part 2)/Issue 1 The procurement of safety critical software in defence equipment. Part 2: guidance. Interim Defence Standard. UK
6. Tierney M (1991) The evolution of Def Stan 00-55 and 00-56. In: An intensification of the formal methods debate in the UK. Research Centre for Social Sciences. University of Edinburgh
7. Gerhart S, Craigen D, Ralston T. (1994). Experience with formal methods in critical systems. IEEE Softw
8. Woodcock J, Larsen PG, Bicarregui J, Fitzgerald J (2009) Formal methods: practice and experience. ACM Comput Surv
9. Kuhn T (1970) The structure of scientific revolutions. University of Chicago Press
10. Bjørner D, Jones C (1978) The vienna development method. The meta language. In: Lecture notes in computer science, vol 61. Springer
11. Bjørner D, Jones C (1982) Formal specification and software development. Prentice Hall International Series in Computer Science

12. O'Regan G (2017) Concise guide to formal methods. Springer
13. Mac An Airchinnigh M (1990) Computation models and computing. PhD Thesis. Department of Computer Science. Trinity College Dublin
14. Polya G (1957) How to solve it. A new aspect of mathematical method. Princeton University Press
15. Lakatos I (1976) Proof and refutations. The logic of mathematical discovery. Cambridge University Press
16. McDonnell E (1994) MSc. Thesis. Department of Computer Science. Trinity College Dublin
17. Hoare JP (1995) Application of the B method to CICS. Appl Formal Methods. Prentice Hall International Series in Computer Science
18. Gries D (1981) The science of programming. Springer, Berlin
19. O'Regan G (2006) Mathematical approaches to software quality. Springer
20. Hoare CAR (1985) Communicating sequential processes. Prentice Hall International Series in Computer Science
21. Robin Milner et al (1989) A Calculus of Mobile Processes. Part 1. LFCS Report Series. ECS-LFCS-89-85. Department of Computer Science. University of Edinburgh

Z Formal Specification Language

17

Key Topics

Sets, relations and functions

Bags and sequences

Data Reification

Refinement

Schema Calculus

Proof in Z

17.1 Introduction

Z is a formal specification language based on Zermelo set theory. It was developed at the Programming Research Group at Oxford University in the early 1980s [1] and became an ISO standard in 2002. Z specifications are mathematical and employ a classical two-valued logic. The use of mathematics ensures precision and allows inconsistencies and gaps in the specification to be identified. Theorem provers may be employed to demonstrate that the software implementation meets its specification.

Z is a '*model-oriented*' approach with an explicit model of the state of an abstract machine given, and operations are defined in terms of this state. Its mathematical notation is used for formal specification, and the schema calculus is used to structure the specifications. The schema calculus is visually striking and consists essentially of boxes, with these boxes or schemas used to describe operations and states. The schemas may be used as building blocks and combined with other

© The Author(s), under exclusive license to Springer Nature Switzerland AG 2023
G. O'Regan, *Mathematical Foundations of Software Engineering*,
Texts in Computer Science, https://doi.org/10.1007/978-3-031-26212-8_17

Fig. 17.1 Specification of
positive square root

$$
\begin{array}{|l}
\text{—}SqRoot\text{———}\\
num?, root! : \mathbb{R}\\
\hline
num? \geq 0\\
root!^2 = num?\\
root! \geq 0\\
\hline
\end{array}
$$

schemas. The simple schema below (Fig. 17.1) is the specification of the positive square root of a real number.

The schema calculus is a powerful means of decomposing a specification into smaller pieces or schemas. This helps to make Z specifications highly readable, as each individual schema is small in size and self-contained. Exception handling is addressed by defining schemas for the exception cases. These are then combined with the original operation schema. Mathematical data types are used to model the data in a system, these data types obey mathematical laws. These laws enable simplification of expressions and are useful with proofs.

Operations are defined in a precondition/postcondition style. A precondition must be true before the operation is executed, and the postcondition must be true after the operation has been executed. The precondition is implicitly defined within the operation. Each operation has an associated proof obligation to ensure that if the precondition is true, then the operation preserves the system invariant. The system invariant is a property of the system that must be true at all times. The initial state itself is, of course, required to satisfy the system invariant.

The precondition for the specification of the square root function above is that $num? \geq 0$; i.e., the function $SqRoot$ may be applied to positive real numbers only. The postcondition for the square root function is $root!^2 = num?$ and $root! \geq 0$. That is, the square root of a number is positive and its square gives the number. Postconditions employ a logical predicate which relates the pre-state to the post-state, with the poststate of a variable being distinguished by priming the variable, e.g., v'.

Z is a typed language and whenever a variable is introduced its type must be given. A type is simply a collection of objects, and there are several standard types in Z. These include the natural numbers $\mathbb{N}$, the integers $\mathbb{Z}$, and the real numbers $\mathbb{R}$. The declaration of a variable x of type X is written $x: X$. It is also possible to create your own types in Z.

Various conventions are employed within Z specification, for example $v?$ indicates that v is an input variable; $v!$ indicates that v is an output variable. The variable $num?$ is an input variable, and $root!$ is an output variable for the square root example above. The notation Ξ in a schema indicates that the operation Op does not affect the state; whereas the notation Δ in the schema indicates that Op is an operation that affects the state.

Many of the data types employed in Z have no counterpart in standard programming languages. It is therefore important to identify and describe the concrete data structures that ultimately will represent the abstract mathematical structures. As the concrete structures may differ from the abstract, the operations on the abstract

Fig. 17.2 Specification of a
library system

$-Library$————
$on\text{-}shelf, missing, borrowed : \mathbb{P}\ Bkd\text{-}Id$
———
$on\text{-}shelf \cap missing = \emptyset$
$on\text{-}shelf \cap borrowed = \emptyset$
$borrowed \cap missing = \emptyset$

Fig. 17.3 Specification of
borrow operation

$-Borrow$————
$\Delta\ Library$
$b? :Bkd\text{-}Id$
———
$b? \in on\text{-}shelf$
$on\text{-}shelf' = on\text{-}shelf \setminus \{b?\}$
$borrowed' = borrowed \cup \{b?\}$

data structures may need to be refined to yield operations on the concrete data
that yield equivalent results. For simple systems, direct refinement (i.e., one step
from abstract specification to implementation) may be possible; in more complex
systems, deferred refinement[1] is employed, where a sequence of increasingly concrete specifications is produced to yield the executable specification. The schema
calculus is employed for combining schemas to make larger specifications and is
discussed later in the chapter.

Example 17.1 The following is a Z specification to borrow a book from a library
system. The library is made up of books that are on the shelf; books that are borrowed;
and books that are missing (Fig. 17.2). The specification models a library with sets
representing books on the shelf, on loan or missing. These are three mutually disjoint
subsets of the set of books *Bkd-Id*.

The system state is defined in the *Library* schema below, and operations such
as *Borrow* and *Return* affect the state. The *Borrow* operation is specified below
(Fig. 17.3).

The notation $\mathbb{P}Bkd\text{-}Id$ is used to represent the power set of *Bkd-Id* (i.e., the set
of all subsets of *Bkd-Id*). The disjointness condition for the library is expressed
by the requirement that the pairwise intersection of the subsets *on-shelf, borrowed,
missing* is the empty set.

The precondition for the *Borrow* operation is that the book must be available
on the shelf to borrow. The postcondition is that the borrowed book is added to
the set of borrowed books and is removed from the books on the shelf.

---

[1] Step-wise refinement involves producing a sequence of increasingly more concrete specifications until eventually the executable code is produced. Each refinement step has associated proof
obligations to prove that it is valid.

Z has been successfully applied in industry including the CICS project at IBM Hursley in the UK.[2] Next, we describe key parts of Z including sets, relations, functions, sequences, and bags.

17.2 Sets

Sets were discussed in Chap. 3, and this section focuses on their use in Z. Sets may be enumerated by listing all of their elements. Thus, the set of all even natural numbers less than or equal to 10 is:

$$\{2, 4, 6, 8, 10\}$$

Sets may be created from other sets using set comprehension: i.e., stating the properties that its members must satisfy. For example, the set of even natural numbers less than 10 is given by set comprehension as:

$$\{n : \mathbb{N} | n \neq 0 \wedge n < 10 \wedge n \bmod 2 = 0 \cdot n\}$$

There are three main parts to the set comprehension above. The first part is the signature of the set and this is given by n: $\mathbb{N}$ above. The first part is separated from the second part by a vertical line. The second part is given by a predicate, and for this example the predicate is $n \neq 0 \wedge n < 10 \wedge n \bmod 2 = 0$. The second part is separated from the third part by a bullet. The third part is a term, and for this example it is simply n. The term is often a more complex expression: e.g., $\log(n^2)$.

In mathematics, there is just one empty set. However, since Z is a typed set theory, there is an empty set for each type of set. Hence, there are an infinite number of empty sets in Z. The empty set is written as $\varnothing$ [X] where X is the type of the empty set. In practice, X is omitted when the type is clear.

Various operations on sets such as union, intersection, set difference, and symmetric difference are employed in Z. The power set of a set X is the set of all subsets of X and is denoted by $\mathbb{P} X$. The set of non-empty subsets of X is denoted by $\mathbb{P}_1 X$ where

$$\mathbb{P}_1 X == \{U : \mathbb{P} X | U \neq \varnothing [X]\}$$

A finite set of elements of type X (denoted by $F X$) is a subset of X that cannot be put into a one-to-one correspondence with a proper subset of itself. This is defined formally as:

$$FX == \{U : \mathbb{P} X | \neg \exists V : \mathbb{P} U \cdot V \neq U \wedge (\exists f : V \leftarrowtail U)\}$$

---

[2] This project claimed a 9% increase in productivity attributed to the use of formal methods.

The expression $f: V \leftarrowtail U$ denotes that f is a bijection from U to V and injective, surjective, and bijective functions were discussed in Chap. 3.

The fact that Z is a typed language means that whenever a variable is introduced (e.g., in quantification with $\forall$ and $\exists$) it is first declared. For example, $\forall j: J \bullet P \Rightarrow Q$. There is also the unique existential quantifier $\exists_1 j: J \mid P$ which states that there is exactly one j of type J that has property P.

17.3 Relations

Relations were discussed in Chap. 3 and are used extensively in Z. A relation R between X and Y is any subset of the Cartesian product of X and Y; i.e., $R \subseteq (X \times Y)$, and a relation in Z is denoted by $R: X \leftrightarrow Y$. The notation $x \mapsto y$ indicates that the pair $(x,y) \in R$.

Consider, the relation *home_owner*: *Person* $\leftrightarrow$ *Home* that exists between people and their homes. An entry *daphne* $\mapsto$ *mandalay* $\in$ *home_owner* if *daphne* is the owner of *mandalay*. It is possible for a person to own more than one home:

$$rebecca \mapsto nirvana \in home\_owner$$
$$rebecca \mapsto trivoli \in home\_owner$$

It is possible for two people to share ownership of a home:

$$rebecca \mapsto nirvana \in home\_owner$$
$$lawrence \mapsto nirvana \in home\_owner$$

There may be some people who do not own a home, and there is no entry for these people in the relation *home_owner*. The type *Person* includes every possible person, and the type *Home* includes every possible home. The domain of the relation *home_owner* is given by:

$$x \in \text{dom } home\_owner \Leftrightarrow \exists h : Home \bullet x \mapsto h \in home\_owner.$$

The range of the relation *home_owner* is given by:

$$h \in \text{ran } home\_owner \Leftrightarrow \exists h : Person \bullet x \mapsto h \in home\_owner.$$

The composition of two relations *home_owner*: *Person* $\leftrightarrow$ *Home* and *home_value*: *Home* $\leftrightarrow$ *Value* yields the relation *owner_wealth*: *Person* $\leftrightarrow$ *Value* and is given by the relational composition *home_owner*; *home_value* where:

$$p \mapsto v \in home\_owner; home\_value \Leftrightarrow$$

$$(\exists h : Home \bullet p \mapsto h \in home\_owner \wedge h \mapsto v \in home\_value)$$

The relational composition may also be expressed as:

$$owner\_wealth = home\_value \text{ o } home\_owner$$

The union of two relations often arises in practice. Suppose a new entry *aisling* $\mapsto$ *muckross* is to be added. Then this is given by

$$home\_owner' = home\_owner \cup \{aisling \mapsto muckross\}$$

Suppose that we are interested in knowing all females who are house owners. Then we restrict the relation *home_owner* so that the first element of all ordered pairs have to be female. Consider *female*: $\mathbb{P}$ *Person* with $\{aisling,\ rebecca\} \subseteq female$.

$$home\_owner = \{aisling \mapsto muckross, rebecca \mapsto nirvana,$$
$$lawrence \mapsto nirvana\}$$

$$female \lhd home\_owner = \{aisling \mapsto muckross,\ rebecca \mapsto nirvana\}$$

That is, *female* $\rhd$ *home_owner* is a relation that is a subset of *home_owner*, and the first element of each ordered pair in the relation is female. The operation $\rhd$ is termed domain restriction, and its fundamental property is:

$$x \mapsto y \in U \lhd R \Leftrightarrow (x \in U \wedge x \mapsto y \in R\}$$

where $R: X \leftrightarrow Y$ and $U: \mathbb{P}\ X$.

There is also a domain anti-restriction (subtraction) operation, and its fundamental property is:

$$x \mapsto y \in U \mathbin{\lhd\!\!\!-} R \Leftrightarrow (x \notin U \wedge x \mapsto y \in R\}$$

where $R: X \leftrightarrow Y$ and $U: \mathbb{P}X$.

There are also range restriction (the $\rhd$ operator) and the range anti-restriction operator (the $\rhd\!\!\!-$ operator). These are discussed in [1].

17.4   Functions

A function [1] is an association between objects of some type X and objects of another type Y such that given an object of type X, there exists only one object in Y associated with that object. A function is a set of ordered pairs where the first element of the ordered pair has at most one element associated with it. A function is therefore a special type of relation, and a function may be *total* or *partial*.

A total function has exactly one element in Y associated with each element of X, whereas a partial function has at most one element of Y associated with each element of X (there may be elements of X that have no element of Y associated with them).

A partial function from X to Y ($f : X \nrightarrow Y$) is a relation $f : X \leftrightarrow Y$ such that:

$$\forall x : X; y, z : Y \cdot (x \mapsto y \in f \wedge x \mapsto z \in f \Rightarrow y = z)$$

The association between x and y is denoted by $f(x) = y$, and this indicates that the value of the partial function f at x is y. A total function from X to Y (denoted $f: X \rightarrow Y$) is a partial function such that every element in X is associated with some value of Y.

$$f : X \rightarrow Y \Leftrightarrow f : X \nrightarrow Y \wedge \mathrm{dom} f = X$$

Clearly, every total function is a partial function but not vice versa.

One operation that arises quite frequently in specifications is the function override operation. Consider the following specification of a temperature map:

```
| − TempMap − − − −
|CityList : ℙCity
|temp : City↛Z
| − − −
|dom temp = CityList
| − − − − − − − − −
```

Suppose the temperature map is given by $temp = \{Cork \mapsto 17, Dublin \mapsto 19, London \mapsto 15\}$. Then consider the problem of updating the temperature map if a new temperature reading is made in Cork: e.g., $\{Cork \mapsto 18\}$. Then the new temperature chart is obtained from the old temperature chart by function override to yield $\{Cork \mapsto 18, Dublin \mapsto 19, London \mapsto 15\}$. This is written as:

$$temp' = temp \oplus \{Cork \mapsto 18\}$$

The function override operation combines two functions of the same type to give a new function of the same type. The effect of the override operation is that the entry $\{Cork \mapsto 17\}$ is removed from the temperature chart and replaced with the entry $\{Cork \mapsto 18\}$.

Suppose $f, g: X \nrightarrow Y$ are partial functions then $f \oplus g$ is defined and indicates that f is overridden by g. It is defined as follows:

$$(f \oplus g)(x) = g(x) \text{ where } x \in \mathrm{dom}\, g$$
$$(f \oplus g)(x) = f(x) \text{ where } x \notin \mathrm{dom}\, g \wedge x \in \mathrm{dom} f$$

This may also be expressed (using domain anti-restriction) as:

$$f \oplus g = ((\mathrm{dom}\, g) \mathbin{\lhd\mkern-9mu -} f) \cup g$$

There is notation in Z for injective, surjective, and bijective functions. An injective function is one to one: i.e.,

$$f(x) = f(y) \Rightarrow x = y.$$

A surjective function is onto: i.e.,

$$\text{Given } y \in Y, \exists x \in X \text{ such that } f(x) = y$$

A bijective function is one to one and onto, and it indicates that the sets X and Y can be put into one-to-one correspondence with one another. Z includes lambda calculus notation to define functions (λ-calculus was discussed in Chap. 12). For example, the function cube $== \lambda x$: $N \cdot x * x * x$. Function composition f; g is similar to relational composition.

17.5   Sequences

The type of all sequences of elements drawn from a set X is denoted by seq X. Sequences are written as $\langle x_1, x_2, \ldots .x_n \rangle$, and the empty sequence is denoted by $\langle \rangle$. Sequences may be used to specify the changing state of a variable over time, with each element of the sequence representing the value of the variable at a discrete time instance.

Sequences are functions and a sequence of elements drawn from a set X is a finite function from the set of natural numbers to X. A partial finite function f from X to Y is denoted by $f: X \nrightarrow Y$. A finite sequence of elements of X is given by a finite function $f: N \nrightarrow X$, and the domain of the function consists of all numbers between 1 and $\#f$ (where $\#f$ is the cardinality of f). It is defined formally as:

$$\text{seq } X == \{f : N \nrightarrow X \mid \text{dom } f = 1..\#f \cdot f\}$$

The sequence $\langle x_1, x_2, \ldots .x_n \rangle$ above is given by:

$$\{1 \mapsto x_1, 2 \mapsto x_2, \ldots n \mapsto x_n\}$$

There are various functions to manipulate sequences. These include the sequence concatenation operation. Suppose $\sigma = \langle x_1, x_2, \ldots x_n \rangle$ and $\tau = \langle y_1, y_2, \ldots y_m \rangle$ then:

$$\sigma ^\frown \tau = \langle x_1, x_2, \ldots x_n, y_1, y_2, \ldots y_m \rangle$$

The head of a non-empty sequence gives the first element of the sequence.

$$\text{head } \sigma = \text{head} \langle x_1, x_2, \ldots .x_n \rangle = x_1$$

The tail of a non-empty sequence is the same sequence except that the first element of the sequence is removed.

$$\text{tail } \sigma = \text{tail}\langle x_1, x_2, \ldots x_n \rangle = \langle x_2, \ldots x_n \rangle$$

Suppose $f: X \to Y$ and a sequence σ: seq X then the function map applies f to each element of σ:

$$\text{map } f \ \sigma = \text{map } \langle x_1, x_2, \ldots x_n \rangle = \langle f(x_1), f(x_2), \ldots f(x_n) \rangle$$

The map function may also be expressed via function composition as:

$$\text{map } f \ \sigma = \sigma; \ f$$

The reverse order of a sequence is given by the rev function:

$$\text{rev } \sigma = \text{rev}\langle x_1, x_2, \ldots x_n \rangle = \langle x_n, \ldots x_2, x_1 \rangle$$

17.6 Bags

A bag is similar to a set except that there may be multiple occurrences of each element in the bag. A bag of elements of type X is defined as a partial function from the type of the elements of the bag to positive whole numbers. The definition of a bag of type X is:

$$\text{bag } X == X \nrightarrow \mathbb{N}_1.$$

For example, a bag of marbles may contain 3 blue marbles, 2 red marbles, and 1 green marble. This is denoted by $B = [b, b, b, g, , r, r]$. The bag of marbles is thus denoted by:

$$\text{bag } Marble == Marble \nrightarrow \mathbb{N}_1.$$

The function count determines the number of occurrences of an element in a bag. For the example above, count $Marble \ b = 3$, and count $Marble \ y = 0$ since there are no yellow marbles in the bag. This is defined formally as:

$$\text{count bag } X \ y = 0 \qquad\qquad y \notin \text{bag } X$$
$$\text{count bag } X \ y = (\text{bag } X)(y) \quad\ y \in \text{bag } X$$

An element y is in bag X if and only if y is in the domain of bag X.

$$y \text{ in bag } X \Leftrightarrow y \in \text{dom (bag } X)$$

Fig. 17.4 Specification of
vending machine using bags

$$-\Delta Vending\ Machine\mbox{------}$$
$stock : bag\ Good$
$price : Good \rightarrow \mathbb{N}_1$

$\mathrm{dom}\ stock \subseteq \mathrm{dom}\ price$

The union of two bags of marbles B1 = [b, b, b, g, , r, r] and B2 = [b, g, , r, y] is given by $B_1 \uplus B_2$ [b, b ,b, b, g, g, r, r, r, y]. It is defined formally as:

$$(B_1 \uplus B_2)(y) = B_2(y) \qquad y \notin \mathrm{dom}\,B_1 \wedge y \in \mathrm{dom}\,B_2$$
$$(B_1 \uplus B_2)(y) = B_1(y) \qquad y \in \mathrm{dom}\ B_1 \wedge y \notin \mathrm{dom}\,B_2$$
$$(B_1 \uplus B_2)(y) = B_1(y) + B_2(y) \quad y \in \mathrm{dom}\,B_1 \wedge y \in \mathrm{dom}\,B_2$$

A bag may be used to record the number of occurrences of each product in a warehouse as part of an inventory system. It may model the number of items remaining for each product in a vending machine (Fig. 17.4).

The operation of a vending machine would require other operations such as identifying the set of acceptable coins, checking that the customer has entered sufficient coins to cover the cost of the good, returning change to the customer, and updating the quantity on hand of each good after a purchase (see [1]).

17.7 Schemas and Schema Composition

The schemas in Z are visually striking, and the specification is presented in two-dimensional graphic boxes. Schemas are used for specifying states and state transitions, and they employ notation to represent the before and after state (e.g., s and s' where s' represents the after state of s). The schemas group all relevant information that belongs to a state description.

There are a number of useful schema operations such as schema inclusion, schema composition, and the use of propositional connectives to link schemas together. The Δ convention indicates that the operation affects the state, whereas the Ξ convention indicates that the state is not affected. These operations and conventions allow complex operations to be specified concisely and assist with the readability of the specification. Schema composition is analogous to relational composition and allows new schemas to be derived from existing schemas.

A schema name S_1 may be included in the declaration part of another schema S_2. The effect of the inclusion is that the declarations in S_1 are now part of S_2, and the predicates of S_1 are S_2 are joined together by conjunction. If the same variable is defined in both S_1 and S_2, then it must be of the same type.

$$
\begin{array}{ll}
|- S_1 --- & |- S_2 --- \\
|x, y : \mathbb{N} & |S_1, z : \mathbb{N} \\
|-- & |-- \\
|x + y > 2 & |z = x + y \\
|--- & |--- \\
\end{array}
$$

The result is that S_2 includes the declarations and predicates of S_1 (Fig. 17.5).

Two schemas may be linked by propositional connectives such as $S_1 \wedge S_2$, $S_1 \vee S_2$, $S_1 \rightarrow S_2$, and $S_1 \leftrightarrow S_2$. The schema $S_1 \vee S_2$ is formed by merging the declaration parts of S_1 and S_2, and then combining their predicates by the logical $\vee$ operator. For example, $S = S_1 \vee S_2$ yields (Fig. 17.6).

Schema inclusion and the linking of schemas use normalization to convert subtypes to maximal types, and predicates are employed to restrict the maximal type to the sub-type. This involves replacing declarations of variables (e.g., $u : 1..35$ with $u : \mathbb{Z}$ and adding the predicate $u > 0$ and $u < 36$ to the predicate part of the schema).

The Δ and Ξ conventions are used extensively, and the notation Δ *TempMap* is used in the specification of schemas that involve a change of state. The notation

$$
\begin{array}{l}
-S_1------ \\
x,y : \mathbb{N} \\
\hline
x + y > 2 \\
\end{array}
\qquad
\begin{array}{l}
-S_2------ \\
S_1 ; z : \mathbb{N} \\
\hline
z = x + y \\
\end{array}
$$

The result is that S_2 includes the declarations and predicates of S_1 (Fig. 17.5):

$$
\begin{array}{l}
-S_2------ \\
x,y : \mathbb{N} \\
z : \mathbb{N} \\
\hline
x + y > 2 \\
z = x + y \\
\end{array}
$$

Fig. 17.5 Schema inclusion

Fig. 17.6 Merging schemas
$(S_1 \vee S_2)$

$$
\begin{array}{l}
-S------ \\
x,y : \mathbb{N} \\
z : \mathbb{N} \\
\hline
x + y > 2 \vee z = x + y \\
\end{array}
$$

Δ *TempMap* represents:

$$\Delta\,TempMap = TempMap \wedge TempMap'$$

The longer form of Δ *TempMap* is written as:

```
| − ΔTempMap
|CityList, CityList' : ℙ City
|temp, temp' : City ⇸ Z
| − −−
|dom temp = CityList
|dom temp' = CityList'
| − − − − − − − −−
```

The notation Ξ *TempMap* is used in the specification of operations that do not involve a change to the state.

```
| − Ξ TempMap − − − −
|ΔTempMap
| − −−
|CityList = CityList'
|temp = temp'
| − − − − − − − −−
```

Schema composition is analogous to relational composition, and it allows new specifications to be built from existing ones. It allows the after-state variables of one schema to be related with the before variables of another schema. The composition of two schemas S and T $(S; T)$ is described in detail in [1] and involves four steps (Table 17.1).

Table 17.1 Schema composition

Step	Procedure
1	Rename all *after*-state variables in S to something new: $S[s^+/s']$
2	Rename all *before* state variables in T to the same new thing: i.e., $T[s^+/s]$
3	Form the conjunction of the two new schemas: $S[s^+/s'] \wedge T[s^+/s]$
4	Hide the variable introduced in steps 1 and 2 $S; T = (S[s^+/s'] \wedge T[s^+/s]) \backslash (s^+)$

$$\begin{array}{|l}
\text{-}S_1\text{------} \\
x,x^+,y? : \mathbb{N} \\
\hline
x^+ = y? - 2
\end{array}
\qquad
\begin{array}{|l}
\text{-}T_1\text{------} \\
x^+,x' : \mathbb{N} \\
\hline
x' = x^+ + 1
\end{array}$$

S_1 and T_1 represent the results of step 1 and step 2, with x' renamed to x^+ in S, and x renamed to x^+ in T. Step 3 and step 4 yield (Fig. 17.7).:

$$\begin{array}{|l}
\text{-}S_1 \wedge T_1\text{------} \\
x,x^+,x',y? : \mathbb{N} \\
\hline
x^+ = y? - 2 \\
x' = x^+ + 1
\end{array}
\qquad
\begin{array}{|l}
\text{-}S ; T\text{------} \\
x, x', y? : \mathbb{N} \\
\hline
\exists x^+ : \mathbb{N} \bullet \\
\quad (x^+ = y? - 2 \\
\quad\; x' = x^+ + 1)
\end{array}$$

Fig. 17.7 Schema composition

The example below should make schema composition clearer. Consider the composition of S and T where S and T are defined as follows:

$$\begin{array}{|l}
\text{-- } S \text{ -- --} \\
x, x', y? : \mathbb{N} \\
\text{-- --} \\
x' = y? - 2 \\
\text{-- --}
\end{array}
\qquad
\begin{array}{|l}
\text{-- } T \text{ -- --} \\
x, x' : \mathbb{N} \\
\text{-- --} \\
x' = x + 1 \\
\text{-- --}
\end{array}$$

$$\begin{array}{|l}
\text{-- } S_1 \text{ -- --} \\
x, x^+, y? : \mathbb{N} \\
\text{-- --} \\
x^+ = y? - 2 \\
\text{-- --}
\end{array}
\qquad
\begin{array}{|l}
\text{-- } T_1 \text{ -- --} \\
x^+, x' : \mathbb{N} \\
\text{-- --} \\
x' = x^+ + 1 \\
\text{-- --}
\end{array}$$

S_1 and T_1 represent the results of step 1 and step 2, with x' renamed to x^+ in S, and x renamed to x^+ in T. Step 3 and step 4 yield (Fig. 17.7).

Schema composition is useful as it allows new specifications to be created from existing ones.

17.8 Reification and Decomposition

A Z specification involves defining the state of the system and then specifying the required operations. The Z specification language employs many constructs that are not part of conventional programming languages, and a Z specification is therefore not directly executable on a computer. A programmer implements the formal specification, and mathematical proof may be employed to prove that a program meets its specification.

Often, there is a need to write an intermediate specification that is between the original Z specification and the eventual program code. This intermediate specification is more algorithmic and uses less abstract data types than the Z specification. The intermediate specification is termed the design and the design needs to be correct with respect to the specification, and the program needs to be correct with respect to the design. The design is a refinement (reification) of the state of the specification, and the operations of the specification have been decomposed into those of the design.

The representation of an abstract data type such as a set by a sequence is termed data reification, and data reification is concerned with the process of transforming an abstract data type into a concrete data type. The abstract and concrete data types are related by the retrieve function, and the retrieve function maps the concrete data type to the abstract data type. There are typically several possible concrete data types for a particular abstract data type (i.e., refinement is a relation), whereas there is one abstract data type for a concrete data type (i.e., retrieval is a function). For example, sets are often reified to unique sequences; and clearly more than one unique sequence can represent a set, whereas a unique sequence represents exactly one set.

The operations defined on the concrete data type are related to the operations defined on the abstract data type. That is, the commuting diagram property is required to hold (Fig. 17.8). That is, for an operation $\boxdot$ on the concrete data type to correctly model the operation $\odot$ on the abstract data type the diagram must commute, and the commuting diagram property requires proof. That is, it is required to prove that:

$$ret \ (\sigma \boxdot \tau) = (ret \ \sigma) \odot (ret \ \tau)$$

In Z, the refinement and decomposition are done with schemas. It is required to prove that the concrete schema is a valid refinement of the abstract schema, and this gives rise to a number of proof obligations. It needs to be proved that the initial states correspond to one another and that each operation in the concrete schema is correct with respect to the operation in the abstract schema, and also that it is applicable (i.e., whenever the abstract operation may be performed the concrete operation may also be performed).

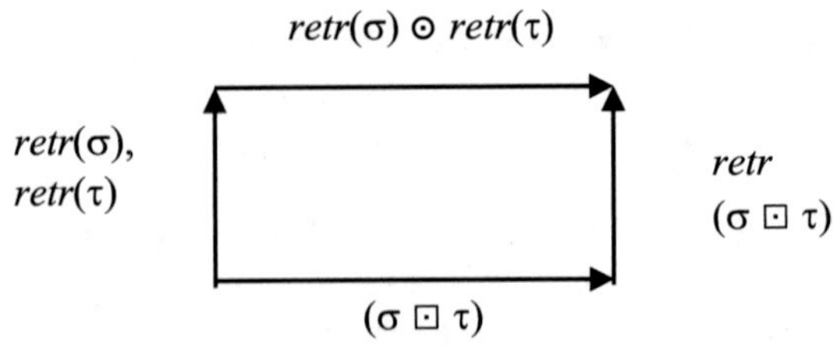

Fig. 17.8 Refinement
commuting diagram

$retr(\sigma) \odot retr(\tau)$

$retr(\sigma),$
$retr(\tau)$

$retr$
$(\sigma \boxdot \tau)$

$(\sigma \boxdot \tau)$

17.9 Proof in *Z*

We discuss the nature of theorem proving in Chap. 19. Mathematicians perform rigorous proof of theorems using technical and natural language, whereas logicians employ formal proofs using propositional and predicate calculus. Formal proofs generally involve a long chain of reasoning with every step of the proof justified, whereas rigorous mathematical proofs involve precise reasoning using a mixture of natural and mathematical language. Rigorous proofs [1] have been described as being analogous to high-level programming languages, whereas formal proofs are analogous to machine language.

A mathematical proof includes natural language and mathematical symbols, and often many of the tedious details of the proof are omitted. Many proofs in formal methods such as Z are concerned with crosschecking on the details of the specification, or on the validity of the refinement step, or proofs that certain properties are satisfied by the specification. There are often many tedious lemmas to be proved, and tool support is essential as proof by hand often contains errors or jumps in reasoning. Machine proofs provide extra confidence as every step in the proof is justified, and the proof of various properties about the programs increases confidence in its correctness.

17.10 Industrial Applications of *Z*

The *Z* specification language is one of the more popular formal methods, and it has been employed for the formal specification and verification of safety critical software. IBM piloted the *Z* formal specification language on the CICS (Customer Information Control System) project at its plant in Hursley, England.

Rolls Royce and Associates (RRA) developed a lifecycle suitable for the development of safety critical software, and the safety critical lifecycle used *Z* for the formal specification and the CADiZ tool provided support for specification, and Ada was the target implementation language.

Logica employed *Z* for the formal verification of a smartcard-based electronic cash system (the Mondex smartcard) in the early 1990s. The smartcard had an 8-bit microprocessor, and the objective was to formally specify both the high-level abstract security policy model and the lower-level concrete architectural design in *Z*, and to provide a formal proof of correspondence between the two.

Computer Management Group (CMG) employed *Z* for modelling data and operations as part of the formal specification of a movable barrier (the MaeslantKering) in the mid-1990s, which is used to protect the port of Rotterdam from flooding. The decisions on opening and closing of the barrier are based on meteorological data provided by the computer system, and the focus of the application of formal methods was to the decision-making subsystem and its interfaces to the environment.

17.11 Review Questions

1. Describe the main features of the Z specification language.
2. Explain the difference between $\mathbb{P}_1\, X$, $\mathbb{P}\, X$ and FX.
3. Give an example of a set derived from another set using set comprehension. Explain the three main parts of set comprehension in Z.
4. Discuss the applications of Z and which areas have benefited most from their use? What problems have arisen?
5. Give examples to illustrate the use of domain and range restriction operators and domain and range anti-restriction operators with relations in Z.
6. Give examples to illustrate relational composition.
7. Explain the difference between a partial and total function and give examples to illustrate function override.
8. Give examples to illustrate the various operations on sequences including concatenation, head, tail, map and reverse operations.
9. Give examples to illustrate the various operations on bags.
10. Discuss the nature of proof in Z and tools to support proof.
11. Explain the process of refining an abstract schema to a more concrete representation, the proof obligations that are generated, and the commuting diagram property.

17.12 Summary

Z is a formal specification language that was developed in the early 1980s at Oxford University in England. It has been employed in both industry and academia, and it was used successfully on the IBM's CICS project. Its specifications are mathematical, and this leads to more rigorous software development. Its mathematical approach allows properties to be proved about the specification, and any gaps or inconsistencies in the specification may be identified.

Z is a 'model-oriented' approach and an explicit model of the state of an abstract machine is given, and the operations are defined in terms of their effect on the state. Its main features include a mathematical notation that is similar to VDM, and the schema calculus. The latter consists essentially of boxes and is used to describe operations and states.

The schema calculus enables schemas to be used as building blocks to form larger specifications. It is a powerful means of decomposing a specification into smaller pieces and helps with the readability of Z specifications, as each schema is small in size and self-contained.

Z is a highly expressive specification language, and it includes notation for sets, functions, relations, bags, sequences, predicate calculus, and schema calculus. Z specifications are not directly executable as many of its data types and constructs are not part of modern programming languages. Therefore, there is a need to refine the Z specification into a more concrete representation and prove that the refinement is valid.

Reference

1. Diller A (1990) An introduction to formal methods. Wiley, England

Model Checking

18

Key Topics

Concurrent Systems

Temporal Logic

State Explosion

Safety and Liveness Properties

Fairness Properties

Linear Temporal Logic

Computational Tree Logic

18.1   Introduction

Model checking is an automated technique such that given a finite-state model of a system and a formal property (expressed in temporal logic), then it systematically checks whether the property is true or false in a given state in the model. It is an effective technique to identify potential design errors, and it increases confidence in the correctness of the system design. Model checking is a highly effective verification technology and is widely used in the hardware and software fields. It has been employed in the verification of microprocessors; in security protocols; in the transportation sector (trains); and in the verification of software in the space sector.

Early work on model checking commenced in the early 1980s (especially in checking the presence of properties such as mutual exclusion and the absence of deadlocks), and the term *"model checking"* was coined by Clarke and Emerson in the early 1980s [1], when they combined the state exploration approach and

© The Author(s), under exclusive license to Springer Nature Switzerland AG 2023
G. O'Regan, *Mathematical Foundations of Software Engineering*,
Texts in Computer Science, https://doi.org/10.1007/978-3-031-26212-8_18

Table 18.1 Model-checking process

Phase	Description
Modelling phase	Model the system under consideration Formalize the property to be checked
Running phase	Run the model checker to determine the validity of the property in the model
Analysis phase	Is the property satisfied? If applicable, check next property If the property is violated then 1. Analyse generated counterexample 2. Refine model, design, or property If out of space try alternative approach (e.g., abstraction of system model)

temporal logic in an efficient manner. Clarke and Emerson received the ACM Turing Award in 2007 for their role in developing model checking into a highly effective verification technology.

Model checking is a formal verification technique based on graph algorithms and formal logic. It allows the desired behaviour (specification) of a system to be verified, and its approach is to employ a suitable model of the system and to carry out a systematic and exhaustive inspection of all states of the model to verify that the desired properties are satisfied. These properties are generally safety properties such as the absence of deadlock, request-response properties, and invariants. The systematic search shows whether a given system model truly satisfies a particular property or not.

The phases in the model-checking process include the modelling, running, and analysis phases (Table 18.1).

The model-based techniques use mathematical models to describe the required system behaviour in precise mathematical language, and the system models have associated algorithms that allow all states of the model to be systematically explored. Model checking is used for formally verifying finite-state concurrent systems (typically modelled by automata), where the specification of the system is expressed in temporal logic, and efficient algorithms are used to traverse the model defined by the system (in its entirety) to check if the specification holds or not. *Of course, any verification using model-based techniques is only as good as the underlying model of the system.*

Model checking is an automated technique such that given a finite-state model of a system and a formal property, then a systematic search may be conducted to determine whether the property holds for a given state in the model. The set of all possible states is called the model's state space, and when a system has a finite-state space it is then feasible to apply model-checking algorithms to automate the demonstration of properties, with a counterexample exhibited if the property is not valid.

The properties to be validated are generally obtained from the system specification, and they may be quite elementary: e.g., a deadlock scenario should never arise (i.e., the system should never be able to reach a situation where no further progress is possible). The formal specification describes what the system should do, whereas the model description (often automatically generated) is an accurate and unambiguous description of how the system actually behaves. The model is often expressed in a finite-state machine consisting of a finite set of states and a finite set of transitions.

Figure 18.1 shows the structure of a typical model-checking system where a preprocessor extracts a state transition graph from a program or circuit. The model-checking engine then takes the state transition graph and a temporal formula P and determines whether the formula is true or not in the model.

The properties need to be expressed precisely and unambiguously (usually in temporal logic) to enable rigorous verification to take place. Model checking extracts a finite model from a system and then checks some property of that model. The model checker performs an exhaustive state search, which involves checking all system states to determine whether they satisfy the desired property or not (Fig. 18.2).

If a state that violates the desired property is determined (i.e., a defect has been found once it is shown that the system does not fulfil one of its specified properties), then the model checker provides a counterexample indicating how the model can reach this undesired state. The system is considered to be correct if it satisfies all of the specified properties. In the cases of where the model is too

Fig. 18.1 Concept of model checking

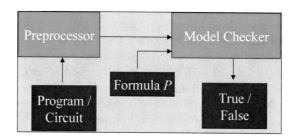

Fig. 18.2 Model checking

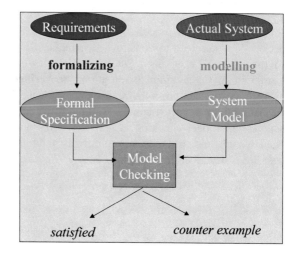

large to fit within the physical memory of the computer (state explosion problem), then other approaches such as abstraction of the system model or probabilistic verification may be employed.

There may be several causes of a state violating the desired property. It may be due to a modelling error (i.e., the model does not reflect the design of the system, and the model may need to be corrected and the model checking restarted). Alternatively, it may be due to a design error with improvements needed to the design, or it may be due to an error in the statement of the property with a modification to the property required and the model checking needs to be restarted.

Model checking is expressed formally by showing that a desired property P (expressed as a temporal logic formula) and a model M with initial state s, that P is always true in any state derivable from s (i.e., $M, s \models P$). We discussed temporal logic briefly in Chap. 11, and model checking is concerned with verifying that linear time properties such as *safety, liveness,* and *fairness* properties are always satisfied, and it employs *linear temporal logic* and *branching temporal logic*. Computational tree logic is a branching temporal logic where the model of time is a tree-like structure, with many different paths in future, one of which might be an actual path which is realized.

One problem with model checking is the state space explosion problem, where the transition graph grows exponentially on the size of the system, which makes the exploration of the state space difficult or impractical. Abstraction is one technique that aims to deal with the state explosion problem, and it involves creating a simplified version of the model (the abstract model). The abstract model may be explored in a reasonable period of time, and the abstract model must respect the original model with respect to key properties such that if the property is valid in the abstract model it is valid in the original model.

Model checking has been applied to areas such as the verification of hardware designs, embedded systems, protocol verification, and software engineering. Its algorithms have improved over the years, and today model checking is a mature technology for verification and debugging with many successful industrial applications.

The advantages of model theory include the fact that the user of the model checker does not need to construct a correctness proof (as in automated theorem proving or proof checking). Essentially, all the user needs to do is to input a description of the program or circuit to be verified and the specification to be checked, and to then press the return key. The checking process is then automatic and fast, and it provides a counterexample if the specification is not satisfied. One weakness of model checking is that it verifies an actual model rather than the actual system, and so its results are only as good as the underlying model. Model checking is described in detail in [2].

18.2   Modelling Concurrent Systems

Concurrency is a form of computing in which multiple computations (processes) are executed during the same period of time. *Parallel computing* allows execution to occur in the same time instant (on separate processors of a multiprocessor machine), whereas *concurrent computing* consists of process lifetimes overlapping and where execution need not happen at the same time instant.

Concurrency employs *interleaving* where the execution steps of each process employ time-sharing slices so that only one process runs at a time, and if it does not complete within its time slice it is paused; another process begins or resumes; and then later the original process is resumed. In other words, only one process is running at a given time instant, whereas multiple processes are part of the way through execution.

It is important to identify concurrency-specific errors such as deadlock and livelock. A *deadlock* is a situation in which the system has reached a state in which no further progress can be made, and at least one process needs to complete its tasks. *Livelock* refers to a situation where the processes in a system are stuck in a repetitive task and are making no progress towards their functional goals.

It is essential that safety properties such as *mutual exclusion* (at most one process is in its critical section at any given time) are not violated. In other words, something bad (e.g., a deadlock situation) should never happen; *liveness properties* (a desired event or something good eventually happens) are satisfied; and *invariants* (properties that are true all the time) are never violated. These behaviour errors may be mechanically detected if the systems are properly modelled and analysed.

Transition systems (Fig. 18.3) are often used as models to describe the behaviour of systems, and these are directed graphs with nodes representing *states* and edges representing *state transitions*. A state describes information about a system at a certain moment of time. For example, the state of a sequential computer consists of the values of all program variables and the current value of the program counter (pointer to next program instruction to be executed).

A transition describes the conditions under which a system moves from one state to another. Transition systems are expressive in that programs are transition systems; communicating processes are transition systems; and hardware circuits are transition systems.

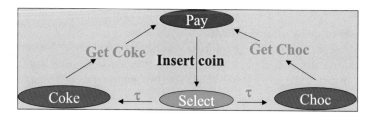

Fig. 18.3 Simple transition system

The transitions are associated with action labels that indicate the actions that cause the transition. For example, in Fig. 18.3 the *Insert coin* is a user action, whereas the *Get coke* and *Get choc* are actions that are performed by the machine. The activity τ represents an internal activity of the vending machine that is not of interest to the modeller. Formally, a transition system TS is a tuple $(S, Act, \rightarrow, I, AP, L)$ such that:

> S is the set of states
> Act is the set of actions
> $\rightarrow S \times Act \times S$ is the transition relation (source state, action and target state)
> $I \subseteq S$ is the set of initial states
> AP is a set of atomic propositions
> $L: S \rightarrow \mathbb{P}\, AP$ (power set of AP) is a labelling function

The transition (s, a, s') is written as $s \xrightarrow{a} s'$
   $L(s)$ are the atomic propositions in AP that are satisfied in state s.
   A concurrent system consists of multiple processes executing concurrently. If a concurrent system consists of n processes where each process $proc_i$ is modelled by a transition system TS_i, then the concurrent system may be modelled by a transition system (|| is the parallel composition operator):

$$TS = TS_1 || TS_2 || \ldots || TS_n$$

There are various operators used in modelling concurrency with transition systems, including operators for interleaving, communication via shared variables, handshaking, and channel systems.

18.3   Linear Temporal Logic

Temporal logic was discussed in Chap. 11 and is concerned with the expression of properties that have time dependencies. The existing temporal logics allow facts about the past, present, and future to be expressed. Temporal logic has been applied to specify temporal properties of natural language, as well as the specification and verification of program and system behaviour. It provides a language to encode temporal knowledge in artificial intelligence applications, and it plays a useful role in the formal specification and verification of temporal properties (e.g., *liveness* and *fairness*) in safety critical systems.

   The statements made in temporal logic can have a truth value that varies over time. In other words, sometimes the statement is true and sometimes it is false, but it is never true or false at the same time. The two main types of temporal logics are *linear time logics* (reason about a single timeline) and *branching time logics* (reason about multiple timelines).

   Linear temporal logic (LTL) is a modal temporal logic that can encode formulae about the future of paths (e.g., a condition that will eventually be true). The

Table 18.2 Basic temporal operators

Operator	Description
Fp	p holds sometime in future
Gp	p holds globally in future
Xp	p holds in next time instant
pUq	p holds until q is true

Fig. 18.4 LTL operators

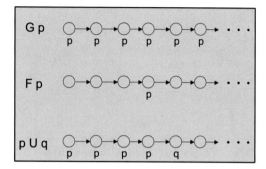

basic linear temporal operators that are often employed (p is an atomic proposition below) are listed in Table 18.2 and illustrated in Fig. 18.4.

For example, consider how the sentence *"This microwave does not heat up until the door is closed"* is expressed in temporal logic. This is naturally expressed with the until operator pUq as follows:

$$\neg \text{Heatup} U \text{DoorClosed}$$

18.4   Computational Tree Logic

In linear logic we look at the execution paths individually, whereas in branching time logics we view the computation as a tree. Computational tree logic (CTL) is a branching time logic, which means that its model of time is a tree-like structure in which the future is not determined, and so there are many paths in future such that any of them could be an actual path that is realized. CTL was first proposed by Clark and Emerson in the early 1980s.

Computational tree logic can express many properties of finite-state concurrent systems. Each operator of the logic has two parts namely the path quantifier (A—"every path", E—"there exists a path"), and the state quantifier (F, P, X, U as explained in Table 18.3). The operators in CTL logic are given by:

For example, the following is a valid CTL formula that states that it is always possible to get to the restart state from any state:

$$AG(EF \text{ restart})$$

Table 18.3 CTL temporal operators

Operator	Description
$A\varphi$ (*all*)	φ holds on *all* paths starting from the current state
$E\varphi$ (*exists*)	φ holds on at least one path starting from the current state
$X\varphi$ (*next*)	φ holds in the next state
$G\varphi$ (*global*)	φ has to hold on the entire subsequent path
$F\varphi$ (*finally*)	φ eventually has to hold (somewhere on the subsequent path)
$\varphi U\psi$ (*until*)	φ has to hold until at some position ψ holds
$\varphi W\psi$ (weak *until*)	φ has to hold until ψ holds (no guarantee ψ will ever hold)

18.5   Tools for Model Checking

There are various tools for model checking including Spin, Bandera, SMV, and UppAal. These tools perform a systematic check on property P in all states and are applicable if the system generates a finite behavioural model. Model-checking tools use a model-based approach rather than a proof rule-based approach, and the goal is to determine whether the concurrent program satisfies a given logical property.

Spin is a popular open-source tool that is used for the verification of distributed software systems (especially concurrent protocols), and it checks finite-state systems with properties specified by linear temporal logic. It generates a counterexample trace if determines that a property is violated.

Spin has its own input specification language (PROMELA), and so the system to be verified needs to be translated into the language of the model checker. The properties are specified using LTL.

Bandera is a tool for model-checking Java source code, and it automates the extraction of a finite-state model from the Java source code. It then translates into an existing model checker's input language. The properties to be verified are specified in the Bandera Specification Language (BSL), which supports pre- and postconditions and temporal properties.

18.6   Industrial Applications of Model Checking

There are many applications of model checking in the hardware and software fields, including the verification of microprocessors and security protocols, as well as applications in the transportation sector (trains) and in the space sector.

The Mars Science Laboratory (MSL) mission used model checking as part of the verification of the critical software for the landing of Curiosity (a large rover) on its mission to Mars. The hardware and software of a spacecraft must be designed for a high degree of reliability, as an error can lead to a loss of the mission. The Spin model checker was employed for the model verification, and the rover was launched in November 2011 and landed safely on Mars in August 2012.

CMG employed formal methods as part of the specification and verification of the software for a movable flood barrier in the mid-1990s. This is used to protect the port of Rotterdam from flooding, Z was employed for modelling data and operations, and Spin/Promela was used for model checking.

Lucent's Pathstar Access Server was developed in the late 1990s, and this system is capable of sending voice and data over the Internet. The automated verification techniques applied to Pathstar consist of generating an abstract model from the implemented C code, and then defining the formal requirements that the application is satisfy. Finally, the model checker is employed to perform the verification.

18.7   Review Questions

1. What is model checking?
2. Explain the state explosion problem.
3. Explain the difference between parallel processing and concurrency.
4. Describe the basic temporal operators.
5. Describe the temporal operators in CTL.
6. Explain the difference between liveness and fairness properties.
7. What is a transition system?
8. Explain the difference between linear temporal logic and branching temporal logic.
9. Investigate tools to support model checking.

18.8   Summary

Model checking is a formal verification technique which allows the desired behaviours of a system to be verified. Its approach is to employ a suitable model of the system and to carry out a systematic inspection of all states of the model to verify the required properties are satisfied in each state. The properties to be validated are generally obtained from the system specification, and a defect is found once it is shown that the system does not fulfil one of its specified properties. The system is considered to be correct if it satisfies all of the specified properties.

The desired behaviour (specification) of the system is verified by employing a suitable model of the system and then carrying out a systematic exhaustive inspection of all states of the model to verify that the desired properties are satisfied. These properties are generally properties such as the absence of deadlock and invariants. The systematic search shows whether a given system model truly satisfies a particular property or not.

The model-based techniques use mathematical models to describe the required system behaviour in precise mathematical language, and the system models have associated algorithms that allow all states of the model to be systematically explored. The specification of the system is expressed in temporal logic, and efficient algorithms are used to traverse the model defined by the system (in its entirety) to check if the specification holds or not. Model-based techniques are only as good as the underlying model of the system.

References

1. Clarke EM, Emerson EA (1981) Design and synthesis of synchronization skeletons using branching time temporal logic. In: Logic of programs: work-shop, Yorktown heights, NY, May 1981, vol 131 of LNCS. Springer
2. Baier C, Katoen JP (2008) Principles of model checking. MIT Press, Cambridge

The Nature of Theorem Proving

19

Key Topics

Mathematical Proof

Formal Proof

Automated Theorem Prover

Interactive Theorem Prover

Logic Theorist

Resolution

Proof Checker

19.1 Introduction

The word *"proof"* is generally interpreted as facts or evidence that support a particular proposition or belief, and such proofs are generally conducted in natural language. Several premises (which are self-evident or already established) are presented, and from these premises (via deductive or inductive reasoning) further propositions are established, until finally the conclusion is established.

The proof of a theorem in mathematics requires additional rigour, and such proofs consist of a mixture of natural language and mathematical argument. It is common to skip over the trivial steps in the proof, and independent mathematicians conduct peer reviews to provide additional confidence in the correctness of the proof and to ensure that no unwarranted assumptions or errors in reasoning have

© The Author(s), under exclusive license to Springer Nature Switzerland AG 2023
G. O'Regan, *Mathematical Foundations of Software Engineering*,
Texts in Computer Science, https://doi.org/10.1007/978-3-031-26212-8_19

been made. Proofs conducted in logic are extremely rigorous with every step in the proof is explicit.[1]

Mathematical proof dates back to the Greeks, and many students are familiar with Euclid's work (*The Elements*) in geometry, where from a small set of axioms and postulates and definitions he derived many of the well-known theorems of geometry. Euclid was a Hellenistic mathematician based in Alexandria around 300BC, and his style of proof was mainly constructive: i.e., in addition to the proof of the existence of an object, he actually constructed the object in the proof. Euclidean geometry remained unchallenged for over 2000 years, until the development of the non-Euclidean geometries in the nineteenth century, and these geometries were based on a rejection of Euclid's controversial 5th postulate (the parallels postulate).

Mathematical proof may employ a *"divide and conquer"* technique; i.e., breaking the conjecture down into subgoals and then attempting to prove each of the subgoals. Another common proof technique is *indirect proof* where we assume the opposite of what we wish to prove, and we show that this results in a contradiction (e.g., see the proof in Chap. 4 that there are an infinite number of primes or the proof that there is no rational number whose square is 2). Other proof techniques used are the method *of mathematical induction*, where involves a proof of the base case and inductive step (see Chap. 6).

Aristotle developed *syllogistic logic* in the fourth century BC, and the rules of reasoning with valid syllogisms remained dominant in logic up to the nineteenth century. Boole develop his mathematical logic in the mid-nineteenth century, and he aimed to develop a calculus of reasoning to verify the correctness of arguments using logical connectives. Predicate logic (including universal and existential quantifiers) was introduced by Frege in the late nineteenth century as part of his efforts to derive mathematics from purely logical principles. Russell and Whitehead continued this attempt in *Principia Mathematica*, and Russell introduced the theory of types to deal with the paradoxes in set theory, which he identified in Frege's system.

The *formalists* introduced extensive axioms in addition to logical principles, and Hilbert's program led to the definition of a *formal mathematical proof* as a sequence of formulae, where each element is either an axiom or derived from a previous element in the series by applying a fixed set of mechanical rules (e.g., *modus ponens*). The last line in the proof is the theorem to be proved, and the formal proof is essentially syntactic following rules with the formulae simply a string of symbols and the meaning of the symbols is unimportant.

The formalists later ran into problems in trying to prove that a formal system powerful enough to include arithmetic was both complete and consistent, and the results of Gödel showed that such a system would be *incomplete* (and one of the

---

[1] Perhaps a good analogy might be that a mathematical proof is like a program written in a high-level language such as C, whereas a formal mathematical proof in logic is like a program written in assembly language.

Fig. 19.1 Idea of automated theorem proving

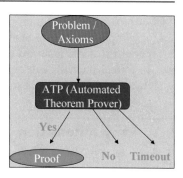

propositions without a proof is that of its own *consistency*). Turing later showed (with his Turing machine) that mathematics is *undecidable*: i.e., there is no algorithm or mechanical procedure that may be applied in a finite number of steps to determine if an arbitrary mathematical proposition is true or false.

The proofs employed in mathematics are rarely formal (in the sense of Hilbert's program), and whereas they involve deductions from a set of axioms, these deductions are rarely expressed as the application of individual rules of logical inference.

The application of formal methods in an industrial environment requires the use of machine-assisted proof, since thousands of proof obligations arise from a formal specification, and theorem provers are essential in resolving these efficiently. Many proofs in formal methods are concerned with crosschecking the details of the specification, checking the validity of the refinement steps, or checking that certain properties are satisfied by the specification. There are often many tedious lemmas to be proved, and theorem provers[2] are essential in dealing with these. Machine proof is explicit, and reliance on some brilliant insight is avoided. Proofs by hand in formal methods are notorious for containing errors or jumps in reasoning, whereas machine proofs are explicit but are often extremely lengthy and essentially unreadable.

Automated theorem proving (ATP) is difficult, as often mathematicians prove a theorem with an initial intuitive feeling that the theorem is true (Fig. 19.1). Human intervention to provide guidance or intuition improves the effectiveness of the theorem prover. There are several tools available to support theorem proving, and these include the Boyer-Moore theorem prover (known as NQTHM); the Isabelle theorem prover; and the HOL system.

The proof of various properties about a program increases confidence in its correctness. However, an absolute proof of correctness[3] is unlikely except for the most

---

[2] Most existing theorem provers are difficult to use and are for specialist use only. There is a need to improve the usability of theorem provers.

[3] This position is controversial with others arguing that if correctness is defined mathematically then the mathematical definition (i.e., formal specification) is a theorem, and the task is to prove that the program satisfies the theorem. They argue that the proofs for non-trivial programs exist and

trivial of programs. A program may consist of legacy software that is assumed to work; a compiler that is assumed to work correctly creates it. Theorem provers are programs that are assumed to function correctly. The best that mathematical proof in formal methods can claim is increased confidence in the correctness of the software, rather than an absolute proof of correctness.

19.2   Early Automation of Proof

Early work on the automation of proof began in the 1950s with the beginning of work in the Artificial Intelligence field, where the early AI practitioners were trying to develop a *"thinking machine"*. One of the earliest programs developed was the *Logic Theorist* (LT), which was presented at the Dartmouth conference on Artificial Intelligence in 1956 [1].

It was developed by Allen Newell and Herbert Simon, and it could prove 38 of the first 52 theorems from Russell and Whitehead's *Principia Mathematica* [2].[4] Russell and Whitehead had attempted to derive all mathematics from axioms and the inference rules of logic, and the LT program conducted proof from a small set of propositional axioms and deduction rules. Its approach was to start with the theorem to be proved, and to then search for relevant axioms and operators to prove the theorem. The Logic Theorist proved theorems in the propositional calculus, but it did not support predicate calculus. It used the five basic axioms of propositional logic and three rules of inference from the *Principii* to prove theorems.[5]

LT demonstrated that computers had the ability to encode knowledge and information, and to perform intelligent operations such as solving theorems in mathematics. The heuristic approach of the LT program tried to emulate human mathematicians but could not guarantee that a proof could be found for every valid theorem.

If no immediate one-step proof could be found, then a set of subgoals was generated (these are formulae from which the theorem may be proved in one step) and proofs of these were then searched for, and so on. The program could use previously proved theorems while developing a proof of a new theorem. Newell and Simon were hoping that the Logic Theorist would do more than just prove theorems in logic, and their goal was that it would attempt to prove theorems in a human-like way and especially with a selective search.

---

that the reason why there are not many examples of such proofs is due to a lack of mathematical specifications.

[4] Russell is said to have remarked that he was delighted to see that the Principia Mathematica could be done by machine, and that if he and Whitehead had known this in advance that they would not have wasted 10 years doing this work by hand in the early twentieth century.

[5] Another possibility (though an inefficient and poor simulation of human intelligence) would be to start with the five axioms of the *Principia*, and to apply the three rules of inference to logically derive all possible sequences of valid deductions. This is known as the British Museum algorithm (as sensible as putting monkeys in front of typewriters to reproduce all of the books of the British Museum).

However, in practice, the Logic Theorist search was not very selective in its approach, and the subproblems were considered in the order in which they were generated, and so there was no actual heuristic procedure (as in human problem solving) to guess at which subproblem was most likely to yield an actual proof. This meant that the Logic Theorist could, in practice, find only very short proofs, since as the number of steps in the proof increased, the amount of search required to find the proof exploded.

The Geometry Machine was developed by Herbert Gelerner at the IBM Research Centre in New York in the late 1950s, with the goal of developing intelligent behaviour in machines. It differed from the Logic Theorist in that it selected only the valid subgoals (i.e., it ignored the invalid ones) and attempted to find a proof of these. The Geometry Machine was successful in finding the solution to a large number of geometry problems taken from high-school text books in plane geometry.

The logicians Hao Wang and Evert Beth (the inventor of semantic tableaux which was discussed in Chap. 10) were critical of the approaches of the AI pioneers and believed that mathematical logic could do a lot more. Wang and others developed a theorem prover for first-order predicate calculus in 1960, but it had serious limitations due to the combinatorial explosion.

Alan Robinson's work on theorem provers in the early 1960s led to a proof procedure termed "*resolution*", which appeared to provide a breakthrough in the automation of predicate calculus theorem provers. A resolution theorem prover is essentially provided with the axioms of the field of mathematics in question, and the negation of the conjecture whose proof is sought. It then proceeds until a contradiction is reached, where there is no possible way for the axioms to be true and for the conjecture to be false.

The initial success of resolution led to excitement in the AI field where pioneers such as John McCarthy (see Chap. 11) believed that human knowledge could be expressed in predicate calculus,[6] and that therefore if resolution was indeed successful for efficient automated theorem provers, then the general problem of Artificial Intelligence was well on the way to a solution. However, while resolution led to improvements with the state explosion problem, it did not eliminate the problem.

This led to a falloff in research into resolution-based approaches to theorem proving, and other heuristic-based techniques were investigated by Bledsoe in the late 1970s. The field of logic programming began in the early 1970s with the development of the Prolog programming language (see Chap. 11). Prolog is in a sense an application of automated theorem proving, where problems are stated in the form of goals (or theorems) that the system tries to prove using a resolution

---

[6] McCarthy's viewpoint that predicate logic was the solution for the AI field was disputed by Minsksy and others (resulting in a civil war between the logicists and the proceduralists). The proceduralists argued that formal logic was an inadequate representation of knowledge for AI and that predicate calculus was an overly rigid and inadequate framework. They argued that an alternative approach such as the procedural representation of knowledge was required.

theorem prover. The theorem prover generally does not need to be very powerful as many Prolog programs require only a very limited search, and a *depth-first* search from the goal backwards to the hypotheses is conducted.

The Argonne Laboratory (based in Chicago in the United States) developed the Aura System in the early 1980s (it was later replaced by Otter), as an improved resolution-based automated theorem prover, and this led to renewed interest in resolution-based approaches to theorem proving. There is a more detailed account of the nature of proof and theorem proving in [1].

19.3   Interactive Theorem Provers

The challenges in developing efficient automated theorem provers led researchers to question whether an effective fully automated theorem prover was possible, and if it made more sense to develop a theorem prover that could be guided by a human in its search for a proof. This led to the concept of *Interactive theorem proving* (ITP) which involves developing formal proofs by man-machine collaboration and is (in a sense) a new way of doing mathematics in front of a computer.

Such a system is potentially useful in mathematical research in formalizing and checking proofs, and it allows the user to concentrate on the creative parts of the proof and relieves the user of the need of carrying out the trivial steps in the proof. It is also a useful way of verifying the correctness of published mathematical proofs by acting as a *proof checker*, where the ITP is provided with a formal proof constructed by a human, which may then be checked for correctness.[7] Such a system is important in *program verification* in showing that the program satisfies its specification, and especially in the safety/security critical field.

A group at Princeton developed a series of systems called Semi-automated mathematics (SAM) in the late 1960s, which combined logic routines with human guidance and control. Their approach placed the mathematician at the heart of the theorem proving, and it was a departure from the existing theorem proving approaches where the computer attempted to find proofs unaided. SAM provided a proof of an unproven conjecture in lattice theory (SAM's lemma), and this is regarded as the first contribution of automated reasoning systems to mathematics [1].

De Bruijn and others at the Technische Hogeschool in Eindhoven in the Netherlands commenced development of the Automath system in the late 1960s. This was a large-scale project for the automated verification of mathematics, and it was tested by treating a full text book. Automath systematically checked the proofs from Landau's text *Grundlagen der Analysis* (this foundations of analysis text was first published in 1930).

---

[7] A formal mathematical proof (of a normal proof) is difficult to write down and can be lengthy. Mathematicians were not really interested in these proof checkers.

The typical components of an Interactive Theorem Prover include an interactive proof editor to allow editing of proofs, formulae and terms in a formal theory of mathematics, and a large library of results which is essential for achieving complex results.

The Gypsy verification environment and its associated theorem prover was developed at the University of Texas in the 1980s, and it achieved early success in program verification with its verification of the encrypted packet interface program (a 4200 line program). It supports the development of software systems and formal mathematical proof of their behaviour.

The Boyer-Moore Theorem prover (NQTHM) was developed in the 1970s/1980s at the University of Texas by Boyer and Moore [3]. It was improved and became known as NQTHM (it has been superseded by ACL2 available from the University of Texas). It supports mathematical induction as a rule of inference, and induction is a useful technique in proving properties of programs. The axioms of Peano arithmetic are built into the theorem prover, and new axioms added to the system need to pass a "correctness test" to prevent the introduction of inconsistencies.

It is far more automated than many other interactive theorem provers, but it requires detailed human guidance (with suggested lemmas) for difficult proofs. The user therefore needs to understand the proof being sought and the internals of the theorem prover.

It has been effective in proving well-known theorems such as Gödel's Incompleteness Theorem, the insolvability of the Halting problem, a formalization of the Motorola MC 68,020 Microprocessor, and many more.

Computational Logic Inc. was a company founded by Boyer and Moore in 1983 to share the benefits of a formal approach to software development with the wider computing community. It was based in Austin, Texas, and provided services in the mathematical modelling of hardware and software systems. This involved the use of mathematics and logic to formally specify microprocessors and other systems. The use of its theorem prover was to formally verify that the implementation meets its specification: i.e., to prove that the microprocessor or other system satisfies its specification.

The HOL system was developed by Michael Gordon and others at Cambridge University in the UK, and it is an environment for interactive theorem proving in a higher-order logic. It has been applied to the formalization of mathematics and to the verification of hardware (including the verification of microprocessor design). It requires skilled human guidance and is one of the most widely used theorem provers. It was originally developed in the early 1980s, and HOL 4 is the latest version. It is an open-source project and is used by academia and industry.

Isabelle is a theorem proving environment developed at Cambridge University by Larry Paulson and Tobias Nipkow of the Technical University of Munich. It allows mathematical formulae to be expressed in a formal language and provides tools for proving those formulae. The main application is the formalization of mathematical proof and proving the correctness of computer hardware or software with respect to its specification and proving properties of computer languages and protocols.

Isabelle is a generic theorem prover in the sense that it has the capacity to accept a variety of formal calculi, whereas most other theorem provers are specific to a specific formal calculus. Isabelle is available free of charge under an open-source licence.

There is a steep learning curve with the theorem provers above, and it generally takes a couple of months for users to become familiar with them. However, automated theorem proving has become a useful tool in the verification of integrated circuit design. Several semiconductor companies use automated theorem proving to demonstrate the correctness of division and other operators on their processors. We present a selection of theorem provers in the next section.

19.4   A Selection of Theorem Provers

Table 19.1 presents a small selection of the available automated and interactive theorem provers.

19.5   Review Questions

1. What is a mathematical proof?
2. What is a formal mathematical proof?
3. What approaches are used to prove a theorem?
4. What is a theorem prover?
5. What role can theorem provers play in software development?
6. What is the difference between an automated theorem prover and an interactive theorem prover?
7. Investigate and give a detailed description of one of the theorem provers in Table 19.1.

19.6   Summary

A mathematical proof includes natural language and mathematical symbols, and often many of the tedious details of the proof are omitted. The proofs in mathematics are rarely formal as such, and many proofs in formal methods are concerned with crosschecking the details of the specification, checking the validity of the refinement steps, or checking that certain properties are satisfied by the specification.

Table 19.1 Selection of theorem provers

Theorem prover	Description
ACL2	A Computational Logic for Applicative Common Lisp (ACL2) is part of the Boyer-Moore family of theorem provers. It is a software system consisting of a programming language (LISP) and an interactive theorem prover. It was developed in the mid-1990s as an industrial strength successor to the Boyer-Moore Theorem prover (NQTHM). It is used in the verification of safety critical hardware and software, and in industrial applications such as the verification of the floating-point module of a microprocessor
OTTER	OTTER is a resolution-style theorem prover for first-order logic developed at the Argonne Laboratory at the University of Chicago (it was the successor to Aura). It has been mainly applied to abstract algebra and formal logic
PVS	The Prototype Verification System (PVS) is a mechanized environment for formal specification and verification. It includes a specification language integrated with support tools and an interactive theorem prover. It was developed by John Rushby and others at SRI in California. The specification language is based on higher-order logic, and the theorem prover is guided by the user in conducting proof. It has been applied to the verification of hardware and software
Theorem Proving System (TPS)	TPS is an automated theorem prover for first-order and higher-order logic (it can also prove theorems interactively). It was developed at Carnegie Mellon University and is used for hardware and software verification
HOL and Isabelle	HOL and Isabelle were developed by the Automated Reasoning Group at the University of Cambridge. The HOL system is an environment for interactive theorem proving in a higher-order logic, and it has been applied to hardware verification. Isabelle is a generic proof assistant which allows mathematical formulae to be expressed in a formal language, and it provides tools for proving those formulae in a logical calculus
Boyer-Moore	The Boyer-Moore Theorem prover (NQTHM) was developed at the University of Texas in the 1970s with the goal of checking the correctness of computer systems. It has been used to verify the correctness of microprocessors, and it has been superseded by ACL2

Machine proof is explicit, and reliance on some brilliant insight is avoided. Proofs by hand often contain errors or jumps in reasoning, while machine proofs are often extremely lengthy and unreadable. The application of formal methods in an industrial environment requires the use of machine-assisted proof, since thousands of proof obligations arise from a formal specification, and theorem provers are essential in resolving these efficiently. The proof of various properties about a program increases confidence in its correctness. However, an absolute proof of correctness is unlikely except for the most trivial of programs.

Automated theorem proving is difficult, as often mathematicians prove a theorem with an initial intuitive feeling that the theorem is true. Human intervention to provide guidance or intuition improves the effectiveness of the theorem prover. Early work on the automation of proof began in the 1950s with the beginning of work in the Artificial Intelligence field, and one of the earliest programs developed was the Logic Theorist, which was presented at the Dartmouth conference on Artificial Intelligence in 1956.

The challenges in developing effective automated theorem provers led researchers to investigate whether it made more sense to develop a theorem prover that could be guided by a human in its search for a proof. This led to the development of Interactive theorem proving which involved developing formal proofs by man-machine collaboration.

The typical components of an interactive Theorem Prover include an interactive proof editor to allow editing of proofs, formulae, and terms in a formal theory of mathematics, and a large library of results which is essential for achieving complex results.

An interactive theorem prover allows the user to concentrate on the creative parts of the proof and relieves the user of the need to carry out and verify the trivial steps in the proof. It is also a useful way of verifying the correctness of published mathematical proofs by acting as a proof checker and is also useful in program verification in showing that the program satisfies its specification, and especially in the safety/security critical fields.

References

1. MacKensie D (1995) The automation of proof. IEEE a historical and sociological exploration. Ann Hist Comput 17(3):7–29
2. Russell B, Whitehead AN (1910) Principia Mathematica. Cambridge University Press, Cambridge
3. Boyer R, Moore JS (1979) A computational logic. The Boyer Moore theorem prover. Academic Press, Cambridge

Cryptography

20

Key Topics

Caesar Cipher

Enigma Codes

Bletchley Park

Turing

Public and Private Keys

Symmetric Keys

Block Ciphers

RSA

20.1 Introduction

Cryptography was originally employed to protect communication of private information between individuals. Today, it consists of mathematical techniques that provide secrecy in the transmission of messages between computers, and its objective is to solve security problems such as privacy and authentication over a communications channel.

It involves enciphering and deciphering messages, and it employs theoretical results from number theory (see Chap. 3) to convert the original message (or plaintext) into cipher text that is then transmitted over a secure channel to the intended recipient. The cipher text is meaningless to anyone other than the intended recipient, and the recipient uses a key to decrypt the received cipher text and to read the original message.

© The Author(s), under exclusive license to Springer Nature Switzerland AG 2023
G. O'Regan, *Mathematical Foundations of Software Engineering*,
Texts in Computer Science, https://doi.org/10.1007/978-3-031-26212-8_20

The origin of the word "cryptography" is from the Greek '*kryptos*' meaning hidden, and '*graphein*' meaning to write. The field of cryptography is concerned with techniques by which information may be concealed in cipher texts and made unintelligible to all but the intended recipient. This ensures the privacy of the information sent, as any information intercepted will be meaningless to anyone other than the authorized recipient.

Julius Caesar developed one of the earliest ciphers on his military campaigns in Gaul (see Chap. 4). His objective was to communicate important messages safely to his generals. His solution is one of the simplest and widely known encryption techniques, and it involves the substitution of each letter in the plaintext (i.e., the original message) by a letter a fixed number of positions down in the alphabet. The Caesar cipher involves a shift of three positions, and this leads to the letter B being replaced by E, the letter C by F, and so on.

The Caesar cipher is easily broken, as the frequency distribution of letters may be employed to determine the mapping. However, the Gaulish tribes were mainly illiterate, and so it is highly likely that the cipher provided good security. The translation of the Roman letters by the Caesar cipher (with a shift key of 3) can be seen in Fig. 4.3.

The process of enciphering a message (i.e., the plaintext) simply involves going through each letter in the plaintext and writing down the corresponding cipher letter, with the reverse process employed in deciphering a cipher message. The encryption and decryption may also be done using modular arithmetic, with the numbers 0–25 used to represent the alphabet letters, and the encryption of a letter x is given by $x + 3 \pmod{26}$ and decryption of a letter y is given by $y - 3 \pmod{26}$.

The Caesar cipher was still in use up to the early twentieth century. However, by then frequency analysis techniques were available to break the cipher. The Vigenère cipher uses a Caesar cipher with a different shift at each position in the text. The value of the shift to be employed with each plaintext letter is defined using a repeating keyword.

20.2   Breaking the Enigma Codes

The Enigma codes were used by the Germans during the Second World War for the secure transmission of naval messages to their submarines. These messages contained top-secret information on German submarine and naval activities in the Atlantic and the threat that they posed to British and Allied shipping.

The codes allowed messages to be passed secretly using encryption, and this meant that any unauthorized interception was meaningless to the Allies. The plaintext (i.e., the original message) was converted by the Enigma machine (Fig. 20.1) into the encrypted text, and these messages were then transmitted by the German military to their submarines in the Atlantic, or to their bases throughout Europe.

The Enigma cipher was invented in 1918, and the Germans believed it to be unbreakable. A letter was typed in German into the machine, and electrical

Fig. 20.1 The Enigma machine

impulses through a series of rotating wheels and wires produced the encrypted letter which was lit up on a panel above the keyboard. The recipient typed the received message into his machine, and the decrypted message was lit up letter by letter above the keyboard. The rotors and wires of the machine could be configured in many different ways, and during the war, the cipher settings were changed at least once a day. The odds against anyone breaking the Enigma machine without knowing the setting were 150×10^{18} to 1.

The British code and cipher school was relocated from London to Bletchley Park at the start of the Second World War (Fig. 20.2). It was located in the town of Bletchley (near Milton Keynes about fifty miles northwest of London). It was commanded by Alistair Dennison and was known as Station X, and several thousands were working there during the second world war. The team at Bletchley Park broke the Enigma codes and therefore made vital contributions to the British and Allied war effort.

Polish cryptanalysts did important work in breaking the Enigma machine in the early 1930s, and they constructed a replica of the machine. They passed their knowledge on to the British and gave them the replica just prior to the German invasion of Poland. The team at Bletchley built upon the Polish work, and the team included Alan Turing[1] (Fig. 20.3), Gordan Welchman,[2] and other mathematicians.

---

[1] Turing made fundamental contributions to computing, including the theoretical Turing machine.
[2] Gordon Welchman was the head of Hut 6 at Bletchley Park, and he made important contributions to code breaking. He invented a method to reduce the time to find the settings of the Enigma machine from days to hours. He also invented a technique known as traffic analysis (later called network analysis/metadata analysis which collected and analysed German messages to determined where and when they were sent.

Fig. 20.2   Bletchley park

Fig. 20.3   Alan Turing

The code-breaking teams worked in various huts in Bletchley park. Hut 6 focussed on air force and army ciphers, and hut 8 focussed on naval ciphers. The deciphered messages were then converted into intelligence reports, with air force and army intelligence reports produced by the team in hut 3, and naval intelligence reports produced by the team in hut 4. The raw material (i.e., the encrypted messages) to be deciphered came from wireless intercept stations dotted around Britain, and from various countries overseas. These stations listened to German radio messages and sent them to Bletchley park to be deciphered and analysed.

Turing devised a machine to assist with breaking the codes (an idea that was originally proposed by the Polish cryptanalysts). This electromechanical machine was known as the bombe (Fig. 20.4), and its goal was to find the right settings of the Enigma machine for that particular day. The machine greatly reduced the odds and the time required to determine the settings on the Enigma machine, and it became the main tool for reading the Enigma traffic during the war. The bombe was first installed in early 1940, and it weighed over a tonne. It was named after a cryptological device designed in 1938 by the Polish cryptologist, Marian Rejewski.

A standard Enigma machine employed a set of rotors, and each rotor could be in any of 26 positions. The bombe tried each possible rotor position and applied a test. The test eliminated almost all of the positions and left a smaller number of

Fig. 20.4 Replica of bombe

cases to be dealt with. The test required the cryptologist to have a suitable "*crib*": i.e., a section of ciphertext for which he could guess the corresponding plaintext.

For each possible setting of the rotors, the bombe employed the crib to perform a chain of logical deductions. The bombe detected when a contradiction had occurred, and it then ruled out that setting and moved onto the next. Most of the possible settings would lead to contradictions and could then be discarded. This would leave only a few settings to be investigated in detail.

The Government Communication Headquarters (GCHQ) was the successor of Bletchley Park, and it relocated to Cheltenham after the war. The site at Bletchley park was then used for training purposes.

The codebreakers who worked at Bletchley Park were required to remain silent about their achievements until the mid-1970s when the wartime information was declassified.[3] The link between British Intelligence and Bletchley Park came to an end in the mid-1980s.

It was decided in the mid-1990s to restore Bletchley Park, and today it is run as a museum by the Bletchley Park Trust.

---

[3] Gordan Welchman published his book 'The Hut Six Story' in 1982 (in the US and UK) describing his wartime experience at Bletchley Park. However, the security services disapproved of its publication and his security clearance was revoked. He was forbidden to speak of his book and wartime work.

Table 20.1 Notation in cryptography

Symbol	Description
M	Represents the message (plaintext)
C	Represents the encrypted message (cipher text)
e_k	Represents the encryption key
d_k	Represents the decryption key
E	Represents the encryption process
D	Represents the decryption process

20.3   Cryptographic Systems

A cryptographic system is a computer system that is concerned with the secure transmission of messages. The message is encrypted prior to its transmission, which ensures that any unauthorized interception and viewing of the message is meaningless to anyone other than the intended recipient. The recipient uses a key to decrypt the cipher text and to retrieve the original message.

There are essentially two different types of cryptographic systems employed, and these are public key cryptosystems and secret key cryptosystems. A *public key cryptosystem* is an asymmetric cryptosystem where two different keys are employed: one for encryption and one for decryption. The fact that a person is able to encrypt a message does not mean that the person is able to decrypt a message.

In a *secret key cryptosystem*, the same key is used for both encryption and decryption. Anyone who has knowledge on how to encrypt messages has sufficient knowledge to decrypt messages, and the sender and receiver need to agree on a shared key prior to any communication. The following notation is employed (Table 20.1).

The encryption and decryption algorithms satisfy the following equation:

$$Dd_k(C) = Dd_k\big(E_{e_k}(M)\big) = M$$

There are two different keys employed in a public key cryptosystem. These are the encryption key e_k and the decryption key d_k with $e_k \neq d_k$. It is called asymmetric since the encryption key differs from the decryption key.

There is just one key employed in a secret key cryptosystem, with the same key e_k is used for both encryption and decryption. It is called *symmetric* since the encryption key is the same as the decryption key: i.e., $e_k = d_k$.

20.4   Symmetric Key Systems

A symmetric key cryptosystem (Fig. 20.5) uses the same secret key for encryption and decryption. The sender and the receiver first need to agree a shared key prior to communication. This needs to be done over a secure channel to ensure that the shared key remains secret. Once this has been done they can begin to encrypt and decrypt messages using the secret key. Anyone who is able to encrypt a message has sufficient information to decrypt the message.

The encryption of a message is in effect a transformation from the space of messages m to the space of cryptosystems $\mathbb{C}$. That is, the encryption of a message with key k is an invertible transformation f such that:

$$f: m \xrightarrow{k} \mathbb{C}$$

The cipher text is given by $C = E_k(M)$ where $M \in m$ and $C \in \mathbb{C}$. The legitimate receiver of the message knows the secret key k (as it will have been transmitted previously over a secure channel), and so the cipher text C can be decrypted by the inverse transformation f^{-1} defined by:

$$f^{-1}: \mathbb{C} \xrightarrow{k} m$$

Therefore, we have that $D_k(C) = D_k(E_k(M)) = M$ the original plaintext message.

The advantages and disadvantages of symmetric key systems include (Table 20.2).

Examples of Symmetric Key Systems

(i) *Caesar Cipher*

The Caesar cipher may be defined using modular arithmetic. It involves a shift of three places for each letter in the plaintext, and the alphabetic letters are represented by the numbers 0–25. The encryption is carried out by addition (module

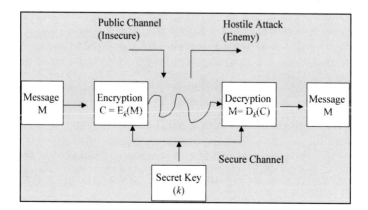

Fig. 20.5 Symmetric key cryptosystem

Table 20.2 Advantages and disadvantages of symmetric key systems

Advantages	Disadvantages
Encryption process is simple (as the same key is used for encryption and decryption)	A shared key must be agreed between two parties
It is faster than public key systems	Key exchange is difficult as there needs to be a secure channel between the two parties (to ensure that the key remains secret)
It uses less computer resources than public key systems	If a user has n trading partners then n secret keys must be maintained (one for each partner)
It uses a different key for communication with every different party	There are problems with the management and security of all of these keys (due to volume of keys that need to be maintained)
	Authenticity of origin or receipt cannot be proved (as key is shared)

26). The encryption of a plaintext letter x to a cipher letter c is given by[4]:

$$c = x + 3 \pmod{26}$$

Similarly, the decryption of a cipher letter c is given by:

$$x = c - 3 \pmod{26}$$

(ii) *Generalized Caesar Cipher*

This is a generalization of the Caesar cipher to a shift of k (the standard Caesar cipher involves a shift of three). This is given by:

$$f_k = E_k(x) \equiv x + k \pmod{26} \quad 0 \leq k \leq 25$$
$$f_k^{-1} = D_k(c) \equiv c - k \pmod{26} \quad 0 \leq k \leq 25$$

(iii) *Affine Transformation*

This is a more general transformation and is defined by:

$$f_{(a,b)} = E_{(a,b)}(x) \equiv ax + b \pmod{26} \quad 0 \leq a, b, x \leq 25 \text{ and } \gcd(a, 26) = 1$$
$$f_{(a,b)}^{-1} = D_{(a,b)}(c) \equiv a^{-1}(c - b) \pmod{26} \quad a^{-1} \text{ is the inverse of } a \bmod 26$$

---

[4] Here x and c are variables rather than the alphabetic characters 'x' and 'c'.

(iv) *Block Ciphers*

Stream ciphers encrypt a single letter at a time and are easy to break. Block ciphers offer greater security, the plaintext is split into groups of letters, and the encryption is performed on the block of letters rather than on a single letter.

The message is split into blocks of n-letters: M_1, M_2, ... M_k where each M_i ($1 \leq i \leq k$) is a block n-letters. The letters in the message are translated into their numerical equivalents, and the cipher text is formed as follows:

$$C_i \equiv AM_i + B \pmod{N} \quad i = 1, 2, \ldots k$$

$$\begin{pmatrix} a_{11} \, a_{12} \, a_{13} \cdots a_{1n} \\ a_{21} \, a_{22} \, a_{23} \cdots a_{2n} \\ a_{31} \, a_{32} \, a_{33} \cdots a_{3n} \\ \cdots \cdots \cdots \cdots \\ \cdots \cdots \cdots \cdots \\ a_{n1} \, a_{n2} \, a_{n3} \cdots a_{nn} \end{pmatrix} \begin{pmatrix} m_1 \\ m_2 \\ m_3 \\ \cdots \\ \cdots \\ m_n \end{pmatrix} + \begin{pmatrix} b_1 \\ b_2 \\ b_3 \\ \cdots \\ \cdots \\ b_n \end{pmatrix} = \begin{pmatrix} c_1 \\ c_2 \\ c_3 \\ \cdots \\ \cdots \\ c_n \end{pmatrix}$$

where (A, B) is the key, A is an invertible $n \times n$ matrix with $\gcd(\det(A), N) = 1$,[5] $M_i = (m_1, m_2, \ldots m_n)^T$, $B = (b_1, b_2, \ldots b_n)^T$, $C_i = (c_1, c_2, \ldots, c_n)^T$. The decryption is performed by:

$$M_i \equiv A^{-1}(C_i - B) \pmod{N} \quad i = 1, 2, \ldots k$$

$$\begin{pmatrix} m_1 \\ m_2 \\ m_3 \\ \cdots \\ \cdots \\ m_n \end{pmatrix} = \begin{pmatrix} a_{11} \, a_{12} \, a_{13} \cdots a_{1n} \\ a_{21} \, a_{22} \, a_{23} \cdots a_{2n} \\ a_{31} \, a_{32} \, a_{33} \cdots a_{3n} \\ \cdots \cdots \cdots \cdots \cdots \\ \cdots \cdots \cdots \cdots \cdots \\ a_{n1} \, a_{n2} \, a_{n3} \cdots a_{nn} \end{pmatrix}^{-1} \begin{pmatrix} c_1 - b_1 \\ c_2 - b_2 \\ c_3 - b_3 \\ \cdots \\ \cdots \\ c_n - b_n \end{pmatrix}$$

(v) Exponential Ciphers

Pohlig and Hellman [1] invented the exponential cipher in 1976. This cipher is less vulnerable to frequency analysis than block ciphers.

Let p be a prime number and let M be the numerical representation of the plaintext, with each letter of the plaintext replaced with its two-digit representation (00–25). That is, $A = 00$, $B = 01$, ..., $Z = 25$.

M is divided into blocks M_i (these are equal size blocks of m letters where the block size is approximately the same number of digits as p). The number of letters

---

[5] This requirement is to ensure that the matrix A is invertible. Matrices are discussed in Chap. 27.

m per block is chosen such that:

$$\underbrace{2525\ldots25}_{m\text{ times}} < p < \underbrace{2525\ldots25}_{m+1\text{ times}}$$

For example, for the prime 8191 a block size of $m = 2$ letters (4 digits) is chosen since:

$$2525 < 8191 < 252{,}525$$

The secret encryption key is chosen to be an integer k such that $0 < k < p$ and $\gcd(k, p - 1) = 1$. Then the encryption of the block M_i is defined by:

$$C_i = E_k(M_i) \equiv M_i^k \pmod{p}$$

The cipher text C_i is an integer such that $0 \le C_i < p$.

The decryption of C_i involves first determining the inverse k^{-1} of the key k (mod $p - 1$), i.e., we determine k^{-1} such that $k.k^{-1} \equiv 1 \pmod{p - 1}$. The secret key k was chosen so that $(k, p - 1) = 1$, and this means that there are integers d and n such that $kd = 1 + n(p - 1)$, and so k^{-1} is d and $kk^{-1} = 1 + n(p - 1)$. Therefore,

$$D_{k^{-1}}(C_i) \equiv C_i^{k^{-1}} \equiv \left(M_i^k\right)^{k^{-1}} \equiv M_i^{1+n(p-1)} \equiv M_i \pmod{p}$$

The fact that $M_i^{1+n(p-1)} \equiv M_i \pmod{p}$ follows from Euler's Theorem and Fermat's Little Theorem (Theorem 3.7 and 3.8), which are discussed in Chap. 3 of [2]. Euler's Theorem states that for two positive integers a and n with $\gcd(a,n) = 1$ that $a^{\phi(n)} \equiv 1 \pmod{n}$.

Clearly, for a prime p we have that $\phi(p) = p - 1$. This allows us to deduce that:

$$M_i^{1+n(p-1)} \equiv M_i^1 M_i^{n(p-1)} \equiv M_i\left(M_i^{(p-1)}\right)^n \equiv M_i(1)^n \equiv M_i \pmod{p}$$

(vi) Data Encryption Standard (DES)

DES is a popular cryptographic system [3] used by governments and private companies around the world. It is based on a symmetric key algorithm and uses a shared secret key that is known only to the sender and receiver. It was designed by IBM and approved by the National Bureau of Standards (NBS[6]) in 1976. It is

---

[6] The NBS is now known as the National Institute of Standards and Technology (NIST).

a block cipher, and a message is split into 64-bit message blocks. The algorithm is employed in reverse to decrypt each cipher text block.

Today, DES is considered to be insecure for many applications as its key size (56 bits) is viewed as being too small, and the cipher has been broken in less than 24 h. This has led to it being withdrawn as a standard and replaced by the Advanced Encryption Standard (AES), which uses a larger key of 128 bits or 256 bits.

The DES algorithm uses the same secret 56-bit key for encryption and decryption. The key consists of 56 bits taken from a 64-bit key that includes 8 parity bits. The parity bits are at position 8, 16, ..., 64, and so every 8th bit of the 64-bit key is discarded leaving behind only the 56-bit key.

The algorithm is then applied to each 64-bit message block, and the plaintext message block is converted into a 64-bit cipher text block. An initial permutation is first applied to M to create M', and M' is divided into a 32-bit left half L_0 and a 32-bit right half R_0. There are then 16 iterations, with the iterations having a left half and a right half:

$$L_i = R_{i-1}$$
$$R_i = L_{i-1} \oplus f(R_{i-1}, K_i)$$

The function f is a function that takes a 32-bit right half and a 48-bit round key K_i (each K_i contains a different subset of the 56-bit key) and produces a 32-bit output. Finally, the pre-cipher text (R_{16}, L_{16}) is permuted to yield the final cipher text C. The function f operates on half a message block (Table 20.3) and involves:

The decryption of the cipher text is similar to the encryption, and it involves running the algorithm in reverse.

DES has been implemented on a microchip. However, it has been superseded in recent years by AES due to security concerns with its small 56-bit key size. The AES uses a key size of 128 bits or 256 bits.

Table 20.3 DES encryption

Step	Description
1	Expansion of the 32-bit half block to 48 bits (by duplicating half of the bits)
2	The 48-bit result is combined with a 48-bit subkey of the secret key using an XOR operation
3	The 48-bit result is broken into 8 * 6 bits and passed through 8 substitution boxes to yield 8 * 4 = 32 bits (This is the core part of the encryption algorithm.)
4	The 32-bit output is re-arranged according to a fixed permutation

20.5   Public Key Systems

A public key cryptosystem (Fig. 20.6) is an asymmetric key system where there is a separate key e_k for encryption and d_k decryption with $e_k \neq d_k$. Martin Hellman and Whitfield Diffie invented it in 1976. The fact that a person is able to encrypt a message does not mean that the person has sufficient information to decrypt messages.

The public key cryptosystem is based on the following (Table 20.4).

The advantages and disadvantages of public key cryptosystems include (Table 20.5).

The implementation of public key cryptosystems is based on *trapdoor one-way functions*. A function $f: X \rightarrow Y$ is a trapdoor one-way function if

- f is easy to computer
- f^{-1} is difficult to compute
- f^{-1} is easy to compute if a trapdoor (secret information associated with the function) becomes available.

A function satisfying just the first two conditions above is termed a *one-way function*.

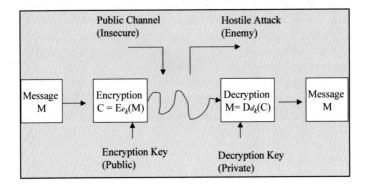

Fig. 20.6   Public key cryptosystem

Table 20.4   Public key encryption system

Item	Description
1	It uses the concept of a key pair (e_k, d_k)
2	One half of the pair can encrypt messages and the other half can decrypt messages
3	One key is private, and one key is public
4	The private key is kept secret, and the public key is published (but associated with trading partner)
5	The key pair is associated with exactly one trading partner

Table 20.5 Advantages and disadvantages of public key cryptosystems

Advantages	Disadvantages
Only the private key needs to be kept secret	Public keys must be authenticated
The distribution of keys for encryption is convenient as everyone publishes their public key and the private key is kept private	It is slow and uses more computer resources
It provides message authentication as it allows the use of digital signatures (which enables the recipient to verify that the message is really from the particular sender)	Security compromise is possible (if private key compromised)
The sender encodes with the private key that is known only to sender. The receiver decodes with the public key and therefore knows that the message is from the sender	Loss of private key may be irreparable (unable to decrypt messages)
Detection of tampering (digital signatures enable the receiver to detect whether message was altered in transit)	
Provides for non-repudiation	

Examples of Trapdoor and One-way Functions

(i) The function $f: pq \rightarrow n$ (where p and q are primes) is a one-way function since it is easy to compute. However, the inverse function f^{-1} is difficult to compute problem for large n since there is no efficient algorithm to factorize a large integer into its prime factors (*integer factorization problem*).

(ii) The function $f_{g, N}: x \rightarrow g^x \pmod{N}$ is a one-way function since it is easy to compute. However, the inverse function f^{-1} is difficult to compute as there is no efficient method to determine x from the knowledge of $g^x \pmod{N}$ and g and N (*the discrete logarithm problem*).

(iii) The function $f_{k, N}: x \rightarrow x^k \pmod{N}$ (where $N = pq$ and p and q are primes) and $kk' \equiv 1 \pmod{\varphi(n)}$ is a trapdoor function. It is easy to compute but the inverse of f (the kth root modulo N) is difficult to compute. However, if the trapdoor k' is given then f can easily be inverted as $(x^k)^{k'} \equiv x \pmod{N}$.

20.5.1 RSA Public Key Cryptosystem

Rivest, Shamir, and Adleman proposed a practical public key cryptosystem (RSA) based on primality testing and integer factorization in the late 1970s. The RSA algorithm was filed as a patent (Patent No. 4405, 829) at the U.S. Patent Office in December 1977. The RSA public key cryptosystem is based on the following assumptions:

- It is straightforward to find two large prime numbers.
- The integer factorization problem is infeasible for large numbers.

The algorithm is based on mod-n arithmetic where n is a product of two large prime numbers.

The encryption of a plaintext message M to produce the cipher text C is given by:

$$C \equiv M^e (\text{mod } n)$$

where e is the public encryption key, M is the plaintext, $\mathbf{C}$ is the cipher text, and n is the product of two large primes p and q. Both e and n are made public, and e is chosen such that $1 < e < \phi(n)$, where $\phi(n)$ is the number of positive integers that are relatively prime to n.

The cipher text C is decrypted by

$$M \equiv C^d (\text{mod } n)$$

where d is the private decryption key that is known only to the receiver, and ed $\equiv$ 1 (mod $\phi(n)$) and d and $\phi(n)$ are kept private.

The calculation of $\phi(n)$ is easy if both p and q are known, as it is given by ϕ $(n) = (p - 1)(q - 1)$. However, its calculation for large n is infeasible if p and q are unknown.

$$\text{ed} \equiv 1(\text{mod } \phi(n))$$
$$\Rightarrow \text{ed} = 1 + k\phi(n) \text{ for some } k \in \mathbb{Z}$$

Euler's Theorem is discussed in Chap. 3 of [2], and this result states that if a and n are positive integers with gcd(a, n) = 1 then $a^{\phi(n)} \equiv 1$ (mod n). Therefore, $M^{\phi(n)} \equiv 1$ (mod n) and $M^{k\phi(n)} \equiv 1$ (mod n). The decryption of the cipher text is given by:

$$C^d (\text{mod } n) \equiv M^{\text{ed}} (\text{mod } n)$$
$$\equiv M^{1+k\phi(n)} (\text{mod } n)$$
$$\equiv M^1 M^{k\phi(n)} (\text{mod } n)$$
$$\equiv M \cdot 1(\text{mod } n)$$
$$\equiv M(\text{mod } n)$$

Table 20.6 Steps for *A* to send secure message and signature to *B*

Step	Description
1	*A* uses *B*'s public key to encrypt the message
2	*A* uses its private key to encrypt its signature
3	*A* sends the message and signature to *B*
4	*B* uses *A*'s public key to decrypt *A*'s signature
5	*B* uses its private key to decrypt *A*'s message

20.5.2 Digital Signatures

The RSA public key cryptography may also be employed to obtain digital signatures. Suppose *A* wishes to send a secure message to *B* as well as a digital signature. This involves signature generation using the private key, and signature verification using the public key. The steps involved are (Table 20.6).

The National Institute of Standards and Technology (NIST) proposed an algorithm for digital signatures in 1991. The algorithm is known as the Digital Signature Algorithm (DSA) and later became the Digital Signature Standard (DSS).

20.6   Review Questions

1. Discuss the Caesar cipher.
2. Describe how the team at Bletchley Park cracked the German Enigma codes.
3. Explain the differences between a public key cryptosystem and a private key cryptosystem.
4. Describe the advantages/disadvantages of symmetric key cryptosystems.
5. Describe the various types of symmetric key systems.
6. What are the advantages and disadvantages of public key cryptosystems?
7. Describe public key cryptosystems.
8. Describe how digital signatures may be generated.

20.7   Summary

Cryptography is the study of mathematical techniques that provide secrecy in the transmission of messages between computers. It was originally employed to protect communication between individuals, and today it is employed to solve security problems such as privacy and authentication over a communications channel.

It involves enciphering and deciphering messages and uses theoretical results from number theory to convert the original messages (or plaintext) into cipher text that is then transmitted over a secure channel to the intended recipient. The cipher text is meaningless to anyone other than the intended recipient, and the received cipher text is then decrypted to allow the recipient to read the message.

A public key cryptosystem is an asymmetric cryptosystem. It has two different encryption and decryption keys, and the fact that a person has knowledge on how to encrypt messages does not mean that the person has sufficient information to decrypt messages.

In a secret key cryptosystem, the same key is used for both encryption and decryption. Anyone who has knowledge on how to encrypt messages has sufficient knowledge to decrypt messages, and it is essential that the key is kept secret between the two parties.

References

1. Pohlig S, Hellman M (1978) An improved algorithm for computing algorithms over GF(p) and its cryptographic significance. IEEE Trans Inf Theory 24:106–110
2. O'Regan G (2021) Guide to discrete mathematics, 2nd edn. Springer, Berlin
3. National Bureau of Standards (1977) Data encryption standard. FIPS-Pub 46. U.S. Department of Commerce

Coding Theory

21

Key Topics

Groups, Rings and Fields

Block Codes

Error Detection and Correction

Generation Matrix

Hamming Codes

21.1 Introduction

Coding theory is a practical branch of mathematics concerned with the reliable transmission of information over communication channels. It allows errors to be detected and corrected, which is essential when messages are transmitted through a noisy communication channel. The channel could be a telephone line, radio link, or satellite link, and coding theory is applicable to mobile communications and satellite communications. It is also applicable to storing information on storage systems such as the compact disc.

It includes theory and practical algorithms for error detection and correction, and it plays an important role in modern communication systems that require reliable and efficient transmission of information.

An error correcting code encodes the data by adding a certain amount of redundancy to the message. This enables the original message to be recovered if a small number of errors have occurred. The extra symbols added are also subject to errors, as accurate transmission cannot be guaranteed in a noisy channel.

© The Author(s), under exclusive license to Springer Nature Switzerland AG 2023
G. O'Regan, *Mathematical Foundations of Software Engineering*,
Texts in Computer Science, https://doi.org/10.1007/978-3-031-26212-8_21

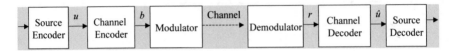

Fig. 21.1 Basic digital communication

  The basic structure of a digital communication system is shown in Fig. 21.1. It includes transmission tasks such as source encoding, channel encoding and modulation; and receiving tasks such as demodulation, channel decoding, and source decoding.

  The modulator generates the signal that is used to transmit the sequence of symbols b across the channel. The transmitted signal may be altered due to the fact that there is noise in the channel, and the signal received is demodulated to yield the sequence of received symbols r.

  The received symbol sequence r may differ from the transmitted symbol sequence b due to the noise in the channel, and therefore a channel code is employed to enable errors to be detected and corrected. The channel encoder introduces redundancy into the information sequence u, and the channel decoder uses the redundancy for error detection and correction. This enables the transmitted symbol sequence $\hat{u}$ to be estimated.

  Shannon [1] showed that it is theoretically possible to produce an information transmission system with an error probability as small as required provided that the information rate is smaller than the channel capacity.

  Coding theory uses several results from pure mathematics, and we discussed several abstract mathematical structures that are used in coding theory in Chap. 5. We briefly discuss its mathematical foundations in the next section.

21.2 Mathematical Foundations of Coding Theory

Coding theory is built from the results of modern algebra, and it uses abstract algebraic structures such as groups, rings, fields, and vector spaces. These abstract structures provide a solid foundation for the discipline, and the main abstract structures used include vector spaces and fields. A *group* is a non-empty set with a single binary operation, whereas *rings* and *fields* are algebraic structures with two binary operations satisfying various laws. A *vector space* consists of vectors over a field.

  The representation of codewords is by n-dimensional vectors over the finite field F_q. A codeword vector v is represented as the n-tuple:

$$v = (a_0, a, \ldots, a_{n-1})$$

where each $a_i \in F_q$. The set of all n-dimensional vectors is the n-dimensional vector space $\mathbf{F}^n_q$ with q^n elements. The addition of two vectors v and w, where $v = (a_0,$

$a_1, ..., a_{n-1})$ and $w = (b_0, b_1, ..., b_{n-1})$ is given by:

$$v + w = (a_0 + b_0, a_1 + b_1, ..., a_{n-1} + b_{n-1})$$

The scalar multiplication of a vector $v = (a_0, a_1, ... a_{n-1}) \in \mathbf{F}^n_q$ by a scalar $\beta \in F_q$ is given by:

$$\beta v = (\beta a_0, \beta a_1, ..., \beta a_{n-1})$$

The set $\mathbf{F}^n_q$ is called the vector space over the finite field F_q if the vector space properties hold. A finite set of vectors $v_1, v_2, ..., v_k$ is said to be *linearly independent* if:

$$\beta_1 v_1 + \beta_2 v_2 + \cdots + \beta_k v_k = 0 \Rightarrow \beta_1 = \beta_2 = \cdots \beta_k = 0$$

Otherwise, the set of vectors $v_1, v_2, ... v_k$ is said to be *linearly dependent*.

The *dimension* (dim W) of a subspace $W \subseteq V$ is k if there are k linearly independent vectors in W but every $k + 1$ vectors are linearly dependent. A subset of a vector space is a *basis* for V if it consists of linearly independent vectors, and its linear span is V (i.e., the basis generates V).

We shall employ the basis of the vector space of codewords to create the generator matrix to simplify the encoding of the information words. The linear span of a set of vectors $v_1, v_2, ..., v_k$ is defined as $\beta_1 v_1 + \beta_2 v_2 + \cdots + \beta_k v_k$.

21.3 Simple Channel Code

We present a simple example to illustrate the concept of an error correcting code, and the example code presented is able to correct a single transmitted error only.

We consider the transmission of binary information over a noisy channel that leads to differences between the transmitted sequence and the received sequence. The differences between the transmitted and received sequence are illustrated by underlining the relevant digits in the example.

$$\text{Sent} \qquad 00\underline{101}110$$
$$\text{Received} \qquad 00\underline{000}110$$

Initially, it is assumed that the transmission is done without channel codes as follows:

$$00101110 \xrightarrow{\text{Channel}} 00000110$$

Next, the use of an encoder is considered and a triple repetition-encoding scheme is employed. That is, the binary symbol 0 is represented by the code word 000, and the binary symbol 1 is represented by the code word 111.

$$00101110 \rightarrow \boxed{\text{Encoder}} \rightarrow 000000111000111111111000$$

Other words, if the symbol 0 is to be transmitted then the encoder emits the codeword 000, and similarly the encoder emits 111 if the symbol 1 is to be transmitted. Assuming that on average one symbol in four is incorrectly transmitted, then transmission with binary triple repetition may result in a received sequence such as:

$$000000111000111111111000 \rightarrow \boxed{\text{Channel}} \rightarrow 0\underline{1}00000\underline{1}10\underline{1}011\underline{1}0\underline{1}0111\underline{0}\underline{1}0$$

The decoder tries to estimate the original sequence by using a *majority decision* on each 3-bit word. Any 3-bit word that contains more zeros than ones is decoded to 0, and similarly if it contains more ones than zero it is decoded to 1. The decoding algorithm yields:

$$0\underline{1}00000\underline{1}10\underline{1}0011\underline{1}0\underline{1}0111\underline{0}\underline{1}0 \rightarrow \boxed{\text{Decoder}} \rightarrow 0010\underline{1}0\underline{1}0$$

In this example, the binary triple repetition code is able to correct a single error within a code word (as the majority decision is two to one). This helps to reduce the number of errors transmitted compared to unprotected transmission. In the first case where an encoder is not employed there are two errors, whereas there is just one error when the encoder is used.

However, there are disadvantages with this approach in that the transmission bandwidth has been significantly reduced. It now takes three times as long to transmit an information symbol with the triple replication code than with standard transmission. Therefore, it is desirable to find more efficient coding schemes.

21.4   Block Codes

There were two code words employed in the simple example above: namely 000 and 111. This is an example of a (n, k) code where the code words are of length $n = 3$, and the information words are of length $k = 1$ (as we were just encoding a single symbol 0 or 1). This is an example of a (3, 1) block code, and the objective of this section is to generalize the simple coding scheme to more efficient and powerful channel codes.

The fundamentals of the q-nary (n, k) block codes (where q is the number of elements in the finite field F_q) involve converting an information block of length k to a codeword of length n. Consider an information sequence $u_0, u_1, u_2, \ldots$ of discrete information symbols where $u_i \in \{0, 1, \ldots, q - 1\} = F_q$. The normal class of channel codes is when we are dealing with binary codes, i.e., $q = 2$. The information sequence is then grouped into blocks of length k as follows:

$$\underbrace{u_0 u_1 u_2 \ldots u_{k-1}} \quad \underbrace{u_k u_{k+1} u_{k+2} \ldots u_{2k-1}} \quad \underbrace{u_{2k} u_{2k+1} u_{2k+2} \ldots u_{3k-1}} \ldots$$

Each block is of length k (i.e., the information words are of length k), and it is then encoded separately into codewords of length n. For example, the information word $u_0 u_1 u_2 \ldots u_{k-1}$ is uniquely mapped to a code word $b_0 b_1 b_2 \ldots b_{n-1}$ of length n where $b_i \in F_q$. Similarly, the information word $u_k u_{k+1} u_{k+2} \ldots u_{2k-1}$ is encoded to the code word $b_n b_{n+1} b_{n+2} \ldots b_{2n-1}$ of length n where $b_i \in F_q$. That is,

$$(u_0 u_1 u_2 \ldots u_{k-1}) \rightarrow \boxed{\text{Encoder}} \rightarrow (b_0 b_1 b_2 \ldots b_{n-1})$$

These code words are then transmitted across the communication channel, and the received words are then decoded. The received word $r = (r_0 r_1 r_2 \ldots r_{n-1})$ is decoded into the information word $\hat{u} = (\hat{u}_0 \hat{u}_1 \hat{u}_2 \ldots \hat{u}_{k-1})$

$$(r_0 r_1 r_2 \ldots r_{n-1}) \rightarrow \boxed{\text{Decoder}} \rightarrow (\hat{u}_0 \hat{u}_1 \hat{u}_2 \ldots \hat{u}_{k-1})$$

Strictly speaking the decoding is done in two steps with the received n-block word r first decoded to the n-block codeword b^*. This is then decoded into the k-block information word $\hat{u}$. The encoding, transmission, and decoding of an (n, k) block may be summarized as follows (Fig. 21.2).

A lookup table may be employed for the encoding to determine the code word b for each information word u. However, the size of the table grows exponentially with increasing information word length k, and so this is inefficient due to the large memory size required. We shall discuss later how a generator matrix may be used to provide an efficient encoding and decoding mechanism.

Notes

(i) The codeword is of length n.
(ii) The information word is of length k.
(iii) The codeword length n is larger than the information word length k.
(iv) A block (n, k) code is a code in which all codewords are of length n and all information words are of length k.

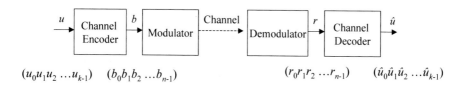

Fig. 21.2 Encoding and decoding of an (n, k) block

(v) The number of possible information words is given by $M = q^k$ (where each information symbol can take one of q possible values and the length of the information word is k).

(vi) The code rate R in which information is transmitted across the channel is given by:

$$R = \frac{k}{n}$$

(vii) The weight of a codeword is $b = (b_0 b_1 b_2 \ldots b_{n-1})$ which is given by the number of non-zero components of b. That is,

$$wt(b) = |\{i : b_i \neq 0, 0 \leq i < n\}|$$

(viii) The distance between two codewords $b = (b_0 b_1 b_2 \ldots b_{n-1})$ and $b' = (b'_0 b'_1 b'_2 \ldots b'_{n-1})$ measures how close the codewords b and b' are to each other. It is given by the Hamming distance:

$$\text{dist}(b, b') = |\{i : b_i \neq b'_i, 0 \leq i < n\}|$$

(ix) The minimum Hamming distance for a code $\mathbf{B}$ consisting of M codewords $b_1, \ldots, b_M$ is given by:

$$d = \min\{\text{dist}(b, b') : \text{where } b \neq b'\}$$

(x) The (n, k) block code $\mathbf{B} = \{b_1, \ldots, b_M\}$ with $M (= q^k)$ codewords of length n and minimum Hamming distance d is denoted by $\mathbf{B}(n, k, d)$.

21.4.1 Error Detection and Correction

The minimum Hamming distance offers a way to assess the error detection and correction capability of a channel code. Consider two codewords b and b' of an (n, k) block code $\mathbf{B}(n, k, d)$.

Then, the distance between these two codewords is greater than or equal to the minimum Hamming distance d, and so errors can be detected as long as the erroneously received word is not equal to a codeword different from the transmitted code word.

That is, the *error detection capability* is guaranteed as long as the number of errors is less than the minimum Hamming distance d, and so the number of detectable errors is $d - 1$.

Fig. 21.3 Error-correcting capability sphere

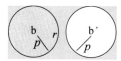

Any two codewords are of distance at least d and so if the number of errors is less than $d/2$ then the received word can be properly decoded to the codeword b. That is, the *error correction capability* is given by:

$$E_{\text{cor}} = \frac{d-1}{2}$$

An error-correcting sphere (Fig. 21.3) may be employed to illustrate the error correction of a received word to the correct codeword b. This may be done when all received words are within the error-correcting sphere with radius p ($< d/2$).

If the received word r is different from b in less than $d/2$ positions, then it is decoded to b as it is more than $d/2$ positions from the next closest codeword. That is, b is the closest codeword to the received word r (provided that the error-correcting radius is less than $d/2$).

21.5 Linear Block Codes

Linear block codes have nice algebraic properties and the codewords in a linear block code are considered to be vectors in the finite vector space $\mathbf{F}^n_q$. The representation of codewords by vectors allows the nice algebraic properties of vector spaces to be used, and this simplifies the encoding of information words as a generator matrix may be employed to create the codewords.

An (n, k) block code $\mathbf{B}(n, k, d)$ with minimum Hamming distance d over the finite field F_q is called *linear* if $\mathbf{B}(n, k, d)$ is a subspace of the vector space $\mathbf{F}^n_q$ of dimension k. The number of codewords is then given by:

$$M = q^k$$

The rate of information (R) through the channel is given by:

$$R = \frac{k}{n}$$

Clearly, since $\mathbf{B}(n, k, d)$ is a subspace of $\mathbf{F}^n_q$ any linear combination of the codewords (vectors) will be a codeword. That is, for the codewords $b_1, b_2, ..., b_r$ we have that:

$$\alpha_1 b_1 + \alpha_2 b_2 + \cdots + \alpha_r b_r \in \mathbf{B}(n, k, d)$$

where $\alpha_1, \alpha_2, \dots, \alpha_r \in F_q$ and $b_1, b_2, ..., b_r \in \mathbf{B}(n, k, d)$.

Clearly, the n-dimensional zero row vector $(0, 0, \ldots, 0)$ is always a codeword, and so $(0, 0, \ldots, 0) \in \mathbf{B}(n, k, d)$. The minimum Hamming distance of a linear block code $\mathbf{B}(n, k, d)$ is equal to the minimum weight of the nonzero codewords: That is,

$$d = \min_{\forall b \neq b'} \left\{ \text{dist}(b'b') \right\} = \min_{\forall b \neq 0} wt(b)$$

In summary, an (n, k) linear block code $\mathbf{B}(n, k, d)$ is:

1. A subspace of $\mathbf{F}^n{}_q$.
2. The number of codewords is $M = q^k$.
3. The minimum Hamming distance d is the minimum weigh of the non-zero codewords.

The encoding of a specific k-dimensional information word $u = (u_0, u_1, \ldots, u_{k-1})$ to a n-dimensional codeword $b = (b_0, b_1, \ldots, b_{n-1})$ may be done efficiently with a generator matrix (matrices are discussed in Chap. 27). First, a basis $\{g_0, g_1, \ldots g_{k-1}\}$ of the k-dimensional subspace spanned by the linear block code is chosen, and this consists of k linearly independent n-dimensional vectors. Each basis element g_i (where $0 \leq i \leq k-1$) is a n-dimensional vector:

$$g_i = \left(g_{i,0}, g_{i,1}, \ldots, g_{i,n-1}\right)$$

The corresponding codeword $b = (b_0, b_1, \ldots, b_{n-1})$ is then a linear combination of the information word with the basis elements. That is,

$$b = u_0 g_0 + u_1 g_1 + \cdots + u_{k-1} g_{k-1}$$

where each information symbol $u_i \in F_q$. The generator matrix G is then constructed from the k linearly independent basis vectors as follows (Fig. 21.4).

The encoding of the k-dimensional information word u to the n-dimensional codeword b involves matrix multiplication (Fig. 21.5).

This may also be written as:

$$b = uG$$

Fig. 21.4 Generator matrix

$$G = \begin{pmatrix} g_{0,} \\ g_{1,} \\ g_{2,} \\ \cdots \\ \cdots \\ \cdots \\ g_{k-1} \end{pmatrix} = \begin{pmatrix} g_{0,0} & g_{0,1} & g_{0,2} & \cdots & g_{0,n-1} \\ g_{1,0} & g_{1,1} & g_{1,2} & \cdots & g_{1,n-1} \\ g_{2,0} & g_{2,1} & g_{2,2} & \cdots & g_{2,n-1} \\ \cdots & \cdots & \cdots & \cdots & \cdots \\ \cdots & \cdots & \cdots & \cdots & \cdots \\ \cdots & \cdots & \cdots & \cdots & \cdots \\ g_{k-1,0} & g_{k-1,1} & g_{k-1,2} & \cdots & g_{k-1,n-1} \end{pmatrix}$$

Fig. 21.5 Generation of codewords

$$(u_0, u_1, ..., u_{k-1}) \begin{pmatrix} g_{0,0} & g_{0,1} & g_{0,2} & \cdots & g_{0,n-1} \\ g_{1,0} & g_{1,1} & g_{1,2} & \cdots & g_{1,n-1} \\ g_{2,0} & g_{2,1} & g_{2,2} & \cdots & g_{2,n-1} \\ \cdots & \cdots & \cdots & \cdots & \cdots \\ \cdots & \cdots & \cdots & \cdots & \cdots \\ \cdots & \cdots & \cdots & \cdots & \cdots \\ g_{k-1,0} & g_{k-1,1} & g_{k-1,2} & \cdots & g_{k-1,n-1} \end{pmatrix} = (b_0, b_1, ..., b_{n-1})$$

Fig. 21.6 Identity matrix $(k \times k)$

$$I_k = \begin{pmatrix} 1 & 0 & 0 & \cdots & 0 \\ 0 & 1 & 0 & \cdots & 0 \\ 0 & 0 & 1 & \cdots & 0 \\ \cdots & \cdots & \cdots & \cdots & \cdots \\ \cdots & \cdots & \cdots & \cdots & \cdots \\ \cdots & \cdots & \cdots & \cdots & \cdots \\ 0 & 0 & 0 & \cdots & 1 \end{pmatrix}$$

Clearly, all $M = q^k$ codewords $b \in \mathbf{B}(n, k, d)$ can be generated according to this rule, and so the matrix G is called the generator matrix. The generator matrix defines the linear block code $\mathbf{B}(n, k, d)$. There is an equivalent $k \times n$ generator matrix for $\mathbf{B}(n, k, d)$ defined as:

$$G = I_k \big| A_{k,n-k}$$

where I_k is the $k \times k$ identity matrix (Fig. 21.6).

The encoding of the information word u yields the codeword b such that the first k symbols b_i of b are the same as the information symbols u_i $0 \le i \le k$.

$$b = uG = \left(u \big| u A_{k,n-k}\right)$$

The remaining $m = n - k$ symbols are generated from $u A_{k,n-k}$ and the last m symbols are the m parity check symbols. These are attached to the information vector u for the purpose of error detection and correction.

21.5.1 Parity Check Matrix

The linear block code $\mathbf{B}(n, k, d)$ with generator matrix $G = (I_k \mid A_{k,n-k})$ may be defined equivalently by the $(n - k) \times n$ parity check matrix H, where this matrix is defined as:

$$H = \left(-A^T_{k,n-k} \big| I_{n-k}\right)$$

The generator matrix G and the parity check matrix H are orthogonal, i.e.,

$$HG^T = 0_{n-k,k}$$

The parity check orthogonality property holds if and only if the vector belongs to the linear block code. That is, for each code vector in $b \in \mathbf{B}(n, k, d)$ we have

$$Hb^T = 0_{n-k,1}$$

and vice versa whenever the property holds for a vector r, then r is a valid codeword in $\mathbf{B}(n, k, d)$. We present an example of a parity check matrix in Example 21.1 below.

21.5.2 Binary Hamming Code

The Hamming code is a linear code that has been employed in dynamic random access memory to detect and correct deteriorated data in memory. The generator matrix for the $\mathbf{B}(7, 4, 3)$ binary Hamming code is given by (Fig. 21.7).

The information words are of length $k = 4$ and the codewords are of length $n = 7$. For example, it can be verified by matrix multiplication that the information word $(0, 0, 1, 1)$ is encoded into the codeword $(0, 0, 1, 1, 0, 0, 1)$.

That is, three parity bits 001 have been added to the information word $(0, 0, 1, 1)$ to yield the codeword $(0, 0, 1, 1, 0, 0, 1)$.

The minimum Hamming distance is $d = 3$, and the Hamming code can detect up to two errors, and it can correct one error.

Example 21.1 (*Parity Check Matrix—Hamming Code*) The objective of this example is to construct the parity check matrix of the binary Hamming code $(7, 4, 3)$ and to show an example of the parity check orthogonality property.

First, we construct the parity check matrix H which is given by $H = (-A^T_{k, n-k} \mid I_{n-k})$ or another words $H = (-A^T_{4,3} \mid I_3)$. We first note that

$$A_{4,3} = \begin{bmatrix} 0 & 1 & 1 \\ 1 & 0 & 1 \\ 1 & 1 & 0 \\ 1 & 1 & 1 \end{bmatrix} \quad A^T_{4,3} = \begin{bmatrix} 0 & 1 & 1 & 1 \\ 1 & 0 & 1 & 1 \\ 1 & 1 & 0 & 1 \end{bmatrix}$$

Therefore, H is given by:

$$H = \begin{bmatrix} 0 & -1 & -1 & -1 & 1 & 0 & 0 \\ -1 & 0 & -1 & -1 & 0 & 1 & 0 \\ -1 & -1 & 0 & -1 & 0 & 0 & 1 \end{bmatrix}$$

Fig. 21.7 Hamming code
$B(7, 4, 3)$ generator matrix

$$G = \begin{pmatrix} 1 & 0 & 0 & 0 & 0 & 1 & 1 \\ 0 & 1 & 0 & 0 & 1 & 0 & 1 \\ 0 & 0 & 1 & 0 & 1 & 1 & 0 \\ 0 & 0 & 0 & 1 & 1 & 1 & 1 \end{pmatrix}$$

We noted that the encoding of the information word $u = (0011)$ yields the code-word $b = (0011001)$. Therefore, the calculation of Hb^{T} yields (recalling that addition is modulo two):

$$
Hb^{\mathrm{T}} = \begin{bmatrix} 0 & -1 & -1 & -1 & 1 & 0 & 0 \\ -1 & 0 & -1 & -1 & 0 & 1 & 0 \\ -1 & -1 & 0 & -1 & 0 & 0 & 1 \end{bmatrix} \begin{pmatrix} 0 \\ 0 \\ 1 \\ 1 \\ 0 \\ 0 \\ 1 \end{pmatrix} = \begin{pmatrix} 0 \\ 0 \\ 0 \end{pmatrix}
$$

21.5.3 Binary Parity Check Code

The binary parity check code is a linear block code over the finite field F_2. The code takes a k-dimensional information word $u = (u_0, u_1, \ldots, u_{k-1})$ and generates the codeword $b = (b_0, b_1, \ldots, b_{k-1}, b_k)$ where $u_i = b_i$ ($0 \le i \le k-1$) and b_k is the parity bit chosen so that the resulting codeword is of even parity. That is,

$$
b_k = u_0 + u_1 + \cdots + u_{k-1} = \sum_{i=0}^{k-1} u_i
$$

21.6 Miscellaneous Codes in Use

There are many examples of codes in use such as repetition codes (such as the triple replication code considered earlier in Sect. 21.3); parity check codes where a parity symbol is attached such as the binary parity check code; Hamming codes such as the (7, 4) code that was discussed in Sect. 21.5.2, which has been applied for error correction of faulty memory.

The Reed–Muller codes form a class of error correcting codes that can correct more than one error. Cyclic codes are special linear block codes with efficient algebraic decoding algorithms. The BCH codes are an important class of cyclic codes, and the Reed–Solomon codes are an example of a BCH code.

Convolution codes have been applied in the telecommunications field, for example, in GSM, UMTS, and in satellite communications. They belong to the class of linear codes, but also employ a memory so that the output depends on the current input symbols and previous input. For more detailed information on coding theory see [2].

21.7   Review Questions

1. Describe the basic structure of a digital communication system.
2. Describe the mathematical structure known as the field. Give examples of fields.
3. Describe the mathematical structure known as the ring and give examples of rings. Give examples of zero divisors in rings.
4. Describe the mathematical structure known as the vector space and give examples
5. Explain the terms linear independence and linear dependence and a basis.
6. Describe the encoding and decoding of an (n, k) block code where an information word of length k is converted to a codeword of length n.
7. Show how the minimum Hamming distance may be employed for error detection and error correction.
8. Describe linear block codes and show how a generator matrix may be employed to generate the codewords from the information words.

21.8   Summary

Coding theory is the branch of mathematics that is concerned with the reliable transmission of information over communication channels. It allows errors to be detected and corrected, and this is extremely useful when messages are transmitted through a noisy communication channel. This branch of mathematics includes theory and practical algorithms for error detection and correction.

The theoretical foundations of coding theory were considered, and its foundations lie in abstract algebra including group theory, ring theory, fields, and vector spaces. The codewords are represented by n-dimensional vectors over a finite field F_q.

An error-correcting code encodes the data by adding a certain amount of redundancy to the message so that the original message can be recovered if a small number of errors have occurred.

The fundamentals of block codes were discussed where an information word is of length k and a codeword is of length n. This led to the linear block codes $\mathbf{B}(n, k, d)$ and a discussion on error detection and error correction capabilities of the codes.

The goal of this chapter was to give a flavour of coding theory, and the reader is referred to more specialized texts (e.g., [2]) for more detailed information.

References

1. Shannon C (1948) A mathematical theory of communication. Bell Syst Tech J 27:379–423
2. Neubauer A, Freunderberger J, Kühn V (2007) Coding theory. Algorithms, architectures and applications. Wiley, New York

Introduction to Statistics

22

Key Topics

Random Sample

Sampling Techniques

Frequency Distribution

Arithmetic Mean, Mode and Median

Bar Chart, Histogram, and Trend Chart

Variance

Standard Deviation

Correlation and Regression

Statistical Inference

Hypothesis Testing

22.1 Introduction

Statistics is an empirical science that is concerned with the collection, organization, analysis, interpretation, and presentation of data. The data collection needs to be planned and this may include surveys and experiments. Statistics are widely used by government and industrial organizations, and they are employed for forecasting as well as for presenting trends. They allow the behaviour of a population to be studied and inferences to be made about the population. These inferences may be tested (*hypothesis testing*) to ensure their validity.

© The Author(s), under exclusive license to Springer Nature Switzerland AG 2023
G. O'Regan, *Mathematical Foundations of Software Engineering*,
Texts in Computer Science, https://doi.org/10.1007/978-3-031-26212-8_22

The analysis of statistical data allows an organization to understand its performance in key areas and to identify problematic areas. Organizations will often examine performance trends over time and will devise appropriate plans and actions to address problematic areas. The effectiveness of the actions taken will be judged by improvements in performance trends over time.

It is often not possible to study the entire population, and instead a representative subset or sample of the population is chosen. This *random sample* is used to make inferences regarding the entire population, and it is essential that the sample chosen is indeed random and representative of the entire population. Otherwise, the inferences made regarding the entire population will be invalid, as a selection bias has occurred. Clearly, a census where every member of the population is sampled is not subject to this type of bias.

A statistical experiment is a causality study that aims to draw a conclusion between the values of a *predictor variable*(s) and a *response variable*(s). For example, a statistical experiment in the medical field may be conducted to determine if there is a causal relationship between the use of a particular drug and the treatment of a medical condition such as lowering of cholesterol in the population. A statistical experiment involves:

- Planning the research
- Designing the experiment
- Performing the experiment
- Analysing the results
- Presenting the results.

22.2 Basic Statistics

The field of statistics is concerned with summarizing, digesting, and extracting information from large quantities of data. It provides a collection of methods for planning an experiment and analysing data to draw accurate conclusions from the experiment. We distinguish between descriptive statistics and inferential statistics:

Descriptive Statistics
This is concerned with describing the information in a set of data elements in graphical format, or by describing its distribution.

Inferential Statistics
This is concerned with making inferences with respect to the population by using information gathered in the sample.

22.2.1 Abuse of Statistics

Statistics are extremely useful in drawing conclusions about a population. However, it is essential that the random sample chosen is actually random, and that the experiment is properly conducted to ensure that valid conclusions are inferred. Some examples of the abuse of statistics include:

- The sample size may be too small to draw conclusions
- It may not be a genuine random sample of the population
- There may be bias introduced from poorly worded questions
- Graphs may be drawn to exaggerate small differences
- Area may be misused in representing proportions
- Misleading percentages may be used.

The quantitative data used in statistics may be discrete or continuous. *Discrete data* is numerical data that has a finite or countable number of possible values, and *continuous data* is numerical data that has an infinite number of possible values.

22.2.2 Statistical Sampling and Data Collection

Statistical sampling is concerned with the methodology for choosing a random sample of a population and the study of the sample with the goal of drawing valid conclusions about the entire population. If a genuine representative random sample of the population is chosen, then a detailed study of the sample will provide insight into the whole population. This helps to avoid a lengthy expensive (and potentially infeasible) study of the entire population.

The sample chosen must be truly random and the sample size sufficiently large to enable valid conclusions to be drawn for the entire population. The probability of being chosen for the random sample should be the same for each member of the population.

Random Sample

A *random sample* is a sample of the population such that each member of the population has an equal chance of being chosen.

A large sample size gives more precise the information about the population. However, little extra information is gained from increasing the sample size above a certain level, and the sample size chosen will depend on factors such as money and time available, the aims of the survey, the degree of precision required, and the number of subsamples required. Table 22.1 summarizes several ways for generating a random sample from the population:

Once the sample is chosen, the next step is to obtain the required information from the sample. The data collection may be done by interviewing each member in the sample; conducting a telephone interview with each member; conducting a postal questionnaire survey, and so on (Table 22.2).

Table 22.1 Sampling techniques

Sampling technique	Description
Systematic sampling	The population is listed and every kth member of the population is sampled. For example, to choose a 2% (1 in 50) sample then every 50th member of the population would be sampled
Stratified sampling	The population is divided into two or more strata and each subpopulation (stratum) is then sampled. Each element in the subpopulation shares the same characteristics (e.g., age groups, gender). The results from the various strata are then combined
Multistage sampling	This approach may be used when the population is spread over a wide geographical area. The area is split up into a number of regions, and a small number of the regions are randomly selected. Each selected region is then sampled. It requires less effort and time, but it may introduce bias if a small number of regions are selected, as it is not a truly random sample
Cluster sampling	A population is divided into clusters, and a few of these clusters are exhaustively sampled (i.e., every element in the cluster is considered). This approach may lead to significant selection bias, as the sampling is not random
Convenience sampling	Sampling is done as convenient, and in this case each person selected may decide whether to participate or not in the sample

Table 22.2 Types of survey

Survey type	Description
Personal interview	Interviews are expensive and time consuming, but allow detailed and accurate information to be collected. Questionnaires are often employed and the interviewers need to be trained in interview techniques. Interviews need to be planned and scheduled, and they are useful in dealing with issues that may arise (e.g., a respondent not fully understanding a question)
Phone survey	This is a reasonably efficient and cost-effective way to gather data. However, refusals or hang-ups may affect the outcome. It also has an in-built bias as only those people with telephones may be contacted and interviewed
Mail questionnaire survey	This involves sending postal questionnaire survey to the participants. The questionnaire needs to be well designed to ensure the respondents understand the questions and answer them correctly. There is a danger of a low response rate that may invalidate the findings
Direct measurement	This may involve a direct measurement of all those in the sample (e.g., the height of all students in a class)
Direct observational study	An observational study allows individuals to be studied, and the variables of interest to be measured
Experiment	An experiment imposes some treatment on individuals in order to study the response

The design of the questionnaire requires careful consideration as a poorly designed questionnaire may lead to invalid results. The questionnaire should be as short as possible, and the questions should be simple and unambiguous. Closed questions where the respondent chooses from simple categories are useful. It is best to pilot the questionnaire prior to carrying out the survey.

22.3 Frequency Distribution and Charts

The data gathered from a statistical study is often raw and may yield little information as it stands. Therefore, the way the data is presented is important, and it is useful to present the information in pictorial form. The advantage of a pictorial presentation is that it allows the data to be presented in an attractive and colourful way, and the reader is not overwhelmed with excessive detail. This enables analysis and conclusions to be drawn. There are several types of charts or graphs that are often employed in the presentation of the data including:

- Bar chart
- Histogram
- Pie chart
- Trend graph.

A *frequency table* is often used to present and summarize data, where a simple frequency distribution consists of a set of data values and the number of items that have that value (i.e., a set of data values and their frequency). The information is then presented pictorially in a bar chart.

The general frequency distribution is employed when dealing with a larger number of data values (e.g., > 20 data values). It involves dividing the data into a set of data classes, and listing the data classes in one column and the frequency of data values in that category in another column. The information is then presented pictorially in a bar chart or histogram.

Figure 22.1 presents the raw data of salaries earned by different people in a company, and Table 22.3 presents the raw data in table format using a frequency table of salaries. Figure 22.2 presents a *bar chart* of the salary data in pictorial form, and it is much easier to read than the raw data presented in Fig. 22.1.

A *histogram* is a way of representing data in bar chart format, and it shows the frequency or relative frequency of various data values or ranges of data values. It is usually employed when there are a large number of data values, and it gives a crisp picture of the spread of the data values, and the centring and variance of the data values from the mean.

Fig. 22.1 Raw salary data

90,000	50,000	50,000	65,000	65,000	45,000	50,000
50,000	50,000	65,000	50,000	50,000	45,000	50,000
65,000						

Table 22.3 Frequency table
of salary data

Salary	Frequency
45,000	2
50,000	8
65,000	4
90,000	1

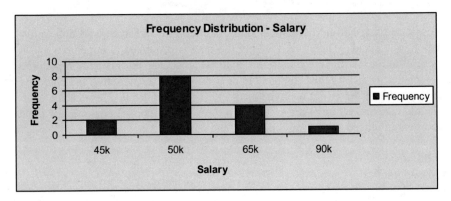

Fig. 22.2   Bar chart of salary data

The data is divided into disjoint intervals where an interval is a certain range of values. The horizontal axis of the histogram contains the intervals (also known as buckets), and the vertical axis shows the frequency (or relative frequency) of each interval.

The bars represent the frequency, and there is no space between the bars. The histogram has an associated shape, e.g., it may be a *normal distribution*, a *bimodal* or *multimodal distribution*, and it may be positively or negatively skewed. The variation and centring refer to the spread of data, and the spread of the data is important as it may indicate whether the entity under study (e.g., a process) is too variable, or whether it is performing within the requirements.

The histogram is termed process centred if its centre coincides with the customer requirements; otherwise, the process is too high or too low. A histogram allows predictions of future performance to be made, where it can be assumed that the future will resemble the past.

The construction of a histogram first requires that a frequency table be constructed, and this requires that the range of the data values be determined. The data are divided into a number of classes (or data buckets), where a bucket is a particular range of data values, and the relative frequency of each bucket is displayed in bar format. The number of class intervals or buckets is determined, and the class intervals are defined. The class intervals are mutually disjoint and span the range of the data values. Each data value belongs to exactly one class interval, and the frequency of each class interval is determined.

Table 22.4 Frequency
table—test results

Mark	Frequency
0–24	3
25–49	10
50–74	15
75–100	2

The results of a class test in mathematics are summarized in Table 22.4. There are 30 students in the class, and each student achieves a score somewhere between 0 and 100. There are four data intervals between 0 and 100 employed to summarize the scores, and the result of each student belongs to exactly one interval. Figure 22.3 is the associated histogram for the frequency data, and it gives a pictorial representation of the marks for the class test.

We may also employ a pie chart as an alternate way to present the class marks. The frequency table is constructed as before, and a visual representation of the percentage in each data class (i.e., the relative frequency) is provided with the pie chart. Each portion of the pie chart represents the percentage of the data values in that interval (Fig. 22.4).

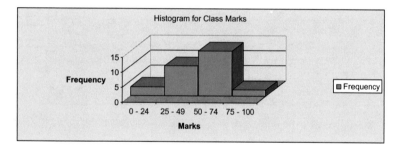

Fig. 22.3   Histogram test results

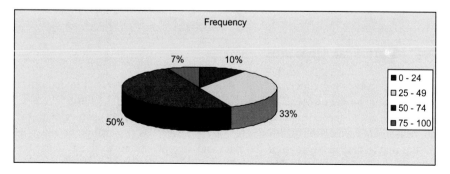

Fig. 22.4   Pie chart test results

Table 22.5 Monthly sales
and profit

	Sales	Profit
Jan	5500	200
Feb	3000	−400
Mar	3500	200
Apr	3000	600
May	4500	−100
Jun	6200	1200
Jul	7350	3200
Aug	4100	100
Sep	9000	3300
Oct	2000	−500
Nov	1100	−800
Dec	3000	300

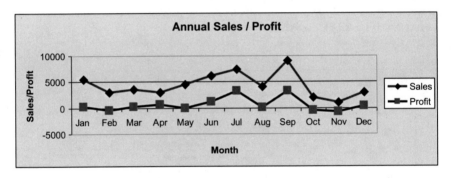

Fig. 22.5 Monthly sales and profit

We present the monthly sales and profit figures for a company in Table 22.5, and Fig. 22.5 gives a pictorial representation of the data in the form of a time series (or trend chart).

22.4 Statistical Measures

Statistical measures are concerned with the basic analysis of the data to determine the average of the data as well as how spread out the data is. The term "average" generally refers to the arithmetic *mean* of a sample, but it may also refer to the statistical *mode* or *median* of the sample. We first discuss the arithmetic mean as it is the mathematical average of the data and is representative of the data. The arithmetic mean is the most widely used average in statistics.

22.4.1 Arithmetic Mean

The *arithmetic mean* (or just mean) of a set of n numbers is defined to be the sum of the data values divided by the number of values. That is, the arithmetic mean of a set of data values $x_1, x_2, \dots, x_n$ (where the sample size is n) is given by:

$$\overline{x} = \frac{\sum_{i=1}^{n} x_i}{n}$$

The arithmetic mean is representative of the data as all values are used in its calculation. The mean of the set of values 5, 11, 9, 4, 16, 9 is given by:

$$m = {}^{5+11+9+4+16+9}/_{6} = 54/6 = 9.$$

The formula for the arithmetic mean of a set of data values given by a frequency table needs to be adjusted.

x_1	x_2	...	...	x_n
f_1	f_2			f_n

$$\overline{x} = \frac{\sum_{i=1}^{n} f_i x_i}{\sum_{i=1}^{n} f_i}$$

The arithmetic mean for the following frequency distribution is calculated by:

x	2	5	7	10	12
f_x	2	4	7	4	2

The mean is given by:

$$m = {}^{(2*2+5*4+7*7+10*4+12*2)}/_{(2+4+7+4+2)}$$
$$= {}^{(4+20+49+40+24)}/_{19} = 137/19 = 7.2 \ .$$

The actual mean of the population is denoted by μ, and it may differ from the sample mean m.

22.4.2 Mode

The *mode* is the most popular element in the sample, i.e., it is the data element that occurs most frequency in the sample. For example, consider a shop that sells mobile phones, then the mode of the annual sales of phones is the most popular phone sold. The mode of the list [1, 4, 1, 2, 7, 4, 3, 2, 4] is 4, whereas the there is no unique mode in the sample [1, 1, 3, 3, 4], and it is said to be bimodal.

The mode of the following frequency distribution is 7, since it occurs the most times in the sample.

x	2	5	7	10	12
f_x	2	4	7	4	2

It is possible that the mode is not unique (i.e., there are at least two elements that occur with the equal highest frequency in the sample), and if this is the case then we are dealing with a bimodal or possibly a multimodel distribution (where there are more than two elements that occur most frequently in the sample).

22.4.3 Median

The *median* of a set of data is the value of the data item that is exactly half way along the set of items, where the data set is arranged in increasing order of magnitude.

If there are an odd number of elements in the sample the median is the middle element. Otherwise, the median is the arithmetic mean of the two middle elements.

The median of 34, 21, 38, 11, 74, 90, 7 is determined by first ordering the set as 7, 11, 21, 34, 38, 74, 90, and the median is then given by the value of the 4th item in the list which is 34.

The median of the list 2, 4, 8, 12, 20, 30 is the mean of the middle two items (as there are an even number of elements and the set is ordered), and so it is given by the arithmetic mean of the 3rd and 4th elements, i.e., $^{8+12}/_2 = 10$.

The calculation of the median of a frequency distribution first requires the calculation of the total number of data elements (i.e., this is given by $N = \Sigma f$), and then determining the value of the middle element in the table, which is the $^{N+1}/_2$ element.

The median for the following frequency distribution is calculated by:

x	2	5	7	10	12
f_x	2	4	7	4	2

The number of elements is given by $N = \Sigma f = 2 + 4 + 7 + 4 + 2 = 19$ and so the middle element is given by the value of the $^{N+1}/_2$ element, i.e., the $^{19+1}/_2$ = the 10th element. From an examination of the table it is clear that the value of the 10th element is 7 and so the median of the frequency distribution is 7.

The final average that we consider is the midrange of the data in the sample, and this is given by the arithmetic mean of the highest and lowest data elements in the sample. That is, $m_{mid} = (x_{max} + x_{min})/2$.

The mean, mode, and median coincide for symmetric frequency distributions but differ for left or right skewed distributions (Fig. 22.6). Skewness describes how non-symmetric the data is.

Dispersion indicates how spread out or scattered the data is, and there are several ways of measuring dispersion including how skewed the distribution is, the range of the data, variance, and the standard deviation.

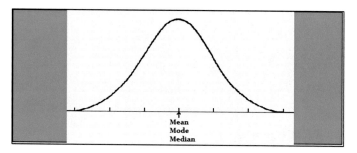

Fig. 22.6   Symmetric distribution

22.5   Variance and Standard Deviation

An important characteristic of a sample is its distribution, and the spread of each element from some measure of central tendency (e.g., the mean). One elementary measure of dispersion is that of the sample *range*, which is defined to be the difference between the maximum and minimum value in the sample. That is, the sample range is defined to be:

$$\text{range} = x_{\max} - x_{\min}$$

The sample range is not a reliable measure of dispersion as just two elements in the sample are used, and so extreme values in the sample may distort the range and make it very large even if most of the elements are quite close to one another.

The standard deviation is the most common way to measure dispersion, and it gives the average distance of each element in the sample from the arithmetic mean. The *sample standard deviation* of a sample $x_1, x_2, \ldots x_n$ is denoted by s, and its calculation first requires the calculation of the sample mean. It is defined by:

$$s = \sqrt{\frac{\sum (x_i - \overline{x})^2}{n-1}} = \sqrt{\frac{\sum x_i^2 - n\overline{x}^2}{n-1}}$$

The *population standard deviation* is denoted by σ and is defined by:

$$\sigma = \sqrt{\frac{\sum (x_i - \mu)^2}{n}} = \sqrt{\frac{\sum x_i^2 - n\mu^2}{n}}$$

Variance is another measure of dispersion, and it is defined as the square of the standard deviation. The *sample variance* s^2 is given by:

$$s^2 = \frac{\sum (x_i - \overline{x})^2}{n-1} = \frac{\sum x_i^2 - n\overline{x}^2}{n-1}$$

The *population variance* σ^2 is given by:

$$\sigma^2 = \frac{\sum (x_i - \mu)^2}{n} = \frac{\sum x_i^2 - n\mu^2}{n}$$

Example 22.1 (*Standard Deviation*) Calculate the standard deviation of the sample 2, 4, 6, 8.

Solution (Standard Deviation)
The sample mean is given by $m = 2 + 4 + 6 + 8/4 = 5$.
   The sample variance is given by:

$$s^2 = (2 - 5)^2 + (4 - 5)^2 + (6 - 5)^2 + (8 - 5)^2/4 - 1$$
$$= 9 + 1 + 1 + 9/3$$
$$= 20/3$$
$$= 6.66$$

The sample standard deviation is given by the square root of the variance and so it is given by:

$$s = \sqrt{6.66}$$
$$= 2.58$$

The formula for the standard deviation and variance may be adjusted for frequency distributions. The standard deviation and mean often go hand in hand, and for normal distributions 68% of the data lies within one standard deviation of the mean; 95% of the data lies within two standard deviations of the mean; and the vast majority (99.7%) of the data lies within three standard deviations of the mean. All data values are used in the calculation of the mean and standard deviation, and so these measures are truly representative of the data.

22.6   Correlation and Regression

The two most common techniques for exploring the relationship between two variables are correlation and linear regression. *Correlation* is concerned with quantifying the strength of the relationship between two variables by measuring the degree of "scatter" of the data values, whereas *regression* expresses the relationship between the variable in the form of an equation (usually a linear equation).
   Correlation quantifies the strength and direction of the relationship between two numeric variables X and Y, and the correlation coefficient may be positive or negative and it lies between -1 and $+1$. If the correlation is positive, then as the value of one variable increases the value of the other variable increases (i.e., the variables move together in the same direction), whereas if the correlation is

negative then as the value of one variable increases the value of the other variable decreases (i.e., the variables move together in the opposite directions). The correlation coefficient r is given by the formula:

$$\text{Corr}(X.Y) = \frac{\overline{XY} - \overline{X}\,\overline{Y}}{\text{Std}(X)\text{Std}(Y)} = \frac{n\Sigma X_i Y_i - \Sigma X_i \Sigma Y_i}{\sqrt{n\Sigma X_i^2 - (\Sigma X_i)^2}\sqrt{n\Sigma Y_i^2 - (\Sigma Y_i)^2}}$$

The sign of the correlation coefficient indicates the direction of the relationship between the two variables. A correlation of $r = +1$ indicates perfect positive correlation, whereas a correlation of $r = -1$ indicates perfect negative correlation. A correlation close to zero indicates no relationship between the two variables; a correlation of $r = -0.3$ indicates a weak negative relationship; whereas a correlation of $r = 0.85$ indicates a strong positive relationship. The extent of the relationship between the two variables may be seen from the following:

- A change in the value of X leads to a change in the value of Y.
- A change in the value of Y leads to a change in the value of X.
- Changes in another variable lead to changes in both X and Y.
- There is no relationship (or correlation) between X and Y.

The relationship (if any) between the two variables can be seen by plotting the values of X and Y in a scatter graph as in Figs. 22.7 and 22.8. The correlation coefficient identifies linear relationships between X and Y, but it does not detect non-linear relationships. It is possible for correlation to exist between two variables but for no causal relationship to exist, i.e., *correlation is not the same as causation*.

Example 22.2 (*Correlation*) The data in Table 22.6 is a summary of the cost of maintenance of eight printers, and it is intended to explore the extent to which the age of the machine is related to the cost of maintenance. It is required to calculate the correlation coefficient.

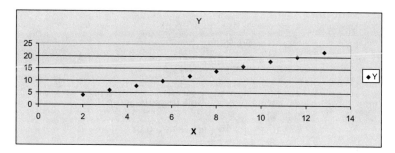

Fig. 22.7 Strong positive correlation

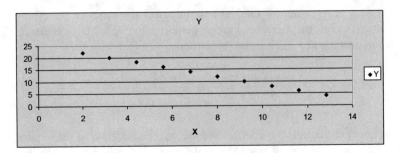

Fig. 22.8 Strong negative correlation

Table 22.6 Cost of maintenance of printers

X (Age)	Y (Cost)	XY	X^2	Y^2
5	50	250	25	2500
12	135	1620	144	18,225
4	60	240	16	3600
20	300	6000	400	90,000
2	25	50	4	625
10	80	800	100	6400
15	200	3000	225	40,000
8	90	720	64	8100
76	940	12,680	978	169,450

Solution (Correlation)

For this example, $n = 8$ (as there are 8 printers) and ΣX_i, ΣY_i, $\Sigma X_i Y_i$, ΣX_i^2, ΣY_i^2 are computed in the last row of the table and so:

$$\Sigma X_i = 76$$
$$\Sigma Y_i = 940$$
$$\Sigma X_i Y_i = 12,680$$
$$\Sigma X_i^2 = 978$$
$$\Sigma Y_i^2 = 169,450$$

We input these values into the correlation formula and get:

$$r = \frac{8 * 12,680 - 76 * 940}{\sqrt{8 * 978 - 76^2}\sqrt{8 * 169,450 - 940^2}}$$
$$= \frac{30,000}{\sqrt{2048}\sqrt{472,000}}$$

$$= \frac{30,000}{45.25 * 687.02}$$
$$= \frac{30,000}{31,087}$$
$$= 0.96$$

Therefore, $r = 0.96$ and so there is strong correlation between the age of the machine and the cost of maintenance of the machine.

22.6.1 Regression

Regression is used to study the relationship (if any) between dependent and independent variables and to predict the dependent variable when the independent variable is known. The prediction capability of regression makes it a more powerful tool than correlation, and regression is useful in identifying which factors impact upon a desired outcome variable.

There are several types of regression that may be employed such as linear or polynomial regression, and this section is concerned with linear regression where the relationship between the dependent and independent variables is expressed by a straight line. More advanced statistical analysis may be conducted with multiple regression models, where there are several independent variables that are believed to affect the value of another variable.

Regression analysis first involves data gathering and plotting the data on a scatter graph. The regression line is the line that best fits the data on the scatter graph (Fig. 22.9), and it is usually determined using the method of least squares or one of the methods summarized in Table 22.7. The regression line is a plot of the expected values of the dependent variable for all values of the independent variable, and the formula (or equation) of the regression line is of the form $y = mx + b$, where the coefficients of a and b are determined.

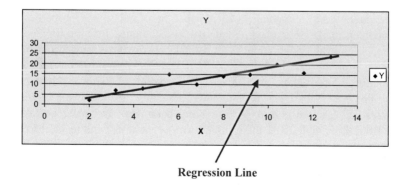

Regression Line

Fig. 22.9 Regression line

Table 22.7 Methods to obtain regression line

Methods	Description
Inspection	This is the simplest method and involves plotting the data in a scatter graph and then drawing a line that best suits the data. (this is subjective and so it is best to draw the mean point, and ensure the regression line passes through this point)
Semi-averages	This involves splitting the data into two equal groups, then finding and drawing the mean point in each group, and joining these points with a straight line (i.e., the regression line)
Least squares	The method of least squares is a mathematical and involves obtaining a regression line where the sum of the squares of the vertical deviations of all the points from the line is minimal

Table 22.8 Hypothesis testing

Action	H_0 true, H_1 false	H_0 false, H_1 true
Fail to reject H_0	Correct	False positive—type 2 error P (accept H_0 I H_0 false) = β
Reject H_0	False negative—type 1 error P (reject H_0 I H_0 true) = α	Correct

The regression line then acts as a model that describes the relationship between the two variables, and the value of the dependent variable may be predicted from the value of the independent variable using the regression line.

22.7   Statistical Inference and Hypothesis Testing

Inferential statistics is concerned with statistical techniques to infer properties of a population from samples taken from the population. Often, it is infeasible or inconvenient to study all members of a population, and so the properties of a representative sample are studied and statistical techniques are used to generalize these properties to the population. A statistical experiment is carried out to gain information from the sample, and it may be repeated as many times as required to gain the desired information. Statistical experiments may be simple or complex.

There are two main types of inferential statistics, and these are *estimating parameters* and *hypothesis testing*. Estimating parameters is concerned taking a statistic from the sample (e.g., the sample mean or variance) and using it to make a statement about the population parameter (i.e., the population mean or variance). Hypothesis testing is concerned with using the sample data to answer research questions such as whether a new drug is effective in the treatment of a particular disease. A sample is not expected to perfectly represent the population, as sampling errors will naturally occur.

A hypothesis is a statement about a particular population whose truth or falsity is unknown. Hypothesis testing is concerned with determining whether the values of the random sample from the population are consistent with the hypothesis. There are two mutually exclusive hypotheses: one of these is the *null hypothesis* H_0 and the other is the *alternate research hypothesis* H_1. The null hypothesis H_0 is what the researcher is hoping to reject, and the research hypothesis H_1 is what the researcher is hoping to accept.

Statistical testing is employed to test the hypothesis, and the result of the test is that we either reject the null hypothesis (and therefore accept the alternative hypothesis), or that we fail to reject it (i.e., we accept) the null hypothesis. The rejection of the null hypothesis means that the null hypothesis is highly unlikely to be true, and that the research hypothesis should be accepted.

Statistical testing is conducted at a certain level of significance, with the probability of the null hypothesis H_0 being rejected when it is true never greater than α. The value α is called the level of significance of the test, with α usually being 0.1, 0.05, or 0.005. A significance level β may also be applied to with respect to accepting the null hypothesis H_0 when H_0 is false. The objective of a statistical test is not to determine whether or not H_0 is actually true, but rather to determine whether its validity is consistent with the observed data. That is, H_0 should only be rejected if the resultant data is very unlikely if H_0 is true (Table 22.8).

The errors that can occur with hypothesis testing include type 1 and type 2 errors. Type 1 errors occur when we reject the null hypothesis when the null hypothesis is actually true. Type 2 errors occur when we accept the null hypothesis when the null hypothesis is false (Fig. 22.7).

For example, an example of a *false positive* is where the results of a blood test come back positive to indicate that a person has a particular disease when in fact the person does not have the disease. Similarly, an example of a *false negative* is where a blood test is negative indicating that a person does not have a particular disease when in fact the person does.

Both errors are potentially very serious, with a false positive generating major stress and distress to the recipient, until further tests are done that show that the person does not have the disease. A false negative is potentially even more serious, as early detection of a serious disease is essential to its treatment, and so a false negative means that valuable time is lost in its detection, which could be very serious.

The terms α and β represent the level of significance that will be accepted, and α may or may not be equal to β. In other words, α is the probability that we will reject the null hypothesis when the null hypothesis is true, and β is the probability that we will accept the null hypothesis when the null hypothesis is false.

Testing a hypothesis at the $\alpha = 0.05$ level is equivalent to establishing a 95% confidence interval. For 99% confidence α will be 0.01, and for 99.999% confidence then α will be 0.00001.

The hypothesis may be concerned with testing a specific statement about the value of an unknown parameter θ of the population. This test is to be done at a certain level of significance, and the unknown parameter may, for example, be the mean or variance of the population. An estimator for the unknown parameter is determined, and the hypothesis that this is an accurate estimate is rejected if the random sample is not consistent with it. Otherwise, it is accepted.

The steps involved in hypothesis testing include:

1. Establish the null and alternative hypothesis
2. Establish error levels (significance)
3. Compute the test statistics (often a *t*-test)
4. Decide on whether to accept or reject the null hypothesis.

The difference between the observed and expected test statistic and whether the difference could be accounted for by normal sampling fluctuations is the key to the acceptance or rejection of the null hypothesis. For more detailed information on statistics see [1, 2].

22.8   Review Questions

1. What is statistics?
2. Explain how statistics may be abused.
3. What is a random sample? How may it be generated?
4. Describe the charts available for the presentation of statistical data.
5. Explain how the average of a sample may be determined.
6. Explain sample variance and sample standard deviation.
7. Explain the difference between correlation and regression.
8. Explain the methods for obtaining the regression line from data.
9. What is hypothesis testing?

22.9   Summary

Statistics is an empirical science that is concerned with the collection, organization, analysis, interpretation, and presentation of data. Statistics are widely used for forecasting as well as for presenting trends. They allow the behaviour of a population to be studied and inferences to be made about the population. These inferences may be tested to ensure their validity.

It is often not possible to study the entire population, and instead a representative subset or sample of the population is chosen. This random sample is used to make inferences regarding the entire population, and it is essential that the sample chosen is indeed random and representative of the entire population. Otherwise, the inferences made regarding the entire population will be invalid due to the introduction of a selection bias.

The data gathered from a statistical study is often raw, and, the way the data is presented is important. It is useful to present the information in pictorial form, as this enables analysis to be done and conclusions to be drawn. Bar charts, histograms, pie chart, and trend graphs may be employed.

Statistical measures are concerned with the basic analysis of the data to determine the average of the data, as well as how spread out the data is. The term "average" generally refers to the arithmetic mean of a sample, but it may also refer to the statistical mode or median of the sample.

An important characteristic of a sample is its distribution and the spread of each element from some measure of central tendency (e.g., the mean). The standard deviation is the most common way to measure dispersion, and it gives the average distance of each element in the sample from the arithmetic mean.

Correlation and linear regression are techniques for exploring the relationship between two variables. Correlation is concerned with quantifying the strength of the relationship between two variables, whereas regression expresses the relationship between the variable in the form of an equation.

Inferential statistics is concerned with statistical techniques to infer properties of a population from samples taken from the population. A hypothesis is a statement about a particular population whose truth or falsity is unknown. Hypothesis testing is concerned with determining whether the values of the random sample from the population are consistent with the hypothesis.

References

1. Dekking FM et al (2010) A modern introduction to probability and statistics. Springer, Berlin
2. Ross SM (1987) Introduction to probability and statistics for engineers and scientists. Wiley Publications, New York

Introduction to Probability Theory

23

Key Topics

Random Variables

Expectation and Variance

Bayes' Formula

Normal Distributions

Binomial Distribution

Poisson Distribution

Unit Normal Distribution

Confidence Intervals and Tests of Significance

23.1 Introduction

Probability is a branch of mathematics that is concerned with measuring uncertainty and random events, and it provides a precise way of expressing the likelihood of a particular event occurring. Probability is also used as part of everyday speech in expressions such as "It is likely to rain in the afternoon", where the corresponding statement expressed mathematically might be "The probability that it will rain in the afternoon is 0.7".

The modern theory of probability theory has its origins in work done on the analysis of games of chance by Cardano in the sixteenth century, and it was developed further in in the seventeenth century by Fermat and Pascal, and refined in the eighteenth century by Laplace. It led to the classical definition of the probability

---

© The Author(s), under exclusive license to Springer Nature Switzerland AG 2023
G. O'Regan, *Mathematical Foundations of Software Engineering*,
Texts in Computer Science, https://doi.org/10.1007/978-3-031-26212-8_23

of an event being:

$$P(\text{Event}) = \frac{\#\text{Favourable Outcomes}}{\#\text{Possible Outcomes}}$$

There are several definitions of probability such as the frequency interpretation and the subjective interpretation of probability. For example, if a geologist sates that "there is a 70% chance of finding gas in a certain region" then this statement is usually interpreted in two ways:

- The geologist is of the view that over the long run 70% of the regions whose environment conditions are very similar to the region under consideration have gas. [*Frequency Interpretation*].
- The geologist is of the view that it is likely that the region contains gas, and that 0.7 is a measure of the geologist's belief in this hypothesis. [*Belief Interpretation*].

That is, according to the frequency interpretation the probability of an event is equal to the long-term frequency of the event's occurrence when the same process is repeated many times.

According to the belief interpretation probability measures the degree of belief about the occurrence of an event or in the truth of a proposition, with a probability of 1 representing the certain belief that something is true and a probability of 0 representing the certain belief that something is false, with a value in between reflecting uncertainty about the belief.

Probabilities may be updated by Bayes' theorem (see Sect. 23.2.2), where the initial belief is the *prior* probability for the event, and this may be updated to a *posterior* probability with availability of new information (see Sect. 23.6 for a short account of Bayesian statistics).

23.2   Basic Probability Theory

Probability theory provides a mathematical indication of the likelihood of an event occurring, and the probability lies between 0 and 1. A probability of 0 indicates that the event cannot occur, whereas a probability of 1 indicates that the event is guaranteed to occur. If the probability of an event is greater than 0.5 then this indicates that the event is more likely to occur than not to occur.

A statistical experiment is conducted to gain certain desired information, and the *sample space* is the set of all possible outcomes of an experiment. The outcomes are all equally likely if no one outcome is more likely to occur than another. An *event E* is a subset of the sample space, and the event is said to have occurred if the outcome of the experiment is in the event E.

For example, the sample space for the experiment of tossing a coin is the set of all possible outcomes of this experiment, i.e., head or tails. The event that the toss results a tail is a subset of the sample space.

$$S = \{h, t\} \quad E = \{t\}$$

Similarly, the sample space for the gender of a newborn baby is the set of outcomes, i.e., the newborn baby is a boy or a girl. The event that the baby is a girl is a subset of the sample space.

$$S = \{b, g\} \quad E = \{g\}$$

For any two events E and F of a sample space S we can also consider the union and intersection of these events. That is,

- $E \cup F$ consists of all outcomes that are in E or F or both.
- $E \cap F$ (usually written as EF) consists of all outcomes that are in both E and F.
- E^c denotes the complement of E with respect to S, and represents the outcomes of S that are not in E.

If $EF = \emptyset$ then there are no outcomes in both E and F, and so the two events E and F are *mutually exclusive*. Events that are mutually exclusive cannot occur at the same time (i.e., they cannot occur together).

Two events are said to be *independent* if the occurrence (or not) of one of the events does not affect the occurrence (or not) of the other. Two mutually exclusive events cannot be independent, since the occurrence of one excludes the occurrence of the other.

The union and intersection of two events can be extended to the union and intersection of a family of events $E_1, E_2, \ldots E_n$ (i.e., $\cup_{i=1}^{n} E_i$ and $\cap_{i=1}^{n} E_i$ E_i).

23.2.1 Laws of Probability

The probability of an event E occurring is given by:

$$P(E) = \frac{\# \text{Outcomes in Event } E}{\# \text{Total Outcomes (in } S)}$$

The laws of probability essentially state that the probability of an event is between 0 and 1, and that the probability of the union of a mutually disjoint set of events is the sum of their individual probabilities. The probability of an event E is zero if E is an impossible event, and the probability of an event E is one if it is a certain event (Table 23.1).

The probability of the union of two events (not necessarily disjoint) is given by:

$$P(E \cup F) = P(E) + P(F) - P(EF)$$

Table 23.1 Axioms of probability

Axiom	Description
1	$P(S) = 1$
2	$P(\emptyset) = 0$
3	$0 \leq P(E) \leq 1$
4	For any sequence of mutually exclusive events $E_1, E_2, \ldots. E_n$. (i.e., $E_i E_j = \emptyset$ where $i \neq j$) then the probability of the union of these events is the sum of their individual probabilities: i.e., $P(\cup_{i=1}^{n} E_i) = \Sigma_{i=1}^{n} P(E_i)$

The complement of an event E is denoted by E^c and denotes that event E does not occur. Clearly, $S = E \cup E^c$ and E and E^c are disjoint and so:

$$P(S) = P(E \cup E^c) = P(E) + P(E^c) = 1$$
$$\Rightarrow P(E^c) = 1 - P(E)$$

The probability of an event E occurring given that an event F has occurred is termed the *conditional probability* (denoted by $P(E|F)$) and is given by:

$$P(E|F) = \frac{P(EF)}{P(F)} \quad \text{where } P(F) > 0$$

This formula allows us to deduce that:

$$P(EF) = P(E|F)P(F)$$

Example 23.1 (Conditional Probability) A family has two children. Find the probability that they are both girls given that they have at least one girl.

Solution (Conditional Probability)
The sample space for a family of two children is $S = \{(g, g), (g, b), b, g), (b, b)\}$. The event E where is at least one girl in the family is given by $E = \{(g, g), (g, b), (b, g)\}$, and the event that G both are girls is $G = \{(g, g)\}$, so we will determine the conditional probability $P(G|E)$ that both children are girls given that there is at least one girl in the family:

$$P(EG) = P(G) = P(g, g) = 1/4$$
$$P(E) = 3/4$$
$$P(G|E) = \frac{P(EG)}{P(E)} = \frac{1/4}{3/4} = 1/3$$

Two events E, F are independent if knowledge that F has occurred does not change the probability that E has occurred. That is, $P(E|F) = P(E)$ and since

$P(E|F) = P(EF)/P(F)$ we have that two events E, F are independent if:

$$P(EF) = P(E)P(F)$$

Two events E and F that are not independent are said to be *dependent*.

23.2.2 Bayes' Formula

Bayes formula enables the probability of an event E to be determined by a weighted average of the conditional probability of E given that the event F has occurred and the conditional probability of E given that F has not occurred:

$$E = E \cap S = E \cap \left(F \cup F^c\right)$$
$$= EF \cup EF^c$$

$$P(E) = P(EF) + P(EF^c) \quad (\text{since } EF \cap EF^c = \emptyset)$$
$$= P(E|F)P(F) + P(E|F^c)P(F^c)$$
$$= P(E|F)P(F) + P(E|F^c)(1 - P(F))$$

We may also get another expression of Bayes' formula from noting that:

$$P(F|E) = \frac{P(FE)}{P(E)} = \frac{P(EF)}{P(E)}$$

Therefore, $P(EF) = P(F|E)P(E) = P(E|F)P(F)$

$$P(E|F) = \frac{P(F|E)P(E)}{P(F)}$$

This version of Bayes' formula allows the probability to be updated where the initial or preconceived belief (i.e., $P(E)$) is the *prior* probability for the event, and this may be updated to a *posterior* probability (i.e., $P(E|F$ is the updated probability), with the new information or evidence (i.e., $P(F)$) and the likelihood that the new information leads to the event (i.e., $P(F|E)$).

Example 23.2 (*Bayes' Formula*) A medical lab is 99% effective in detecting a certain disease when it is actually present, and it yields a false positive for 1% of healthy people tested. If 0.25% of the population actually have the disease what is the probability that a person has the disease if the patient's blood test is positive.

Solution (Bayes' Formula)
Let T be the event that that the patient's test result is positive, and D the event that the tested person has the disease. Then the desired probability is $P(D|T)$ and is given by:

$$P(D|T) = \frac{P(DT)}{P(T)} = \frac{P(T|D)P(D)}{P(T|D)P(D) + P(T|D^c)P(D^c)}$$
$$= \frac{0.99 * 0.0025}{0.99 * 0.0025 + 0.01 * 0.9975} = 0.1988$$

The reason that only approximately 20% of the population whose test results are positive actually have the disease may seem surprising, but is explained by the low incidence of the disease, just one person out of every 400 tested will have the disease and the test will correctly confirm that 0.99 have the disease, but the test will also state that $399 * 0.01 = 3.99$ have the disease and so the proportion of time that the test is correct is $^{0.99}/_{0.99+3.99} = 0.1988$.

23.3   Random Variables

Often, some numerical quantity determined by the result of the experiment is of interest rather than the result of the experiment itself. These numerical quantities are termed *random variables*. A random variable is termed *discrete* if it can take on a finite or countable number of values, and otherwise it is termed *continuous*.

The *distribution function* (denoted by $F(x)$) of a random variable is the probability that the random variable X takes on a value less than or equal to x. It is given by:

$$F(x) = P\{X \leq x\}$$

All probability questions about X can be answered in terms of its distribution function F. For example, the computation of $P\{a < X < b\}$ is given by:

$$P\{a < X < b\} = P\{X \leq b\} - P\{X \leq a\}$$
$$= F(b) - F(a)$$

The *probability mass function* for a discrete random variable X (denoted by $p(a)$) is the probability that the random variable is a certain value. It is given by:

$$p(a) = P\{X = a\}$$

Further, $F(a)$ can also be expressed in terms of the probability mass function

$$F(a) = P\{X \leq a\} = \sum_{\forall x \leq a} p(x)$$

X is a continuous random variable if there exists a non-negative function $f(x)$ (termed the *probability density function*) defined for all $x \in (-\infty, \infty)$ such that

$$P\{X \in B\} = \int_B f(x)dx$$

All probability statements about X can be answered in terms of its density function $f(x)$. For example:

$$P\{a \leq X \leq b\} = \int_a^b f(x)dx$$

$$P\{X \in (-\infty, \infty)\} = 1 = \int_{-\infty}^{\infty} f(x)dx$$

The function $f(x)$ is termed the probability density function, and the probability distribution function $F(a)$ is defined by:

$$F(a) = P\{X \leq a\} = \int_{-\infty}^{a} f(x)dx$$

Further, the first derivative of the probability distribution function yields the probability density function. That is,

$$d/da\ F(a) = f(a).$$

The expected value (i.e., the *mean*) of a discrete random variable X (denoted $E[X]$) is given by the weighted average of the possible values of X:

$$E[X] = \begin{cases} \sum_i x_i P\{X = x_i\} & \text{Discrete Random variable} \\ \int_{-\infty}^{\infty} xf(x)dx & \text{Continuous Random variable} \end{cases}$$

Further, the expected value of a function of a random variable is given by $E[g(X)]$ and is defined for the discrete and continuous case respectively.

$$E[g(X)] = \begin{cases} \Sigma_i g(x_i) P\{X = x_i\} & \text{Discrete Random variable} \\ \int_{-\infty}^{\infty} g(x)f(x)dx & \text{Continuous Random variable} \end{cases}$$

The *variance* of a random variable is a measure of the spread of values from the mean and is defined by:

$$\text{Var}(X) = E[X^2] - (E[X])^2$$

The standard deviation σ is given by the square root of the variance. That is,

$$\sigma = \sqrt{\text{Var}(X)}$$

The *covariance* of two random variables is a measure of the relationship between two random variables X and Y, and indicates the extent to which they both change (in either similar or opposite ways) together. It is defined by:

$$\text{Cov}(X, Y) = E[XY] - E[X].E[Y].$$

It follows that the covariance of two independent random variables is zero. Variance is a special case of covariance (when the two random variables are identical). This follows since $\text{Cov}(X, X) = E[X.X] - (E[X])(E[X]) = E[X^2] - (E[X])^2 = \text{Var}(X)$.

A positive covariance $(\text{Cov}(X, Y) \geq 0)$ indicates that Y tends to increase as X does, whereas a negative covariance indicates that Y tends to decrease as X increases.

The *correlation* of two random variables is an indication of the relationship between two variables X and Y (we discussed correlation and regression in Sect. 22.6). If the correlation is negative and close to -1 then Y tends to decrease as X increases, and if it is positive and close to 1 then Y tends to increase as X increases. A correlation close to zero indicates no relationship between the two variables; a correlation of $r = -0.4$ indicates a weak negative relationship; whereas a correlation of $r = 0.8$ indicates a strong positive relationship. The correlation coefficient is between ± 1 and is defined by:

$$\text{Corr}(X, Y) = \frac{\text{Cov}(X, Y)}{\sqrt{\text{Var}(X)\text{Var}(Y)}}$$

Once the correlation between two variables has been calculated the probability that the observed correlation was due to chance can be computed. This is to ensure that the observed correlation is a real one and not due to a chance occurrence.

23.4 Binomial and Poisson Distributions

The binomial and Poisson distributions are two important distributions in statistics, and the Poisson distribution may be used as an approximation for the binomial. The binomial distribution was first used in games of chance, and it has the following characteristics:

- The existence of a trial of an experiment, which is, defined in terms of two states namely success or failure.
- The identical trials may be repeated a number of times yielding several successes and failures.

- The probability of success (or failure) is the same for each trial.

A *Bernoulli trial* is where there are just two possible outcomes of an experiment, i.e., success or failure. The probability of success and failure is given by:

$$P\{X = 1\} = p$$
$$P\{X = 0\} = 1 - p$$

The mean of the Bernoulli distribution is given by p (since $E[X] = 1.p + 0.(1 - p) = p$), and the variance is given by $p(1 - p)$ (since $E[X^2] - E[X]^2 = p - p^2 = p(1 - p)$).

The *Binomial distribution* involves n Bernoulli trials, where each trial is independent and results in either success (with probability p) or failure (with probability $1 - p$). The binomial random variable X with parameters n and p represents the number of successes in n independent trials, where X_i is the result of the ith trial and X is represented as:

$$X = \sum_{i=1}^{n} X_i$$

$$X_i = \begin{cases} 1 \text{ if the } i\text{th trial is a success} \\ 0 \text{ otherwise} \end{cases}$$

The probability of i successes from n independent trials is then given by the binomial theorem:

$$P\{X = i\} = \binom{n}{i} p^i (1 - p)^{n-i} \quad i = 0, 1, \ldots n$$

Clearly, $E[X_i] = p$ and $\text{Var}(X_i) = p(1 - p)$ (since X_i is an independent Bernoulli random variable). The mean of the binomial distribution $E[X]$ is the sum of the mean of the $E[X_i]$, i.e., $\Sigma_1{}^n E[X_i] = np$, and the variance $\text{Var}(X)$ is the sum of the $\text{Var}(X_i)$ (since the X_i are independent random variables) and so $\text{Var}(X) = np(1 - p)$. The binomial distribution is symmetric when $p = 0.5$, and the distribution is skewed to the left or right when $p \neq 0.5$ (Fig. 23.1).

Example 23.3 (*Binomial Distribution*) The probability that a printer will need correcting adjustments during a day is 0.2. If there are five printers running on a particular day determine the probability of:

1. No printers need correcting
2. One printer needs correcting
3. Two printers require correcting
4. More than two printers require adjusting.

Binomial Distribution (n=10, p=0.5)

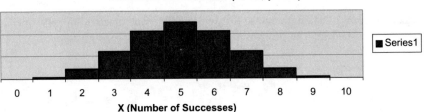

Fig. 23.1 Binomial distribution

Solution (Binomial Distribution)

There are five trials (with $n = 5$, $p = 0.2$, and the success of a trial is a printer needing adjustments). And so,

(1) This is given by $P(X = 0) = (^5_0)\, 0.2^0 * 0.8^5 = 0.3277$
(2) This is given by $P(X = 1) = (^5_1)\, 0.2^1 * 0.8^4 = 0.4096$
(3) This is given by $P(X = 2) = (^5_2)\, 0.2^2 * 0.8^3 = 0.205$
(4) This is given by $1 - P(2$ or fewer printers need correcting)

$$= 1 - [P(X = 0) + P(X = 1) + P(X = 2)$$
$$= 1 - [0.3277 + 0.4096 + 0.205]$$
$$= 1 - 0.9423$$
$$= 0.0577$$

The *Poisson distribution* may be used as an approximation to the binomial distribution when n is large (e.g., $n > 30$) and p is small (e.g., $p < 0.1$). The characteristics of the Poisson distribution are:

- The existence of events that occur at random and may be rare (e.g., road accidents).
- An interval of time is defined in which events may occur.

The probability of i successes (where $i = 0, 1, 2 \ldots$) is given by:

$$P(X = i) = \frac{e^{-\lambda}\lambda^i}{i!}$$

The mean and variance of the Poisson distribution are given by λ.

Example 23.4 (*Poisson Distribution*) Customers arrive randomly at a supermarket at an average rate of 2.5 customers per minute, where the customer arrivals form a Poisson distribution. Determine the probability that:

1. No customers arrive in any particular minute
2. Exactly one customer arrives in any particular minute
3. Two or more customers arrive in any particular minute
4. One or more customers arrive in any 30 s period.

Solution (Poisson Distribution)
The mean λ is 2.5/min for parts 1–3, and λ is 1.25 for part 4.

1. $P(X = 0) = e^{-2.5} * 2.5^0/0! = 0.0821$
2. $P(X = 1) = e^{-2.5} * 2.5^1/1! = 0.2052$
3. $P(2 \text{ or more}) = 1 - P(X = 0 \text{ or } X = 1)] = 1 - [P(X = 0) + P(X = 1)] = 0.7127$
4. $P(1 \text{ or more}) = 1 - P(X = 0) = 1 - e^{-1.25} * 1.25^0/0! = 1 - 0.2865 = 0.7134.$

23.5 The Normal Distribution

The *normal distribution* is the most important distribution in statistics, and it occurs frequently in practice. It is shaped like a bell, and it is popularly known as the bell-shaped distribution, and the curve is symmetric about the mean of the distribution. The empirical frequencies of many populations naturally exhibit a bell-shaped (normal) curve, such as the frequencies of the height and weight of people. The largest frequencies cluster around the mean and taper away symmetrically on either side of the mean. The German mathematician, Gauss (Fig. 23.2), originally studied the normal distribution, and it is also known as the Gaussian distribution.

The normal distribution is a continuous distribution, and it has two parameters, namely the mean μ and the standard deviation σ. It is a continuous distribution, and so it is not possible to find the probability of individual values, and thus it is only possible to find the probabilities of ranges of values. The normal distribution has the important properties that 68.2% of the values lie within one standard deviation of the mean, with 95% of the values within two standard deviations; and

Fig. 23.2 Carl Friedrich Gauss

Fig. 23.3 Standard normal
bell curve (Gaussian
distribution)

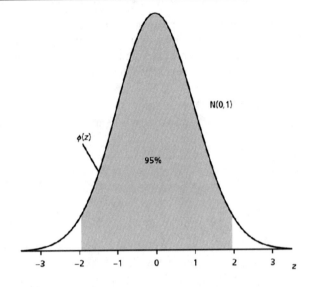

$\phi(z)$

95%

N(0, 1)

99.7% of values are within three standard deviations of the mean. Another words, the value of the normal distribution is practically zero when the value of x is more than three standard deviations from the mean. The shaded area under the curve in Fig. 23.3 represents two standard deviations of the mean and comprises 95% of the population.

The *normal distribution* N has mean μ and standard deviation σ. Its density function $f(x)$ where (where $-\infty < x < \infty$) is given by:

$$f(x) = \frac{1}{\sqrt{2\pi}\sigma} e^{-(x-\mu)^2/2\sigma^2}$$

23.5.1 Unit Normal Distribution

The *unit* (or *standard*) normal distribution $Z(0,1)$ has mean 0 and standard deviation of 1. Every normal distribution may be converted to the unit normal distribution by $Z = (X - \mu)/\sigma$, and every probability statement about X has an equivalent probability statement about Z. The unit normal density function is given by:

$$f(y) = \frac{1}{\sqrt{2\pi}} e^{-\frac{1}{2}y^2}$$

The standard normal distribution is symmetric about 0 (as the mean is zero), and the process of converting a normal distribution with mean μ and standard deviation σ is termed standardizing the x-value. There are tables of values that give the probability of a Z score between zero and the one specified.

Example 23.5 (*Normal Distribution*) Weights of bags of oranges are normally distributed with mean 3 lbs and standard deviation 0.2 lb. The delivery to a supermarket is 350 bags at a time. Determine the following:

1. Standardize to a unit normal distribution.
2. What is the probability that a standard bag will weigh more than 3.5 lbs?
3. How many bags from a single delivery would be expected to weigh more than 3.5 lbs.

Solution (Normal Distribution)

1. $Z = X - \mu/\sigma = X - 3/0.2$.
2. Therefore, when $X = 3.5$ we have $Z = 3.5 - 3/0.2 = 2.5$.

For $Z = 2.5$ we have from the unit normal tables that

$$P(Z \le 2.50) = 0.9938 = P(X \le 3.5)$$

Therefore, $P(X > 3.5) = 1 - P(X \le 3.5) = 1 - 0.9938 = 0.0062$

3. The proportion of all bags that have a weight greater than 3.5 lbs is 0.0062, and so it would be expected that there are $350 * 0.0062 = 2.17$ bags with a weight > 3.5, and so in practical terms we would expect two bags to weigh more than 3.5 lbs.

The normal distribution may be used as an approximation to the binomial when n is large (e.g., $n > 30$), and when p is not too small or large. This is discussed in the next section, where the mean of the normal distribution is np and the standard deviation is $\sqrt{np(1 - p)}$.

23.5.2 Confidence Intervals and Tests of Significance

The study of normal distributions helps in the process of estimating or specifying a range of values, where certain population parameters (such as the mean) lie from the results of small samples. Further, the estimate may be stated with a certain degree of confidence, such as there is 95% or 99% confidence that the mean value lies between 4.5 and 5.5. That is, *confidence intervals* (also known as confidence limits) specify a range of values within which some unknown population parameter lies with a stated degree of confidence, and it is based on the results of the sample.

The confidence interval for an unknown population mean where the sample mean, sample variance, the sample size, and desired confidence level are known is given by:

$$\bar{x} \pm z \frac{s}{\sqrt{n}}$$

In the formula $\bar{x}$ is the sample mean, s is the sample standard deviation, n is the sample size, and z is the confidence factor (for a 90% confidence interval $z = 1.64$, for the more common 95% confidence interval $z = 1.96$, and $z = 2.58$ for the 99% confidence interval).

Example 23.6 (*Confidence Intervals*) Suppose a new motor fuel has been tested on 30 similar cars, and the fuel consumption was 44.1 mpg with a standard deviation of 2.9 mpg. Calculate a 95% confidence interval for the fuel consumption of this model of car.

Solution (Confidence Intervals)

The sample mean is 44.1, the sample standard deviation is 2.9, the sample size is 30, and the confidence factor is 1.96, and so the 95% confidence interval is:

$$\bar{x} \pm z \frac{s}{\sqrt{n}} = 44.1 \pm 1.96 * \frac{2.9}{\sqrt{30}}$$
$$= 44.1 \pm 1.96 * 0.5295$$
$$= 44.1 \pm 1.0378$$
$$= (43.0622, 45.1378)$$

That is, we can say with 95% confidence that the fuel consumption for this model of car is between 43.0 and 45.1 mpg.

The confidence interval for an unknown population mean where the sample proportion and sample size are known is given by:

$$p \pm z \sqrt{\frac{p(1-p)}{n}}$$

In the formula p is the sample proportion, n is the sample size, and z is the confidence factor.

Example 23.7 (*Confidence Intervals*) Suppose 3 faulty components are identified in a random sample of 20 products taken from a production line. What statement can be made about the defect rate of all finished products?

Solution (Confidence Intervals)

The proportion of defective products in the sample is $p = {}^3/_{20} = 0.15$, and the sample size is $n = 20$. Therefore, the 95% confidence interval for the population mean is given by:

$$p \pm z \sqrt{\frac{p(1-p)}{n}} = 0.15 \pm 1.96 \sqrt{\frac{0.15(1-0.15)}{20}}$$
$$= 0.15 \pm 1.96 * 0.0798$$
$$= 0.15 \pm 0.1565$$

$$= (-0.0065, 0.3065)$$

That is, we can say with 95% confidence that the defective rate of finished products lies between 0 and 0.3065.

Tests of Significance for the Mean

Tests of significance are related to confidence intervals and use the concepts from the normal distribution. To test whether a sample of size n, with sample mean $\overline{x}$ and sample standard deviation s could be considered as having been drawn from a population with mean μ the test statistic must lie in the range -1.96 to 1.96.

$$z = \frac{\overline{x} - \mu}{\left[\frac{s}{\sqrt{n}}\right]}$$

That is, the test is looking for evidence of a significant difference between the sample mean $\overline{x}$ and the population mean μ, and evidence is found if z lies outside of the stated limits, whereas if z lies within the limits then there is no evidence that the sample mean is different from the population mean.

Example 23.8 (*Tests of Significance*) A new machine has been introduced, and management is questioning whether it is more productive than the previous one. Management takes 15 samples of this week's hourly output to test whether it is less productive, and the average production per hour is 1250 items with a standard deviation of 50. The output per hour of the previous machine was 1275 items per hour. Determine with a test of significance whether the new machine is less productive.

Solution (Tests of Significance)

The sample mean is 1250, the population mean is 1275, the sample standard deviation is 50, and the sample size is 15.

$$z = \frac{\overline{x} - \mu}{\left[\frac{s}{\sqrt{n}}\right]} = \frac{1250 - 1275}{\left[\frac{50}{\sqrt{15}}\right]} = \frac{-25}{12.91} = -0.1936$$

This lies within the range -1.96 to 1.96 and so there is no evidence of any significant difference between the sample mean and the population mean, and so management is unable to make any statement on differences in productivity.

23.5.3 The Central Limit Theorem

A fundamental result in probability theory is the *central limit theorem*, which essentially states that the sum of a large number of independent and identically distributed random variables has a distribution that is approximately normal. That is, suppose X_1, X_2, ..., X_n is a sequence of independent random variables each with mean μ and variance σ^2. Then for large n the distribution of

$$\frac{X_1 + X_2 + \cdots + X_n - n\mu}{\sigma\sqrt{n}}$$

is approximately that of a unit normal variable Z. One application of the central limit theorem is in relation to the binomial random variables, where a binomial random variable with parameters (n, p) represents the number of successes of n independent trials, where each trial has a probability of p of success. This may be expressed as:

$$X = X_1 + X_2 + \cdots + X_n$$

where $X_i = 1$ if the ith trial is a success and is 0 otherwise. The mean of the Bernoulli trial $E(X_i) = p$, and its variance is $\text{Var}(X_i) = p(1 - p)$. (The mean of the Binomial distribution with n Bernoulli trials is np and the variance is $np(1 - p)$). By applying the central limit theorem it follows that for large n that

$$\frac{X - np}{\sqrt{np(1 - p)}}$$

will be approximately a unit normal variable (which becomes more normal as n becomes larger).

The sum of independent normal random variables is normally distributed, and it can be shown that the sample average of X_1, X_2, ... X_n is normal, with a mean equal to the population mean but with a variance reduced by a factor of $1/n$.

$$E(\overline{X}) = \sum_{i=1}^{n} \frac{E(X_i)}{n} = \mu$$

$$\text{Var}(\overline{X}) = \frac{1}{n^2} \sum_{i=1}^{n} \text{Var}(X_i) = \frac{\sigma^2}{n}$$

It follows that from this that the following is a unit normal random variable.

$$\sqrt{n}\frac{(X - \mu)}{\sigma}$$

The term *six-sigma* (6σ) is a methodology concerned with continuous process improvement to improve business performance, and it aims to develop very high quality close to perfection. It was developed by Bill Smith at Motorola in the early 1980s, and it was later used by leading companies such as General Electric. A 6σ process is one in which 99.9996% of the products are expected to be free from defects (3.4 defects per million) [1].

Table 23.2 Probability distributions

Distribution name	Density function	Mean/variance
Hypergeometric	$P\{X = i\} = \binom{N}{i}\binom{M}{n-i}/\binom{N+M}{n}$	$nN/N + M, np(1-p)[1-(n-1)/N + M - 1]$
Uniform	$f(x) = 1/(\beta - \alpha)\ \alpha \le x \le \beta, 0$	$(\alpha + \beta)/2, (\beta - \alpha)^2/12$
Exponential	$f(x) = \lambda e^{-\lambda x}$	$1/\lambda, 1/\lambda^2$
Gamma	$f(x) = \lambda e^{-\lambda x}(\lambda x)^{\alpha-1}/\Gamma(\alpha)$	$\alpha/\lambda, \alpha/\lambda^2$

There are many other well-known distributions such as the *hypergeometric distribution* that describes the probability of *i* successes in *n* draws from a finite population without replacement; the *uniform distribution*; the *exponential distribution*; and the *gamma* distribution. The mean and variance of these distributions are summarized in Table 23.2.

23.6 Bayesian Statistics

Bayesian statistics is named after Thomas Bayes who was an 18th-century English theologian and statistician, and it differs from the frequency interpretation of probability in that it considers the probability of an event to be a measure of one's personal belief in the event. According to the frequentist approach only repeatable events such as the result from flipping a coin have probabilities, where the probability of an event is the long-term frequency of its occurrence. Bayesians view probability in a more general way and probabilities may be used to represent the uncertainty of an event or hypothesis. It is perfectly acceptable in the Bayesian view of the world to assign probabilities to non-repeatable events, whereas a strict frequentist would claim that such probabilities do not make sense, as they are not repeatable.

Bayesian thinking provides a way of dealing rationally with randomness and risk in daily life, and it is very useful when the more common frequency interpretation is unavailable or has limited information. It interprets probability as a measure of one's personal belief in a proposition or outcome, and it is essential to first use all your available prior knowledge to form an initial estimate of the probability of the event or hypothesis. Further, when reliable frequency data becomes available the measure of personal belief would be updated accordingly to equal the probability calculated by the frequency calculation. Further, the probabilities must be updated in the light of new information that becomes available, as probabilities may change significantly from new information and knowledge. Finally, no matter how much the odds move in your favour there is eventually one final outcome (which may or may not be the desired event).

Often, in an unreliable and uncertain world we base our decision-making on a mixture of reflection and our gut instinct (which can be wrong). Often, we encounter several constantly changing random events and so it is natural to wonder on the extent to which rational methods may be applied to risk assessment and decision making in an uncertain world.

An initial estimate is made of the belief in the proposition, and if you always rely on the most reliable and objective probability estimates while keeping track of possible uncertainties and updating probabilities in line with new data then the final probability number computed will be the best possible.

We illustrate the idea of probabilities being updated with an adapted excerpt from a children's story called "Fortunately", which was written by Remy Charlip in the 1960s [2]:

- A lady went on a hot air balloon trip
- Unfortunately she fell out
- Fortunately she had a parachute on
- Unfortunately the parachute did not open
- Fortunately there was a haystack directly below
- Unfortunately there was a pitchfork sticking out at the top of haystack
- Fortunately she missed the pitchforks
- Unfortunately she missed the haystack.

The story illustrates how probabilities can change dramatically based on new information, and despite all the changes to the probabilities during the fall the final outcome is a single result (i.e., either life or death). Let p be the probability of survival then the value of p changes as she falls through sky based on new information at each step. Table 23.3 illustrates an estimate of what the probabilities might be:

However, even if probability calculations become irrelevant after the event they still give the best chances over the long term. Over our lives we make many thousands of decisions about where and how to travel, what diet we should have and so on, and though the impact of each of these decisions on our life expectancy is very

Table 23.3 Probability of survival

Step	Prob. survival
A lady went on a hot air balloon trip	$p = 0.999998$
Unfortunately she fell out	$p = 0.000001$
Fortunately she had a parachute on	$p = 0.999999$
Unfortunately the parachute did not open	$p = 0.000001$
Fortunately there was a haystack directly below	$p = 0.5$
Unfortunately there was a pitchfork sticking out at the top of haystack	$p = 0.000,001$
Fortunately she missed the pitchforks	$p = 0.5$
Unfortunately she missed the haystack	$p = 0.000001$

small, their combined effect is potentially significant. Clearly, careful analysis is needed for major decisions rather than just deciding based on gut instinct.

For the example above we could estimate probabilities for the various steps based on the expectation of probability of survival on falling without a parachute, the expectation of probability of survival on falling onto a haystack without a parachute and we would see wildly changing probabilities from the changing circumstances.

We discussed Bayes' formula in Sect. 22.2.2, which allows the probability to be updated where the initial or preconceived belief (i.e., $P(E)$ is the *prior* probability for the event), and this may be updated to a *posterior* probability (i.e., $P(E|F$ is the updated probability), with the new information or evidence (i.e., $P(F)$) and the likelihood that the new information leads to the event (i.e., $P(F|E)$). The reader is referred to [3] for a more detailed account of probability and statistics.

23.7 Review Questions

1. What is probability?
2. Explain the laws of probability.
3. What is a sample space? What is an event?
4. Prove Boole's inequality $P(\cup_{i=1}^{n} E_i) \leq \Sigma_{i=1}^{n} P(E_i)$ where the E_i are not necessarily disjoint.
5. A couple has 2 children. What is the probability that both are girls if the eldest is a girl?
6. What is a random variable?
7. Explain the difference between the probability mass function and the probability density function (for both discrete and continuous random variables).
8. Explain variance, covariance, and correlation.
9. What is the binomial distribution and what is its mean and variance?
10. What is the Poisson distribution and what is its mean and variance?
11. What is the normal distribution and what is its mean and variance?
12. What is the unit normal distribution and what is its mean and variance?
13. Explain the significance of the central limit theorem.
14. What is Bayes' theorem? Explain the importance of Bayesian thinking.

23.8 Summary

Probability is a branch of mathematics that is concerned with measuring uncertainty and random events, and it provides a precise way of expressing the likelihood of a particular event occurring, and the probability is a numerical value between 0 and 1. A probability of 0 indicates that the event cannot occur, whereas a probability of 1 indicates that the event is guaranteed to occur. If the probability of an event is greater than 0.5, then this indicates that the event is more likely to occur than not.

A sample space is the set of all possible outcomes of an experiment, and an event E is a subset of the sample space, and the event is said to have occurred if the outcome of the experiment is in the event E. Bayes formula enables the probability of an event E to be determined by a weighted average of the conditional probability of E given that the event F occurred and the conditional probability of E given that F has not occurred.

Often, some numerical quantity from an experiment is of interest rather than the result of the experiment itself. These numerical quantities are termed random variables. The distribution function of a random variable is the probability that the random variable X takes on a value less than or equal to x.

The binomial and Poisson distributions are important distributions in statistics, and the Poisson distribution may be used as an approximation for the binomial. The Binomial distribution involves n Bernoulli trials, where each trial is independent and results in either success or failure. The mean of the Bernoulli distribution is given by p and the variance by $p(1 - p)$.

The normal distribution is popularly known as the bell-shaped distribution. It is a continuous distribution, and the curve is symmetric about the mean of the distribution. It has two parameters, namely the mean μ and the standard deviation σ. Every normal distribution may be converted to the unit normal distribution by $Z = (X - \mu)/\sigma$, and every probability statement about X has an equivalent probability statement about the unit distribution Z.

The central limit theorem essentially states that the sum of a large number of independent and identically distributed random variables has a distribution that is approximately normal. Bayesian statistics provides a way of dealing rationally with randomness and risk in daily life, and it interprets probability as a measure of one's personal belief in a proposition or outcome.

References

1. O'Regan G (2014) Introduction to software quality. Springer, Berlin
2. Charlip R (1993) Fortunately. Simon and Schuster, New York
3. Dekking FM et al (2010) A modern introduction to probability and statistics. Springer, Berlin

Introduction to Data Science

24

Key Topics

Data Science

Data Scientist

GDPR

Privacy

Security

AI

Internet of Things

Social Media

24.1 Introduction

Information is power in the digital age, and the collection, processing, and use of information need to be regulated. Data science involves the extraction of knowledge from data sets that consist of structured and unstructured data, and data scientists have a responsibility to ensure that this knowledge is used wisely and not abused. Data science may be regarded as a branch of statistics as it uses many concepts from the field, and in order to prevent errors occurring during data analysis it is essential that both the data and models are valid.

The question of ownership of the data is important; as if, for example, I take a picture of another individual does the picture belong to me (as owner of the camera and the collector of the data)? Or does it belong to the individual who is the subject of the image? Most reasonable people would say that the image is my

---

© The Author(s), under exclusive license to Springer Nature Switzerland AG 2023
G. O'Regan, *Mathematical Foundations of Software Engineering*,
Texts in Computer Science, https://doi.org/10.1007/978-3-031-26212-8_24

property, and, if so, what responsibilities or obligations do I have (if any) to the other individual? That is, although I may technically be the owner of the image, the fact that it contains the personal data (or image) of another should indicate that I have an ethical responsibility or obligation to ensure that the image (or personal data) is not misused in any way to harm that individual. Further, if I misuse the image in any way then I may be open to a lawsuit from the individual.

Personal data is collected about individuals from their use of computer resources such as their use of email, their Google searches, their Internet, and social media use to build up revealing profiles of the user that may be targeted to advertisers. Modern technology has allowed governments to conduct mass surveillance on its citizens, with face recognition software allowing citizens to be recognized at demonstrations or other mass assemblies.

Further, smartphones provide location data that allows the location of the user to be tracked and may be used for mass surveillance. Many online service providers give customer data to the security and intelligence agencies (e.g., NSA and CIA), and these agencies often have the ability to hack into electronic devices. It is important that such surveillance technologies are regulated and not abused by the state. Privacy has become more important in the information age, and it is the way in which we separate ourselves from other people and is the right to be left alone. The European GDPR law has become an important protector of privacy and personal data, and it has been adopted by many countries around the world.

Companies collect lots of personal data about individuals, and so the question is how should a company respond to a request for personal information on particular users? Does it have a policy to deal with that situation? What happens to the personal data that a bankrupt company has gathered? Is the personal data part of the assets of the bankrupt company and sold on with the remainder of the company? How does this affect privacy information agreements and compliance to them or does the agreement cease on termination of business activities?

The consequence of an error in data collection or processing could result in harm to an individual, and so the data collection and processing needs to be accurate. Decisions may be made on the basis of public and private data, and often individuals are unaware as to what data was collected about them, whether the data is accurate, and whether it is possible to correct errors in the data.

Further, the conclusions from the analysis may be invalid due to errors in incorrect or biased algorithms, and so a reasonable question is how to keep algorithmically driven systems from harming people? Data scientists have a responsibility to ensure that the algorithm is fit for purpose and uses the right training data, and as far as practical to detect and eliminate unintentional discrimination in algorithms against individuals or groups.

That is, problems may arise when the algorithm uses criteria tuned to fit the majority, as this may be unfair to minorities. Another words, the results are correct, but presented in an over simplistic manner. This could involve presenting the correct aggregate outcome but ignoring the differences within the population, and so leading to the suppression of diversity, and discriminating against the minority group. Another problem is where the data may be correct but presented in a misleading way (e.g., the scales of the axis may be used to present the results visually in an exaggerated way).

24.2 Ethics of Data Science

There has been a phenomenal growth in the use of digital data in information technology, with vast amounts of data collected, processed, and used, and so the ethics of data science has become important. There are social consequences to the use of data, and the ethics of data science aims to investigate what is fair and ethical in data science, and what should or should not be done with data.

A fundamental principle of ethics in data science refers to *informed consent*, and this has its origins in the ethics of medical experiments on individuals. The concept of informed consent in medical ethics is where the individual is informed about the experiment and gives their *consent voluntarily*. The individual has the right to withdraw consent at any time during the experiment. Such experiments are generally conducted to benefit society, and often there is a board that approves the study and oversees it to ensure that all participants have given their informed consent, and attempts to balance the benefits to society with any potential harm to individuals. Once individuals have given their informed consent data may be collected about them.

The principle of informed consent is part of information technology, in the sense that individuals accept the terms and conditions before they may use software applications, and these terms state that data may be collected, processed, and shared. However, it is important to note that generally users do not give informed consent in the sense of medical experiments, as the details of the data collection and processing is hidden in the small print of the terms and condition, and this is generally a long and largely unreadable document. Further, the consent is not given voluntarily, in the sense that if a user wishes to use the software, then he or she has no choice but to click acceptance of the terms and conditions of use for the site. Otherwise, they are unable to access the site, and so for many software applications (apps) consent is essentially coerced rather than freely given.

There was some early research done on user behaviour by Facebook in 2012, where they conducted an experiment to determine if they could influence the mood of users by posing happy or sad stories to their news feed. The experiment was done without the consent of the users, and while the study indicated that happy or sad stories did influence the user's mood and postings, it led to controversy and major dissatisfaction with Facebook when users became aware that they were the *subject of a psychological experiment without their consent*.

The dating site OKCupid uses an algorithm to find compatibility matches for its users based on their profiles, and two people are assigned a match rating based on the extent to which the algorithm judges them to be compatible. OKCupid conducted psychological experiments on its users without their knowledge, with the first experiment being a "love is blind" day where all images were removed from the site, and so compatibilities were determined without the use of images.

Another experiment was very controversial and unethical, as the site lied to the users on their match ratings (e.g., two people with a compatibility rating of 90% were given a rating of 30%, and vice versa). The site was trying to determine the extent that two people would get along irrespective of the rating that they were given, and it showed that two people talked more when falsely told that the algorithm matched them, and vice versa. The controversy arose once users became aware of the deception by the company, and it provides a case study on the *socially unacceptable manipulation of user data* by an Internet company.

Data collection is not a new phenomenon as devices such as cameras and telephones have been around for some time. People have reasonable expectations on privacy and do not expect their phone calls to be monitored and eavesdropped on by others, or they do not expect to be recorded in a changing room or in their home. Individuals will wish to avoid the harm that could occur due to data about them being collected, processed, and shared. The question is whether reasonable rules can be defined and agreed, and whether tradeoffs may be made to balance the conflicting rights and to protect the individual as far as is possible. Some relevant questions on data collection and ownership are considered in Table 24.1.

24.2.1 Data Science and Data Scientists

Data science is a multidisciplinary field that extracts knowledge from data sets that consist of structured and unstructured data, and large data sets (*big data*[1]) may be analysed to extract useful information. The field has great power to harm and to help, and data scientists have a responsibility to use this power wisely. Data science may be regarded as a branch of statistics as it uses many concepts from the field, and in order to prevent errors occurring during data analysis it is essential that both the data and models are valid.

The consequence of an error in the data analysis or with the analysis method could result in harm to the individual. There are many sources of error such as the sample chosen, which may not be representative of the entire population. Other problems arise with knowledge acquisition by machine learning, where the learning algorithm has used incomplete training data for pattern (or other knowledge) recognition. Training data may also be incomplete if the future population differs from the past population.

The data collection needs to decide on the data and attributes to be collected, and often the attributes chosen are limited to what is available, and the data scientist will also need to decide what to do with missing attributes. Often errors arise in data processing tasks such as analysing text information or recognizing faces

---

[1] Big data involves combining data from lots of sources such as bar codes, cctv, shopping data, driver's license, and so on.

Table 24.1 Some reasons for data collection

Question	Answers
Who owns the data?	A user's personal information may legally belong to the data collector, but the data subject may have some control as the data is about him/her – The author of the biography of an individual owns the copyright not the individual – The photographer of a (legally taken) photo owns the image not the subject – Recording of audio/video is similar – May be a need to acknowledge copyright (if applicable) – May be limits in rights as to how data is collected and used (e.g., privacy of phone calls) – The data subject may have some control of the data collected
What are the expected responsibilities of the collector	The collector of the data is expected to: – Collect only required data – Collect legal and ethical data only – Preserve confidentiality/integrity of collected personal data – Not misuse the data (e.g., alter image) – Use data only for purpose gathered – Share data only with user consent
What is the purpose of the data collection?	The purpose may be to: – Carry out service for a user – Improve user experience – Understand users – Build up profile of user behaviour – Exploit user data for commercial purposes
How is user consent to data collection given?	User consent may be given in various ways – User informed of purpose of data collection – User consents to use of data – May be hidden in terms and conditions of site
User control	This refers to the ability of the user to control the way that their personal data is being collected/used: – Ability of user to modify their personal data – Ability of user to delete their personal data

from photos. There may be human errors in the data (e.g., spelling errors or where the data field was misunderstood), and errors may lead to poor results and possible harm to the user. The problem with such errors is that often decisions are made on the basis of public and private data, and often individuals are unaware as to what data was collected and whether there is a method to correct it.

Even with perfect data the conclusions from the analysis may be invalid due to errors in the model, and there are many ways in which the model may be incorrect.

Many machine-learning algorithms just estimate parameters to fit a pre-determined model, without knowing whether the model is appropriate or not (e.g., the model may be attempting to fit a linear model to a non-linear reality). This becomes problematic when estimating (or extrapolating) values outside of the given data unless there is confidence in the correctness of the model.

Further, care is required before assigning results to an individual from an analysis of group data, as there may be other explanations (e.g., Simpson's paradox in probability/statistics is where a trend that appears in several groups of data disappears or reverses when these groups are combined). It is important to think about the population that you are studying, and to make sure that you are collecting data on the right population, and whether to segment it into population groups, as well as how best to do the segmentation.

It may seem reasonable to assume that data-driven analysis is fair and neutral, but unfortunately the problem is that humans may unintentionally introduce bias, as they set the boundary conditions. The bias may be through their choice of the model, the use of training data that may not be representative of the population, or the past population may not be representative of the future population, and so on. This may potentially lead to algorithmic decisions that are unfair (e.g., the case of the Amazon hiring algorithm that was biased towards the hiring of males), and so the question is how to be confident that the algorithms are fair and unbiased. Data scientists have a responsibility to ensure that the algorithm is fit for purpose and uses the right training data, and as far as practical to detect and eliminate unintentional discrimination (individual or target group).

Another problem that may arise is data that is correct but presented in a misleading way. One simple way to do this is to manipulate the scales of the axis to present the results visually in an exaggerated way. Another example is where the results are correct, but presented in an over simplistic manner (e.g., there may be two or more groups in the population with distinct behaviour where one group is the dominant), where the correct aggregate outcome is presented but this is misleading due to the differences within the population, and by suppressing diversity there may be discrimination against the minority group. In other words, the algorithm may use criteria tuned to fit the majority and may be unfair to minorities.

Exploration is the first phase in data analysis, and a hypothesis may be devised to fit the observed data (this is the opposite of traditional approaches where the starting point is the hypothesis, and the data is used to confirm or reject the hypothesis based on the data from the control and target groups, and so this approach needs to be used carefully to ensure the validity of the results).

24.2.2 Data Science and Society

Data science has consequences for society with one problem being that algorithms tend to learn and codify the current state of the world, and it is therefore harder to change the algorithm to reflect the reality of a changing world. The impact of innovative technologies affects the different cohorts and social groups in society in different ways, and there may also be differences between how different groups view privacy. Data scientists tend to be focused on getting the algorithm to perform correctly to do the right processing, and so often may not consider the wide societal impacts of the technology.

Algorithms may be unfair to individuals in that an individual may be classified as being a member of a group in view of the value of a particular attribute, and so the individual could be typecast due to their perceived membership of the group. Another words, the individual may be assigned opinions or properties of the group, and this means that there is a danger of developing a stereotype view of the individual. Further, it may be difficult for individuals to break out of these stereotypes, as these biases become embedded within the algorithm thereby helping to maintain the status quo.

There are further dangers when predications are made, as predictions are probabilistic and may be wrong, and only suggest a greater likelihood of occurrence of an event. Predictive techniques have been applied to predictive policing and to the prediction of uprisings, but there are dangers of false positives and false negatives (see type I and type II errors in probability/statistics in Chap. 9).

It is important that the societal consequences of algorithms are fully considered by companies, in order to ensure that the benefits of data science are achieved, and harm to individuals is avoided.

24.3 What Is Data Analytics?

Data analytics is the science of handling data collection by computer-driven systems, where the goal is to generate insights that will improve decision making. It involves the overlap of several disciplines such as statistics, information technology, and domain knowledge. It is widely used in social media, e-commerce, the Internet of Things, recommendation engines, gaming, and may potentially be applied to other fields such as information security, logistics, and so on.

Data analytics involves the analysis of data to create structure, order, meaning, and patterns from the data. It uses the collected data to produce information as well as generating insights from the data for decision makers. This is essential in making informed decisions to meet current and future business needs. Data analytics may involve machine learning, or it could be quick and simple if the data set is ready, and the goal is to perform just simple descriptive analysis. There are four types of data analytics (Table 24.2).

Descriptive analysis is a data analysis method that is used to give summary of what is going on and nothing more. It provides information as to what happened, and it allows the data collected by the system to be used to identify what went wrong. This type of data is often used to summarize large data sets, and to describe a summary of the outcomes to the stakeholders. The most relevant metrics produced include the key performance indicators (KPIs).

Table 24.2 Types of data analytics

Type	Description
Descriptive	These metrics describe what happened in the past and gives a summary of what is going on
Diagnostic	These are concerned with why it happened and involve analysis to determine why something has happened
Predictive	These are concerned with what is likely to happen in the future
Prescriptive	These are concerned with analysis to make better decisions, and it may involve considering several factors to suggest a course of action for the business. It may involve the use of AI techniques such as machine learning, and the goal is to make progress and avoid problems in the future

Diagnostic analysis is concerned with the analysis of the descriptive metrics to solve problems, and to identify what the issue could potentially be, and to understand why something has happened.

Predictive analysis involves predicting what is likely to happen in the future based on data from the past, i.e., it is attempting to predict the future based on actions in the past, and it may involve the use of statistics and modeling to predict future performance, based on current and historical data. Other techniques employed include neural networks, regression, and decision trees.

Prescriptive analysis is used to help business to make better decisions through the analysis of data and is effective when the organization knows the right questions to ask and responds appropriately to the answers. It often uses AI techniques such as machine learning to process a vast amount of data, to find patterns, and to recommend a course of action that will resolve or improve the situation. The recommended course of action is based on past events and outcomes, and the use of machine learning strategies builds upon the predictive analysis of what is likely to happen to recommend a future course of action.

Prescriptive analytics may be used to automate prices based on several factors such as demand, weather, and commodity prices. These algorithms may automatically raise or lower prices at a much faster rate than human intervention.

Companies may use data analytics to create and sell useful products by drilling down into customer data to determine what they are looking for. This includes understanding the features desired of the product and the price that they are willing to pay, and so data analytics has a role to play in new product design. They may be used by the business to improve customer loyalty and retention, and this may be done by gathering data (e.g., the opinions of customers from social media, email, and phone calls) to ensure that the voice of the customer is heard and acted upon appropriately.

Marketing groups often use data analytics to determine how successful their marketing campaign has been and to make changes where required. The marketing team may use the analytics to run targeted marketing and advertisement campaigns to segmented audiences (i.e., subsets of the population based on their unique characteristics such as demographics, interests, needs, and location). Market segmentation is useful in getting to know the customers, and determining what is needed in their market segment, and to determine how best to meet their needs.

Big data analytics may be used for targeted advertisements. For example, Netflix collects data on its customers including their searches and viewing history, and this data provides an insight into the specific interests of the customer, which is then used to send suggestions to the customer on the next movie that they should watch.

Big data analytics involves examining large amounts of data to identify the hidden patterns and correlations, and to give insights to enable the right business decisions to be made. Big data analytics is often done with sophisticated software systems that provide fast analytic procedures, where the use of big data allows the business to identify patterns and trends. It enables the business to collect as much data it requires to understand the customers and to derive critical insights to maintain customers.

24.3.1 Business Analytics and Business Intelligence

Business analytics involves converting business data into useful business information through the use of statistical techniques and advanced software. It includes a set of analytical methods for solving problems and assisting decision making, especially in the context of vast quantities of data. The combination of analysis with intuition allows useful insights into business organizations to be provided and helps them to achieve their objectives. Many organizations use the principles and practice of business analytics.

Business intelligence (BI) processes all the data generated by a business, and uses it to generate clear reports (e.g., a dashboard report of the key metrics), as well as the key trends and performance measures that are used by management in decision making. That is, business intelligence is data analytics with insight that allows managers to make informed decisions, and so it is focused on the decision-making part of the process. It may employ data mining, performance benchmarking, process analysis, and descriptive analytics. That is, business analytics allows management issues to be explored and solved.

The effectiveness of management decision making is influenced by the accuracy and completeness of the information that managers have, with inaccurate or incomplete information leading to poorer decisions. Companies often have data that is unstructured or in diverse formats, and such data is generally more difficult to gather and analyse. This has led software firms to offer business intelligence solutions to organizations that wish to make better use of their data, and to optimize the information gathered from the data. There are several software applications designed to unify a company's data and analytics.

24.3.2 Big Data and Data Mining

The term *"Big data"* refers to the large, diverse sets of data that arrives at ever-increasing rates and volumes. It encompasses the volume of data, the velocity or speed at which it is created and collected, and the variety or scope of the data points being covered (these are generally referred to as the three V's of big data). There has been an explosion in the volume of big data with approximately 40 zettabytes[2] (ZB) of data employed globally.

Big data often comes from *data mining*, where data mining involves exploring and analysing large blocks of data to gather meaningful patterns and trends. Data is gathered and loaded into data warehouses by organizations (i.e., the data is centralized into a single database or program), and then stored either on in-house servers or on the cloud. The user decides how to organize the data, and application software sorts the data accordingly, and the data is presented in an easy-to-read format such as a graph or report.

The data may be internal or external. It may be structured or unstructured, where structured data is often already managed in the organization's databases or spreadsheets and may be numeric and easily formatted. Unstructured data uses data that may be unformatted, and so it does not fall into a predetermined format (i.e., it is free form), and it may come from search engines or from forum discussions on social media.

Big data may be collected in various ways such as from publicly shared comments on social media, or gathered from personal electronics or apps, through questionnaires, product purchases, and so on. Big data is generally stored in databases and is analysed with software that is designed to handle large and complex data sets (usually software as a service, SaaS).

24.3.3 Data Analytics for Social Media

Data analytics provides a quantitative insight into human behaviour on a social media website and is a way to understand users and how to communicate with them better. It enables the business to understand its audience better, to improve the user experience, and to create content that will be of interest to them. Data analytics consist of a collection of data that says something about the social media conversation, and it involves the collection, monitoring, analysis, summarization, and a graph to visualize the behaviour of users.

Another words, *data analytics* involves learning to read a social media community through data, and the interpretations of the quantifiable data (or metrics) gives information on the activities, events, and conversations. This includes what users like when they are online, but other important information such as their

---

[2] A zettabyte is 1 sextillion bytes $= 2^{70}$ bytes (approximately a billion terabytes or 1000 exabytes or a trillion gigabytes).

opinions and emotions need to be gathered through *social listening*. Social listening involves monitoring keywords and mentions in social media conversations in the target audience and industry, to understand and analyse what the audience is saying about the business and allows the business to engage with its audience.

Social media companies use data analytics to gain an insight into customers, and elementary data such as the number of likes, the number of followers, the number of times a video is played on YouTube, and so on are gathered to obtained a quantified understanding of a conversation. This data is valuable in judging the effectiveness of a social media campaign, where the focus is to determine how effective the campaign has been in meeting its goals. The goals may be to increase the number of users or to build a brand, and data analytics combined with social listening help in understanding how people are interacting, as well as what they are interacting about and how successful the interactions has been.

Facebook and Twitter maintain a comprehensive set of measurements for data analytics, with Facebook maintaining several metrics such as the number of page views and the number of likes and reach of posts (i.e., the number of people who saw posts at least once). Twitter includes a dashboard view to summarize how successful tweet activity has been, as well as the interests and locations of the user's followers. Social listening considers user opinions, emotions, views, evaluations, and attitude, and social media data contains rich collection of human emotions.

The design of a social media campaign is often an iterative process, with the first step being to determine the objective of the campaign and designing the campaign to meet the requirements. The effectiveness of a campaign is judged by a combination of social media analytics and social listening, with the campaign refined appropriately to meet its goals and the cycle repeating. The key performance indicators (KPI) may include increased followers/subscribers or an increase in the content shared, and so on.

24.3.4 Sources of Data

The collected data is commercially valuable, especially when data about individuals are linked from several sources. *Data brokers* are companies that aggregate and link information from multiple sources to create more complete and valuable information products (i.e., profiles of individuals) that may then be sold on to interested parties. Meta data (i.e., data about the data such as the time of a phone call or who the call is made to) also provides useful information that may be collected and shared (Table 24.3).

For example, suppose the probability of an individual buying a pair of hiking books is very low (say 1 in 5000 probability). Next, that individual starts scanning a website (say Amazon) for boots then that individual is now viewed as being more likely to buy a pair of hiking boots (say a 1 in 100 probability). This large increase in probability will mean that the individual is now of interest to advertisers and sellers, and various targeted (popup) advertisements will appear to the individual

Table 24.3 Sources of data

Source	Answers
Data collected by merchants and service providers	This includes personal data entered for the purchase of products and services such as name, address, date of birth, products and services purchased, etc.
Activity tracking	This involves monitoring the user's activity on the site (or app), and recording the user's searches, and the products browsed and purchased It may involve recording the user's interests, their activities, and their interactions and communications with others on the site
Search profile	The history of a person's searches over a period of time on a search engine such as Google reveals information about the individual and their interests
Sensors from devices	There are many sensors in the world around us such as personal devices as part of the Internet of Things that may record information such as health data or what the individual is eating. Third-party devices such as security cameras may be conducting public or private surveillance. GPS technology on smart phones may be tracking the user's location

advertising different hiking boots. This may become quite tedious and annoying to the individual, who may have been just browsing, and is now subject to an invasion of advertisements, but many apps are free and often the source of their revenue is from advertisements, and so they gather data about the user that is then sold on to advertisers.

Users should be in control of how their data is used, and most user agreements are "all-or-nothing" in the sense that a user must give up control of their data to use the application, and so essentially the user has no control. That is, a user must click acceptance of the terms and conditions in order to use the services of a web application. Clearly, users would be happier and feel that they are in control if they were offered graduated choices by the vendor, to allow them to make tradeoffs, and to choose a level of privacy that they are comfortable with.

24.4 Mathematics Used in Data Science

Mathematics is employed in data science and analytics and includes areas such as (Table 24.4).

Other areas of mathematics that may arise in data analytics include graph theory (see Chap. 7), operations research (see Chap. 31), and discrete mathematics (see [1]).

Table 24.4 Mathematics in data analytics

Type	Description
Probability	An introduction to some of the concepts in probability theory such as basic probability, expectation, conditional probability, Bayes' theorem, and probability density functions was discussed in Chap. 23
Statistics	Statistics is a vast area and an introduction to some of the important concepts in the field including descriptive statistics; measures of central tendency such as the mean, mode and median, variance and covariance, and correlation was discussed in Chap. 22
Linear algebra	This includes topics such as matrix theory and Gaussian elimination as discussed in Chap. 27, as well as basic algebra as discussed in Chap. 5
Calculus	This includes the study of differentiation and integration and includes topics such as limits, continuity, rules of differentiation, Taylor's series, and area and volume as discussed in Chaps. 25 and 26

24.5 Review Questions

1. What is data science?
2. What is the role of the data scientist?
3. What are the main sources of personal data collected on line?
4. What are the main risks to an individual using social media?
6. What mathematics are employed in data science?
7. What is data analytics?

24.6 Summary

Companies collect lots of personal data about individuals from their use of computer resources such as email, search engines, their Internet, and social media use, and the data is processed to build up revealing profiles of the user that may be targeted to advertisers. Modern technology allows mass surveillance to be conducted by governments on its citizens, with face recognition software allowing citizens to be recognized at demonstrations or other mass assemblies.

Modern technology allows the location of the user to be tracked, and privacy is important in the information age, and it is the way in which we separate ourselves from other people, and is the right to be left alone. The European GDPR law has become an important protector of privacy and personal data, and both European and other countries have adapted it.

Data analytics is the science of handling data collection by computer-driven systems, where the goal is to generate insights that will improve decision making. It involves the overlap of several disciplines such as statistics, information technology, and domain knowledge. It is widely used in social media, e-commerce, the Internet of Things, recommendation engines, gaming, and may potentially be applied to other fields such as information security, logistics, and so on.

Reference

1. O'Regan G (2021) Guide to discrete mathematics, 2nd edn. Springer, Berlin

Calculus I

25

Key Topics

Limit of a function

Continuity

Mean value theorem

Taylor's theorem

Differentiation

Maxima and minima

Integration

25.1 Introduction

Newton and Leibniz independently developed calculus in the late seventeenth century.[1] Calculus plays a key role in describing how rapidly things change, and it may be employed to determine the velocity and acceleration of moving bodies as well as calculating the area of a region under a curve or between two curves. It may be used to determine the volumes of solids, computing the length of a curve,

---

[1] The question of who invented the calculus led to a bitter controversy between Newton and Leibniz with the latter accused of plagiarising Newton's work. Newton, an English mathematician and physicist was the giant of the late seventeenth century, and Leibnitz was a German mathematician and philosopher. Today, both Newton and Leibniz are credited with the independent development of the calculus.

© The Author(s), under exclusive license to Springer Nature Switzerland AG 2023
G. O'Regan, *Mathematical Foundations of Software Engineering*,
Texts in Computer Science, https://doi.org/10.1007/978-3-031-26212-8_25

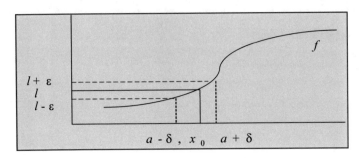

Fig. 25.1 Limit of a function

and in finding the tangent to a curve. It is an important branch of mathematics concerned with limits, continuity, derivatives, and integrals of functions.

The concept of a *limit* is fundamental in the calculus. Let f be a function defined on the set of real numbers, then the limit of f at a is l (written as $\lim_{x \to a} f(x) = l$) if given any real number $\varepsilon > 0$ then there exists a real number $\delta > 0$ such that $|f(x) - l| < \varepsilon$ whenever $|x - a| < \delta$. The idea of a limit can be seen in Fig. 25.1.

The function f defined on the real numbers is *continuous* at a if $\lim_{x \to a} f(x) = f(a)$. The set of all continuous functions on the closed interval $[a, b]$ is denoted by $C[a, b]$.

If f is a function defined on an open interval containing x_0 then f is said to be *differentiable* at x_0 if the limit

$$\lim_{x \to x_0} \frac{f(x) - f(x_0)}{x - x_0}$$

exists. Whenever this limit exists it is denoted by $f'(x_0)$ and is called the *derivative* of f at x_0. Differential calculus is concerned with the properties of the derivative of a function. The derivative of f at x_0 is the slope of the tangent line to the graph of f at $(x_0, f(x_0))$ (Fig. 25.2).

It is easy to see that if a function f is differentiable at x_0 then f is continuous at x_0.

Fig. 25.2 Derivative as a tangent to curve

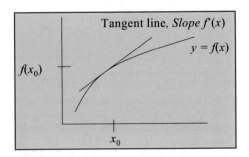

Rolle's Theorem

Suppose $f \in C[a, b]$ and f is differentiable on (a, b). If $f(a) = f(b)$ then there exists c such that $a < c < b$ with $f'(c) = 0$.

Mean Value Theorem

Suppose $f \in C[a, b]$ and f is differentiable on (a, b). Then there exists c such that $a < c < b$ with

$$f'(c) = \frac{f(b) - f(a)}{b - a}$$

Proof The mean value theorem is a special case of Rolle's theorem, and the proof involves defining the function $g(x) = f(x) - rx$ where $r = (f(b) - f(a))/(b - a)$.

It is easy to verify that $g(a) = g(b)$. Clearly, g is differentiable on (a, b) and so by Rolle's theorem there is a c in (a, b) such that $g'(c) = 0$. Therefore, $f'(c) - r = 0$ and so $f'(c) = r = f(b) - f(a)/(b - a)$ as required.

Interpretation of the Mean Value Theorem

The mean value theorem essentially states that there is at least one point c on the curve $f'(x)$ between a and b such that slope of the chord is the same as the tangent $f(c)$ (Fig. 25.3).

Intermediate Value Theorem

Suppose $f \in C[a, b]$ and K is any real number between $f(a)$ and $f(b)$. Then there exists c in (a, b) for which $f(c) = K$.

Fig. 25.3 Interpretation of mean value theorem

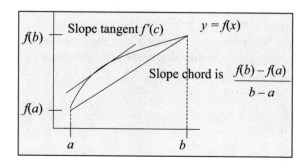

Fig. 25.4 Interpretation of intermediate value theorem

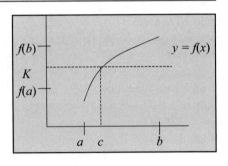

Proof The proof of this relies on the completeness property of the real numbers. It involves considering the set S in $[a, b]$ such that $f(x) \leq K$ and noting that this set is non-empty since $a \in S$ and bounded above by b. Therefore the supremum[2] $\sup S = c$ exists, and it is straightforward to show (using ε and δ arguments and the fact that f is continuous) that $f(c) = K$ (Fig. 25.4).

L'Hôpital's Rule

Suppose that $f(a) = g(a) = 0$ and that $f'(a)$ and $g'(a)$ exist and that $g'a) \neq 0$. Then L'Hopital's rule states that:

$$\lim_{x \to a} \frac{f(x)}{g(x)} = \frac{f'(a)}{g'(a)}$$

Proof

$$\lim_{x \to a} \frac{f(x)}{g(x)} = \lim_{x \to a} \frac{f(x) - f(a)}{g(x) - g(a)}$$

$$= \lim_{x \to a} \frac{\frac{f(x) - f(a)}{x - a}}{\frac{g(x) - g(a)}{x - a}}$$

$$= \frac{\lim_{x \to a} \frac{f(x) - f(a)}{x - a}}{\lim_{x \to a} \frac{g(x) - g(a)}{x - a}}$$

$$= \frac{f'(a)}{g'(a)}$$

Taylor's Theorem

The Taylor series is concerned with the approximation to values of the function f near x_0. The approximation employs a polynomial (or power series) in powers of $(x - x_0)$ as well as the derivatives of f at $x = x_0$. There is an error term (or remainder) associated with the approximation.

---

[2] The supremum is the least upper bound and the infimum is the greatest lower bound.

Suppose $f \in C^n[a, b]$ and f^{n+1} exists on (a, b). Let $x_0 \in (a, b)$ then for every $x \in (a, b)$ there exists $\xi(x)$ between x_0 and x with

$$f(x) = P_n(x) + R_n(x)$$

where $P_n(x)$ is the nth Taylor polynomial for f about x_0 and $R_n(x)$ is the called the remainder term associated with $P_n(x)$. The infinite series obtained by taking the limit of $P_n(x)$ as $n \to \infty$ is termed the Taylor series for f about x_0.

$$P_n(x) = f(x_0) + f'(x_0)(x - x_0) + \frac{f''(x_0)}{2!}(x - x_0)^2 + \cdots + \frac{f^n(x_0)}{n!}(x - x_0)^n$$

The remainder term is given by:

$$R_n(x) = \frac{f^{n+1}(\xi(x))(x - x_0)^{n+1}}{(n + 1)!}$$

25.2 Differentiation

Mathematicians of the seventeenth century were working on various problems concerned with motion. These included problems such as determining the motion or velocity of objects on or near the earth, as well as the motion of the planets. They were also interested in changes of motion, i.e., in the acceleration of these moving bodies.

Velocity is the rate at which distance changes with respect to time, and the average speed during a journey is the distance travelled divided by the elapsed time. However, since the speed of an object may be variable over a period of time, there is a need to be able to determine its velocity at a specific time instance. That is, there is a need to determine the rate of change of distance with respect to time at any time instant.

The direction in which an object is moving at any instant of its flight was also studied. For example, the direction in which a projectile is fired determines the horizontal and vertical components of its velocity. The direction in which an object is moving can vary from one instant to another.

The problem of finding the maximum and minimum values of a function was also studied, e.g., the problem of determining the maximum height that a bullet reaches when it is fired. Other problems studied include problems to determine the lengths of paths, the areas of figures, and the volume of objects.

Newton and Leibnitz (Figs. 25.5 and 25.6) showed that these problems could be solved by means of the concept of the derivative of a function, i.e., the rate of change of one variable with respect to another.

Fig. 25.5 Isaac Newton

Fig. 25.6 Wilhelm Gottfried
Leibniz

Rate of Change

The average rate of change and instantaneous rate of change are of practical interest.
For example, if a motorist drives 200 miles in four hours, then the average speed is
50 miles per hour, i.e., the distance travelled divided by the elapsed time. The actual
speed during the journey may vary as if the driver stops for lunch, then the actual
speed is zero for the duration of lunch.

The actual speed is the instantaneous rate of change of distance with respect to
time. This has practical implications as motorists are required to observe speed limits,
and a speed camera may record the actual speed of a vehicle with the driver subject
to a fine if the permitted speed limit has been exceeded. The actual speed is relevant
in a car crash as speed is a major factor in road fatalities.

In calculus, the term Δx means a change in x and Δy means the corresponding
change in y. The derivative of f at x is the instantaneous rate of change of f, and f is
said to be differentiable at x. It is defined as:

$$\frac{dy}{dx} = \lim_{\Delta x \to 0} \frac{\Delta y}{\Delta x} = \lim_{\Delta x \to 0} \frac{f(x + \Delta x) - f(x)}{\Delta x}$$

In the formula, Δy is the increment $f(x + \Delta x) - f(x)$

The average velocity of a body moving along a line in the time interval t to $t + \Delta t$ where the body moves from position $s = f(t)$ to position $s + \Delta s$ is given by:

$$V_{av} = \frac{\text{displacement}}{\text{Time travelled}} = \frac{\Delta s}{\Delta t} = \frac{f(t + \Delta t) - f(t)}{\Delta t}$$

The instantaneous velocity of a body moving along a line is the derivative of its position $s = f(t)$ with respect to t. It is given by:

$$v = \frac{ds}{dt} = \lim_{\Delta t \to 0} \frac{\Delta s}{\Delta t} = f'(t)$$

25.2.1 Rules of Differentiation

Table 25.1 presents several rules of differentiation.

Derivatives of Well-Known Functions
The following are the derivatives of some well-known functions including basic trigonometric functions, the exponential function, and the natural logarithm function.

(i) $d/dx \operatorname{Sin} x = \operatorname{Cos} x$

Table 25.1 Rules of differentiation

No.	Rule	Definition
1	Constant	The derivative of a constant is 0. That is, for $y = f(x) = c$ (a constant value) we have $dy/dx = 0$
2	Sum	$d/dx\,(f + g) = df/dx + dg/dx$
3	Power	The derivative of $y = f(x) = x^n$ is given by $dy/dx = nx^{n-1}$
4	Scalar	If c is a constant and u is a differentiable function of x then $dy/dx = c\,d\,u/dx$ where $y = cu(x)$
5	Product	The product of two differentiable functions u and v is differentiable and $\frac{d}{dx}(uv) = v\frac{du}{dx} + u\frac{dv}{dx}$
6	Quotient	The quotient of two differentiable functions u, v is differentiable (where $v \neq 0$) and $\frac{d}{dx}\left[\frac{u}{v}\right] = \frac{v\frac{du}{dx} - u\frac{dv}{dx}}{v^2}$
7	Chain rule	*Chain Rule.* Suppose $h = g \circ f$ is the composite of two differentiable functions $y = g(x)$ and $x = f(t)$. Then h is a differentiable function of t whose derivative at each value of t is: $h'(t) = (g \circ f)'(t) = g'(f(t))f'(t)$ This may also be written as: $\frac{dy}{dt} = \frac{dy}{dx}\frac{dx}{dt}$

(ii) $d/dx \, \mathrm{Cos} x = -\mathrm{Sin} x$
(iii) $d/dx \, \mathrm{Tan} x = \mathrm{Sec}^2 x$
(iv) $d/dx \, e^x = e^x$
(v) $d/dx \, \ln x = 1/x$ (where $x > 0$)
(vi) $d/dx \, a^x = \ln(a) a^x$
(vii) $d/dx \, \log_a x = 1/x \ln(a)$
(viii) $d/dx \, \arcsin x = 1/\sqrt{(1 - x^2)}$
(ix) $d/dx \, \arccos x = -1/\sqrt{(1 - x^2)}$
(x) $d/dx \, \arctan x = 1/(1 + x^2)$

Increasing and Decreasing Functions

Suppose that a function f has a derivative at every point x of an interval I. Then

(i) f increases on I if $f'(x) > 0$ for all x in I.
(ii) f decreases on I if $f'(x) < 0$ for all x in I.

The geometric interpretation of the first derivative test is that it states that differentiable functions increase on intervals where their graphs have positive slopes and decrease on intervals where their graphs have negative slopes.

If f' changes from positive to negative values as x passes from left to right through point c then the value of f at c is a *local maximum* value of f. Similarly, if f' changes from negative to positive values as x passes from left to right through point c then the value of f at c is a *local minimum* value of f (Fig. 25.7).

The graph of a differentiable function $y = f(x)$ is concave down in an interval where f' decreases and concave up in an interval where f' increases. This may be defined by the second interval test for concavity. Other words the graph of $y = f(x)$ are concave down in an interval where $f'' < 0$ and concave up in an interval where $f'' > 0$.

A point on the curve where the concavity changes from concave up to concave down or vice versa is termed a point of inflection. That is, at a *point of inflection c* we have that f' is positive on one side and negative on the other side. At the point of inflection c we have the value of the second derivative is zero, i.e., $f''(c) = 0$, or in other words f' goes through a local maximum or minimum.

Fig. 25.7 Local minima and maxima

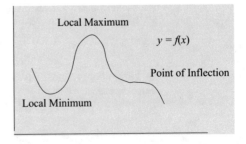

25.3 Integration

The derivative is a functional operator that takes a function as an argument and returns a function as a result. The inverse operation involves determining the original function from the known derivative, and integral calculus is the branch of the calculus concerned with this problem. The integral of a function consists of all those functions that have it as a derivative.

Integration is applicable to problems involving area and volume. It is the mathematical process that allows the area of a region with curved boundaries to be determined, and it also allows the volume of a solid to be determined.

The problem of finding functions whose derivatives is known involves finding a function $y = F(x)$ whose derivative is given by the differential equation:

$$\frac{dy}{dx} = f(x)$$

The solution to this differentiable equation over the interval I is F if F is differentiable at every point of I and for every x in I we have:

$$\frac{d}{dx} F(x) = f(x)$$

Clearly, if $F(x)$ is a particular solution to d/dx $F(x) = f(x)$ then the general solution is given by:

$$y = \int f(x)dx = F(x) + k$$

since

$\frac{d}{dx}(F(x) + k) = f(x) + 0 = f(x)$.

Rules of Integration
The following are rules of integration as well as the integrals of some well-known functions such as basic trigonometric functions and the exponential function. Table 25.2 presents several rules of integration.

It is easy to check that the integration has been carried out correctly. This is done by computing the derivative of the function obtained and checking that it is the same as the function to be integrated.

Often, the goal may be to determine a particular solution satisfying certain conditions rather than the general solution. The general solution is first determined, and then the constant k that satisfies the particular solution is calculated.

The *substitution method* is a useful method that is often employed in performing integration, and its effect is to potentially change an unfamiliar integral into one that is easier to evaluate. The procedure to evaluate $\int f(g(x))g'(x)dx$ where f', g' are continuous functions is as follows:

Table 25.2   Rules of integration

No.	Rule	Definition
1	Constant	$\int u'(x)dx = u(x) + k$
2	Sum	$\int (u(x) + v(x))dx = \int u(x)dx + \int v(x)dx$
3	Scalar	$\int au(x)dx = a\int u(x)dx$ (where a is a constant)
4	Power	$\int x^n dx = \frac{x^{n+1}}{n+1} + k$ (where $n \neq -1$)
5	Cos	$\int \cos x dx = \sin x + k$
6	Sin	$\int \sin x dx = -\cos x + k$
7	$\sec^2 x$	$\int \sec^2 x dx = \tan x + k$
8	e^x	$\int e^x dx = e^x + k$
9	Logarithm	$\int 1/x dx = \ln x + k$

1. Substitute $u = g(x)$ and $du = g'(x)dx$ to obtain $\int f(u)du$
2. Integrate with respect to u.
3. Replace u by $g(x)$ in the result.

The method of *integration by parts* is a rule of integration that transforms the integral of a product of functions into simpler integrals. It is a consequence of the product rule for differentiation.

$$\int udv = uv - \int vdu$$

$$\int f(x)g'(x)dx = f(x)g(x) - \int f'(x)g(x)dx$$

25.3.1 Definite Integrals

A definite integral defines the area under the curve $y = f(x)$, and the area of the region between the graph of a non-negative continuous function $y = f(x)$ for the interval $a \leq x \leq b$ of the x-axis is given by the definite integral.

The sum of the area of the rectangles approximates the area under the curve and the more rectangles that are used the better the approximation (Fig. 25.8).

The definition of the area of the region beneath the graph of $y = f(x)$ from a to b is defined to be the limit of the sum of the rectangle areas as the width of the rectangles become smaller and smaller, and the number of rectangles used approaches infinity. The limit of the sum of the rectangle areas exists for any continuous function.

The approximation of the area under the graph $y = f(x)$ between $x = a$ and $x = b$ is done by dividing the region into n strips with each strip of uniform width given by $\Delta x = (b - a)/n$ and drawing lines perpendicular to the x-axis (Fig. 25.9). Each

Fig. 25.8 Area under the curve

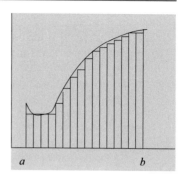

strip is approximated with an inscribed rectangle where the base of the rectangle is on the x-axis to the to the lowest point on the curve above (lower Riemann sum). We let c_k be a point in which f takes on its minimum value in the interval from x_{k-1} to x_k and the height of the rectangle is $f(c_k)$. The sum of these areas is the approximation of the area under the curve and is given by:

$$S_n = f(c_1)\Delta x + f(c_2)\Delta x + \cdots + f(c_n)\Delta x$$

The area under the graph of a nonnegative continuous function f over the interval $[a, b]$ is the limit of the sums of the areas of inscribed rectangles of equal base length as n approaches infinity.

$$\begin{aligned} A &= \lim_{n\to\infty} S_n \\ &= \lim_{n\to\infty} f(c_1)\Delta x + f(c_2)\Delta x + \cdots + f(c_n)\Delta x \\ &= \lim_{n\to\infty} \sum_{k=1}^{n} f(c_k)\Delta x \end{aligned}$$

It is not essential that the division of $[a, b]$ into $a, x_1, x_2, \ldots x_{n-1}, b$ gives equal subintervals $\Delta x_1 = x_1 - a$, $\Delta x_2 = x_2 - x_1$, $\ldots \Delta x_n = b - x_{n-1}$. The *norm* of the subdivision is the largest interval length.

Fig. 25.9 Area under the curves—lower sum

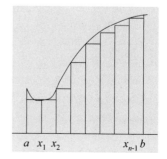

The lower Riemann sum L and the upper sum U may be formed, and the more finely divided that $[a, b]$ is the closer the values of the lower and upper sum U and L. The upper and lower sums may be written as:

$$L = \min_{1} \Delta x_1 + \min_{2} \Delta x_2 + \cdots + \min_{n} \Delta x_n$$

$$U = \max_{1} \Delta x_1 + \max_{2} \Delta x_2 + \cdots + \max_{n} \Delta x_n$$

$$\lim_{\text{norm} x \to 0} U - L = 0 \text{ (i.e., } \lim_{\text{norm} \to 0} U = \lim_{\text{norm} \to 0} L \text{)}$$

Further, if $S = \Sigma f(c_k) \Delta x_k$ (where c_k is any point in the subinterval and $\min_k \le f(c_k) \le \max_k$) we have:

$$\lim_{\text{norm} \to 0} L = \lim_{\text{norm} \to 0} S = \lim_{\text{norm} \to 0} U$$

Integral Existence Theorem (Riemann Integral)
If f is continuous on $[a, b]$ then.

$$\int_{a}^{b} f(x) dx = \lim_{\text{norm} \to 0} \sum f(c_x) \Delta x_k$$

exists and is the same number for any choice of the numbers c_k.

Properties of Definite Integrals
Table 25.3 presents some algebraic properties of the definite integral.

Table 25.3 Properties of definite integral

Properties of definite integral
(i) $\int_a^a f(x)dx = 0$
(ii) $\int_b^a f(x)dx = -\int_a^b f(x)dx$
(iii) $\int_a^b kf(x)dx = k \int_a^b f(x)dx$
(iv) $\int_a^b f(x)dx \ge 0$ if $f(x) \ge 0$ on $[a, b]$
(v) $\int_a^b f(x)dx \le \int_a^b g(x)dx$ if $f(x) \le g(x)$ on $[a, b]$
(vi) $\int_a^b f(x)dx + \int_b^c f(x)dx = \int_a^c f(x)dx$
(vii) $\int_a^b \{f(x) + g(x)\}dx = \int_a^b f(x)dx + \int_a^b g(x)dx$
(viii) $\int_a^b \{f(x) - g(x)\}dx = \int_a^b f(x)dx - \int_a^b g(x)dx$

Table 25.4 Fundamental theorems of integral calculus

Theorem
First fundamental theorem: (existence of anti-derivative) If f is continuous on $[a, b]$ then $F(x)$ is differentiable at every point x in $[a, b]$ where $F(x)$ is given by: $F(x) = \int_a^x f(t)at$ If f is continuous on $[a, b]$ then there exists a function $F(x)$ whose derivative on $[a, b]$ is f $\frac{dF}{dx} = \frac{d}{dx} \int_a^x f(t)dt = f(x)$
Second fundamental theorem: (integral evaluation theorem) If f is continuous on $[a, b]$ and F is any anti-derivative of f on $[a, b]$ then: $\int_a^b f(x)dx = F(b) - F(a)$

25.3.2 Fundamental Theorems of Integral Calculus

Table 25.4 presents two fundamental theorems of integral calculus.

That is, the procedure to calculate the definite integral of f over $[a, b]$ involves just two steps:

(i) Find an antiderivative F of f
(ii) Calculate $F(b) - F(a)$

For a more detailed account of integral and differential calculus the reader is referred to Finney [1].

25.4   Review Questions

1. Explain the concept of the limit of a function.
2. Explain the concept of continuity.
3. Explain the difference between average velocity and instantaneous velocity, and explain the concept of the derivative of a function.
4. Determine the following
   a. $\lim_{x \to 0} \text{Sin } x$
   b. $\lim_{x \to 0} x \text{ Cos } x$
   c. $\lim_{x \to -\infty} |x|$
5. Determine the derivative of the following functions
   a. $y = x^3 + 2x + 1$
   b. $y = x^2 + 1, x = (t + 1)^2$
   c. $y = \text{Cos } x^2$
6. Determine the integral of the following functions
   a. $\int (x^2 - 6x) \, dx$

b. $\int \sqrt{(x-6)}\, dx$
c. $\int (x^2 - 4)^2\, 3x^3 dx$
7. Explain the significance of the fundamental theorems of the calculus.

25.5   Summary

This chapter provided a brief introduction to the calculus including limits, continuity, differentiation, and integration. Newton and Leibniz developed the calculus independently in the late seventeenth century. It plays a key role in describing how rapidly things change and may be employed to calculate areas of regions under curves, volumes of figures, and in finding tangents to curves.

In calculus, the term Δx means a small change in x and Δy means the corresponding change in y. The derivative of f at x is the instantaneous rate of change of f, and is defined as

$$\frac{dy}{dx} = \lim_{\Delta x \to 0} \frac{\Delta y}{\Delta x} = \lim_{\Delta x \to 0} \frac{f(x + \Delta x) - f(x)}{\Delta x}$$

Integration is the inverse operation of differentiation and involves determining the original function from the known derivative. The integral of a function consists of all those functions that have the function as a derivative.

Integration is applicable to problems involving area and volume, and it allows the area of a region with curved boundaries to be determined.

Reference

1. Finney T (1988) Calculus and analytic geometry, 7th edn. Addison Wesleys, Boston

Calculus II

26

Key Topics

Applications of calculus

Velocity and acceleration

Area and volume

Length of curve

Trapezoidal rule

Simpsons rule

Fourier series

Laplace transforms

Differential equations

26.1 Introduction

This chapter considers several applications of calculus including the use of differentiation to deal with problems involving the rate of change, and we show how velocity, speed, and acceleration may be determined, as well as solving maxima/minima problems. We show how integration may be used to solve problems involving area, volume, and length of a curve.

© The Author(s), under exclusive license to Springer Nature Switzerland AG 2023
G. O'Regan, *Mathematical Foundations of Software Engineering*,
Texts in Computer Science, https://doi.org/10.1007/978-3-031-26212-8_26

The definite integral may be used to determine the area under a curve as well as computing the area bounded by two curves. We show how the volume of a solid of known cross-sectional area may be determined and show how to compute the volume of a solid generated by rotating a region around the x- or y-axis.

We show how the length of a curve may be determined, and we present the formulae for the Trapezoidal Rule and Simpson's Rule which are used to approximate definite integrals.

Finally, we introduce Fourier series, Laplace transforms, and differential equations. A Fourier series consists of the sum of a possibly infinite set of sine and cosine functions. The Laplace transform is an integral transform which takes a function f and transforms it to another function F by means of an integral. An equation that contains one or more derivatives of an unknown function is termed a differential equation, and many important problems in engineering and physics involve determining a solution to these equations.

26.2   Applications of Calculus

There are rich applications of the calculus in science and engineering, and we consider several applications of differentiation and integration. This includes a discussion of problems involving velocity and acceleration of moving objects, problems to determine the rate at which one variable changes from the rate at which another variable is known to change, and maxima and minima problems that are solved with differentiation. We then solve problems involving area and volume that are solved by integration.

Differentiation may be used to determine the speed, velocity, and acceleration of bodies. Velocity is the rate of change of position with respect to time, and acceleration is given by the rate of change of velocity with respect to time. The speed is the magnitude of the velocity. This is expressed mathematically by letting $s(t)$ be a function giving the position of an object at time t, and the velocity, speed, and acceleration are then given by:

$$\text{Velocity of object at time } t = y(t) = s'(t)$$
$$\text{Acceleration of object at time } t = a(t) = v'(t) = s''(t)$$
$$\text{Speed of object at time } t = |v(t)|$$

Example 26.1 A ball is dropped from a height of 64 ft (Imperial system), and its height above the ground after t seconds is given by the equation $s(t) = -16t^2 + 64$. Determine.

(a) The velocity when it hits the ground.
(b) The average velocity during its fall.

Solution

The ball hits the ground when $s(t) = 0$: i.e.,

$$0 = -16t^2 + 64$$
$$16t^2 = 64$$
$$t^2 = 4$$
$$t = 2$$

The velocity is given by $s'(t) = -32t$. and so the velocity when the ball hits the ground is equal to $-32 * 2 = -64$ ft/s.

The average velocity is given by distance travelled/time $= (s(2) - s(0))/2 - 0 = -64/2 = -32$ ft/s.

Occasionally, problems arise where one variable is known to change at a certain rate and the problem is to determine the rate of change on another variable. For example, how fast does the height of the water level drop when a cylindrical tank is drained at a certain rate.

Example 26.2 How fast does the water level in a cylindrical tank drop when water is removed at the rate of 3 L/s?

Solution

The radius of the tank is r (a constant) and the height is h and the volume of water V (which are both changing over time and so V and h are considered differentiable functions of time).

$$dV/dt = -3 \ (3 \text{ L/sec. is being removed})$$

We wish to determine $\frac{dh}{dt}$ and we can determine this using the formula for the volume of a cylinder.

$$V = \pi r^2 h$$

Then

$$dV/dt = \pi r^2 \, dh/dt = -3 \quad (\text{since } r \text{ is a constant})$$
$$dh/dt = -3/\pi r^2$$

That is, the water level is dropping at the constant rate of $3/\pi r^2$ L/s.

Maxima and minima problems refer to problems where the goal is to maximize or minimize a function on a particular interval, where the function is continuous and differentiable on the interval, and the function does not attain its maximum or minimum at the endpoints of the interval. Then we know that the maximum or minimum is at an interior point of the interval where the derivative is zero.

Example 26.3 Find two positive integers whose sum is 20 and whose product is as large as possible.

Solution
Let x be one of the numbers then the other number is $20 - x$ and so the product of both numbers is given by:

$$f(x) = x(20 - x) = 20x - x^2$$

The objective is to determine the value of x that will maximize the product (i.e., the value of $f(x)$ in the interval $0 \le x \le 20$). The function $f(x)$ is continuous and differentiable and attains a local maximum where its derivative is zero. The derivative is given by:

$$f'(x) = 20 - 2x$$

The derivative is 0 when $20 - 2x = 0$ or when $x = 10$, and the maximum value of $f(x)$ is $200 - 100 = 100$.

Problems involving area and volume may be solved with integration. The definite integral may be applied to problems to determine the area below the curve as may be seen in the following example:

Example 26.4 Find the area under the curve $x^2 - 4$ and the x-axis from $x = -2$ to $x = 2$.

Solution
The area under the curve $y = f(x)$ between $x = a$ and $x = b$ is given by:

$$A = \int_a^b f(x)dx$$

And so the area of the curve is $y = f(x)$ between $x = -2$ to $x = 2$ is given by:

$$\int_{-2}^{2} (x^2 - 4)dx$$

$$= \frac{x^3}{3} - 4x \Big|_{-2}^{2}$$

$$= \left[\frac{8}{3} - 8\right] - \left[\frac{-8}{3} + 8\right]$$

$$= -\frac{32}{3}$$

The area between two curves $y = f(x)$ and $y = g(x)$ where $f(x) \geq g(x)$ on the interval $[a, b]$ is given by:

$$A = \int_a^b (f(x) - g(x))dx$$

Example 26.5 Find the area of the region bounded by the curve $y = 2 - x^2$ and the line $y = -x$ on the interval $[-1, 2]$.

Solution
We take $f(x) = 2 - x^2$ and $g(x) = -x$, and it can be seen by drawing both curves that $f(x) \geq g(x)$ on the interval $[-1, 2]$. Therefore, the area between both curves is given by:

$$A = \int_{-1}^2 ((2 - x^2) - (-x))dx$$

$$= \int_{-1}^2 ((2 + x - x^2)dx$$

$$= 2x + 1/2x^2 - 1/3x^3 \Big|_{-1}^2 - 1$$
$$= 10/3 + 7/6$$
$$= 27/6$$
$$= 4.5$$

The *volume* of a solid of known *cross functional area* $A(x)$ from $x = a$ to $x = b$ is given by:

$$V = \int_a^b A(x)dx$$

Example 26.6 Find the volume of the pyramid that has a square base that is 3 m on a side and is 3 m high. The area of a cross section of the pyramid is given by $A(x) = x^2$. Find the volume of the pyramid.

Solution

The volume is given by:

$$V = \int_0^3 x^2 dx$$

$$= 1/3x^3\big|^3 0$$

$$= 9\,m^2$$

The volume of a solid created by *revolving the region* bounded by $y = f(x)$ and $x = a$ to $x = b$ about the x-axis is given by:

$$V = \int_a^b \pi(f(x))^2 dx$$

Example 26.7 Find the volume of the sphere generated by rotating the semi-circle $y = \sqrt{(a^2 - x^2)}$ about the x-axis (between $x = -a$ and $x = a$).

Solution

The volume is given by:

$$V = \int_{-a}^a \pi(a^2 - x^2) dx$$

$$= \pi\left(a^2 x - x^3/3\right)\big|_{-a}^a$$

$$= 4/3\pi a^3$$

The *length* of a *curve* $y = f(x)$ from $x = a$ to $x = b$ is given by:

$$L = \int_a^b \sqrt{1 + (f'(x))^2}\, dx$$

Example 26.8

Find the length of the curve $y = f(x) = x^{3/2}$ from $(0, 0)$ to $(4, 8)$.

Solution

The derivative $f'(x)$ is given by: $f'(x) = {}^3/_2\, x^{1/2}$ and so $(f'(x))^2 = {}^9/_4\, x$

The length is then given by:

$$L = \int_0^4 \sqrt{1 + 9/4x}\, dx$$
$$= 2/3\,4/9(1 + 9/4x)^{3/2}\big|_0^4$$
$$= 8/27\left(10^{3/2} - 1\right)$$

There are various rules for approximating definite integrals including the Trapezoidal Rule and Simpson's Rule. The *Trapezoidal Rule* approximates short stretches of the curve with line segments, and the sum of the areas of the trapezoids under the curve is calculated and used as the approximate to the definite integral. *Simpson's Rule* approximates short stretches of the curve with parabolas, and the sum of the areas under the parabolic arcs is calculated and used as the approximate to the definite integral.

The approximation of the Trapezoidal rule to the definite integral is given by the formula:

$$\int_a^b f(x)dx \approx \frac{h}{2}(y_0 + 2y_1 + 2y_2 + \cdots + 2y_{n-1} + y_n)$$

where there are n subintervals and $h = (b - a)/n$.

Example 26.9 Determine an approximation to the definite integral $\int_1^2 x^2\, dx$ using the Trapezoidal rule with $n = 4$ and compare to the exact value of the integral.

Solution
The approximate value is given by:

$$1/4/2(1 + 2*1.5625 + 2*2.25 + 2*3.0625 + 4)$$
$$= 1/8, (18.75)$$
$$= 2.34$$

The exact value is given by

$$x^3/3\big|_1^2 = 2.33$$

The approximation of Simpson's rule to the definite integral is given by the formula:

$$\int_a^b f(x)dx \approx \frac{h}{3}(y_0 + 4y_1 + 2y_2 + 4y_3 \ldots + 2y_{n-2} + 4y_{n-1} + y_n)$$

where there are n subintervals (n is even) and $h = (b - a)/n$.

26.3   Fourier Series

Fourier series are named after Joseph Fourier, a 19th-century French mathematician, and are used to solve practical problems in physics. A Fourier series consists of the sum of a possibly infinite set of sine and cosine functions. The Fourier series for f on the interval $0 \leq x \leq l$ defines a function f whose value at each point is the sum of the series for that value of x.

$$f(x) = \frac{a_0}{2} + \sum_{m=1}^{\infty} \left[a_m \cos \frac{m\pi x}{l} + b_m \sin \frac{m\pi x}{l} \right]$$

The sine and cosine functions are periodic functions

Note 1: (Period of Function)
A function f is periodic with period $T > 0$ if $f(x + T) = f(x)$ for every value of x. The sine and cosine functions are periodic with period 2π: i.e., $\sin(x + 2\pi) = \sin(x)$ and $\cos(x + 2\pi) = \cos(x)$. The functions $\sin m\pi x/l$ and $\cos m\pi x/l$ have period $T = 2l/m$.

Note 2: (Orthogonality)
Two functions f and g are said to be orthogonal on $a \leq x \leq b$ if:

$$\int_a^b f(x)g(x)\mathrm{d}x = 0$$

A set of functions is said to be mutually orthogonal if each distinct pair in the set is orthogonal. The functions $\sin m\pi x/l$ and $\cos m\pi x/l$ where $m = 1, 2, \ldots$ form a mutually orthogonal set of functions on the interval $-l \leq x \leq l$, and they satisfy the following orthogonal relations as specified in Table 26.1.

The orthogonality property of the set of sine and cosine functions allows the coefficients of the Fourier series to be determined. Thus, the coefficients a_n, b_n for

Table 26.1 Orthogonality properties of sine and cosine

Orthogonality properties of sine and cosine
$\int_{-l}^{l} \cos \frac{m\pi x}{l} \sin \frac{n\pi x}{l} \mathrm{d}x = 0$ all m, n
$\int_{-l}^{l} \cos \frac{m\pi x}{l} \cos \frac{n\pi x}{l} \mathrm{d}x = \begin{cases} 0 & m \neq n \\ l & m = n \end{cases}$
$\int_{-l}^{l} \sin \frac{m\pi x}{l} \sin \frac{n\pi x}{l} \mathrm{d}x = \begin{cases} 0 & m \neq n \\ l & m = n \end{cases}$

the convergent Fourier series $f(x)$ are given by:

$$a_n = \frac{1}{l} \int_{-l}^{l} f(x) \cos \frac{n\pi x}{l} dx \quad n = 0, 1, 2, \ldots$$

$$b_n = \frac{1}{l} \int_{-l}^{l} f(x) \sin \frac{n\pi x}{l} dx \quad n = 1, 2, \ldots.$$

The values of the coefficients a_n and b_n are determined from the integrals, and the ease of computation depends on the particular function f involved.

$$f(x) = \frac{a_0}{2} + \sum_{m=1}^{\infty} \left[a_m \cos \frac{m\pi x}{l} + b_m \sin \frac{m\pi x}{l} \right]$$

The values of a_n and b_n depend only on the value of $f(x)$ in the interval $-l \le x \le l$. The terms in the Fourier series are periodic with period $2l$, and the function converges for all x whenever it converges on $-l \le x \le l$. Further, its sum is a periodic function with period $2l$, and therefore, $f(x)$ is determined for all x by its values in the interval $-l \le x \le l$.

26.4 The Laplace Transform

An integral transform takes a function f and transforms it to another function F by means of an integral. Often, the objective is to transform a problem for f into a simpler problem and then to recover the desired function from its transform F. Integral transforms are useful in solving differential equations, and an integral transform is a relation of the form:

$$F(s) = \int_{\alpha}^{\beta} K(s, t) f(t) dt$$

The function F is said to be the transform of f, and the function K is called the kernel of the transformation.

The Laplace transform is named after the well-known 18th-century French mathematician and astronomer, Pierre Laplace. The Laplace transform of f (denoted by $\mathscr{L}\{f(t)\}$ or $F(s)$) is given by:

$$\mathcal{L}\{f(t)\} = F(s) = \int_{0}^{\infty} e^{-st} f(t) dt$$

The kernel $K(s, t)$ of the transformation is e^{-st}, and the Laplace transform is defined over an integral from zero to infinity. This is defined as a limit of integrals over finite intervals as follows:

$$\int_a^\infty f(t)dt = \lim_{A\to\infty} \int_a^A f(t)dt$$

Theorem (Sufficient Condition for Existence of Laplace Transform)
Suppose that f is a piecewise continuous function on the interval $0 \le x \le A$ for any positive A and $|f(t)| \le Ke^{at}$ when $t \ge M$ where a, K, and M are constants and K, M > 0 then the Laplace transform $\mathcal{L}\{f(t)\} = F(s)$ exists for $s > a$.

The following examples are Laplace transforms of some well-known elementary functions.

$$L\{1\} = \int_0^\infty e^{-st}dt = \frac{1}{s}, \quad s > 0$$

$$L\{e^{at}\} = \int_0^\infty e^{-st} e^{at}dt = \frac{1}{s - a} \quad s > a$$

$$L\{\sin at\} = \int_0^\infty e^{-st} \sin at\, dt = \frac{a}{s^2 + a^2} \quad s > 0$$

26.5   Differential Equations

Many important problems in engineering and physics involve determining a solution to an equation that contains one or more derivatives of the unknown function. Such an equation is termed a differential equation, and the study of these equations began with the development of the calculus by Newton and Leibnitz.

Differential equations are classified as ordinary or partial based on whether the unknown function depends on a single independent variable or on several independent variables. In the first case only ordinary derivatives appear in the differential equation and it is said to be an *ordinary differential equation*. In the second case the derivatives are partial, and the equation is termed *a partial differential equation*.

For example, Newton's second law of motion ($F = ma$) expresses the relationship between the force exerted on an object of mass m and the acceleration of the object. The force vector is in the same direction as the acceleration vector. It is given by the ordinary differential equation:

$$m\frac{d^2x(t)}{dt^2} = F(x(t))$$

The next example is that of a second-order partial differential equation. It is the wave equation and is used for the description of waves (e.g., sound, light, and water waves) as they occur in physics. It is given by:

$$a^2 \frac{\partial^2 u(x,t)}{\partial x^2} = \frac{\partial^2 u(x,t)}{\partial t^2}$$

There are several fundamental questions with respect to a given differential equation. First, there is the question as to the existence of a solution to the differential equation. Second, if it does have a solution then is this solution unique. A third question is to how to determine a solution to a particular differential equation.

Differential equations are classified as to whether they are linear or nonlinear. The ordinary differential equation $F(x, y, y', \ldots, y^{(n)}) = 0$ is said to be *linear* if F is a linear function of the variables $y, y', \ldots, y^{(n)}$. The general ordinary differential equation is of the form:

$$a_0(x)y^{(n)} + a_1(x)y^{(n-1)} + \ldots a_n(x)y = g(x)$$

A similar definition applies to partial differential equations, and an equation is *nonlinear* if it is not linear.

26.6 Review Questions

1. What is the difference between velocity and acceleration?
2. How fast does the radius of a spherical soap bubble change when air is blown into it at the rate of 10 cm^3/s?
3. Find the area under the curve $y = x^3 - 4x$ and the x-axis between $x = -2$ to 0.
4. Find the area between the curves $y = x - 2x$ and $y = x^{1/2}$ between $x = 2$ to 4.
5. Determine the volume of the figure generated by revolving the line $x + y = 2$ about the x-axis bounded by $x = 0$ and $y = 0$.
6. Determine the length of the curve $y = \frac{1}{3} (x^2 + 2)^{3/2}$ from $x = 0$ to $x = 3$.
7. What is a periodic function and give examples?
8. Describe applications of Fourier series, Laplace transforms, and differential equations.

26.7   Summary

This chapter provided a short account of applications of the calculus to calculating the velocity and acceleration of moving bodies and problems involving rates of change and maxima/minima problems.

We showed that integration allows the area under a curve to be calculated and the area of the region between two curves to be computed to numerical analysis, Fourier series, Laplace transforms, and differential equations.

Numerical analysis is concerned with devising methods for approximating solutions to mathematical problems. Often an exact formula is not available for solving a particular problem, numerical analysis provides techniques to approximate the solution in an efficient manner. We discussed the Trapezoidal and Simpson's rule which provide an approximation to the definite integral.

A Fourier series consists of the sum of a possibly infinite set of sine and cosine functions. A differential equation is an equation that contains one or more derivatives of the unknown function.

This chapter has sketched some important results in the calculus, and the reader is referred to Finney [1] for more detailed information.

Reference

1. Finney T (1988) Calculus and analytic geometry, 7th edn. Addison Wesley, Boston

Matrix Theory

27

Key Topics

Matrix

Matrix operations

Inverse of a matrix

Determinant

Eigen vectors and values

Cayley Hamilton theorem

Cramer's rule

27.1 Introduction

A *matrix* is a rectangular array of numbers that consists of horizontal rows and vertical columns. A matrix with m rows and n columns is termed an $m \times n$ matrix, where m and n are its dimensions. A matrix with an equal number of rows and columns (e.g., n rows and n columns) is termed a *square* matrix. Figure 27.1 is an example of a square matrix with four rows and four columns.

The entry in the ith row and the jth column of a matrix A is denoted by $A[i, j]$, $A_{i,j}$, or a_{ij}, and the matrix A may be denoted by the formula for its (i, j)th entry: i.e., (a_{ij}) where i ranges from 1 to m and j ranges from 1 to n.

An $m \times 1$ matrix is termed a *column vector*, and a $1 \times n$ matrix is termed a *row vector*. Any row or column of a $m \times n$ matrix determines a row or column vector which is obtained by removing the other rows (respectively columns) from the

---

Fig. 27.1 Example of a 4 ×
4 square matrix

$$\begin{pmatrix} 6 & 0 & -2 & 3 \\ 4 & 2 & 3 & 7 \\ 11 & -5 & 5 & 3 \\ 3 & -5 & -8 & 1 \end{pmatrix}$$

matrix. For example, the row vector $(11, -5, 5, 3)$ is obtained from the matrix example by removing rows 1, 2, and 4 of the matrix.

Two matrices A and B are equal if they are both of the same dimensions, and if $a_{ij} = b_{ij}$ for each $i = 1, 2..., m$ and each $j = 1, 2, ..., n$.

Matrices be added and multiplied (provided certain conditions are satisfied). There are identity matrices under the addition and multiplication binary operations such that the addition of the (additive) identity matrix to any matrix A yields A and similarly for the multiplicative identity. Square matrices have inverses (provided that their determinant is nonzero), and every square matrix satisfies its characteristic polynomial.

It is possible to consider matrices with infinite rows and columns, and although it is not possible to write down such matrices explicitly, it is still possible to add, subtract, and multiply by a scalar provided; there is a well-defined entry in each (i, j)th element of the matrix.

Matrices are an example of an algebraic structure known as an *algebra*. Chapter 5 discussed several algebraic structures such as groups, rings, fields, and vector spaces. The matrix algebra for $m \times n$ matrices A, B, C and scalars λ, μ satisfies the following properties (there are additional multiplicative properties for square matrices).

1. $A + B = B + A$
2. $A + (B + C) = (A + B) + C$
3. $A + 0 = 0 + A = A$
4. $A + (-A) = (-A) + A = 0$
5. $\lambda(A + B) = \lambda A + \lambda B$
6. $(\lambda + \mu)A = \lambda A + \mu B$
7. $\lambda(\mu A) = (\lambda \mu)A$
8. $1A = A$

Matrices have many applications including their use in graph theory to keep track of the distance between pairs of vertices in the graph; a rotation matrix may be employed to represent the rotation of a vector in three-dimensional space. The product of two matrices represents the composition of two linear transformations, and matrices may be employed to determine the solution to a set of linear equations. They also arise in computer graphics and may be employed to project a three-dimensional image onto a two-dimensional screen. It is essential to employ efficient algorithms for matrix computation, and this is an active area of research in the field of numerical analysis.

27.2   Two × Two Matrices

Matrices arose in practice as a means of solving a set of linear equations. One of the earliest examples of their use is in a Chinese text dating from between 300BC and 200AD. The Chinese text showed how matrices could be employed to solve simultaneous equations. Consider the set of equations:

$$ax + by = r$$
$$cx + dy = s$$

Then the coefficients of the linear equations in x and y above may be represented by the matrix A, where A is given by:

$$A = \begin{bmatrix} a & b \\ c & d \end{bmatrix}$$

The linear equations may be represented as the multiplication of the matrix A and a vector $\underline{x}$ resulting in a vector $\underline{v}$:

$$A\underline{x} = \underline{v}.$$

The matrix representation of the linear equations and its solution are as follows:

$$\begin{bmatrix} a & b \\ c & d \end{bmatrix} \begin{bmatrix} x \\ y \end{bmatrix} = \begin{bmatrix} r \\ s \end{bmatrix}$$

The vector $\underline{x}$ may be calculated by determining the inverse of the matrix A (provided that its inverse exists). The vector $\underline{x}$ is then given by:

$$\underline{x} = A^{-1}\underline{v}$$

The solution to the set of linear equations is then given by:

$$\begin{bmatrix} x \\ y \end{bmatrix} = \begin{bmatrix} a & b \\ c & d \end{bmatrix}^{-1} \begin{bmatrix} r \\ s \end{bmatrix}$$

The inverse of a matrix A exists if and only if its *determinant* is nonzero, and if this is the case the vector $\underline{x}$ is given by:

$$\begin{bmatrix} x \\ y \end{bmatrix} = \frac{1}{\det\ A} \begin{bmatrix} d & -b \\ -c & a \end{bmatrix} \begin{bmatrix} r \\ s \end{bmatrix}$$

The determinant of a 2×2 matrix A is given by:

$$\det A = ad - cb.$$

The determinant of a 2×2 matrix is denoted by:

$$\begin{vmatrix} a & b \\ c & d \end{vmatrix}$$

A key property of determinants is that

$$\det(AB) = \det(A).\det(B)$$

The transpose of a 2×2 matrix A (denoted by A^{T}) involves exchanging rows and columns and is given by:

$$A^{\mathrm{T}} = \begin{bmatrix} a & c \\ b & d \end{bmatrix}$$

The inverse of the matrix A (denoted by A^{-1}) is given by:

$$A^{-1} = \frac{1}{\det A} \begin{bmatrix} d & -b \\ -c & a \end{bmatrix}$$

Further, $A \cdot A^{-1} = A^{-1} \cdot A = I$ where I is the identity matrix of the algebra of 2×2 matrices under multiplication. That is:

$$AA^{-1} = A^{-1}A = \begin{bmatrix} 1 & 0 \\ 0 & 1 \end{bmatrix}$$

The addition of two 2×2 matrices A and B is given by a matrix whose entries are the addition of the individual components of A and B. The addition of two matrices is commutative and we have:

$$A + B = B + A = \begin{bmatrix} a+p & b+q \\ c+r & d+s \end{bmatrix}$$

where A and B are given by:

$$A = \begin{bmatrix} a & b \\ c & d \end{bmatrix} \quad B = \begin{bmatrix} p & q \\ r & s \end{bmatrix}$$

The identity matrix under addition is given by the matrix whose entries are all 0, and it has the property that $A + 0 = 0 + A = A$.

$$\begin{bmatrix} 0 & 0 \\ 0 & 0 \end{bmatrix}$$

The multiplication of two 2×2 matrices is given by:

$$AB = \begin{bmatrix} ap + br & aq + bs \\ cp + dr & cq + ds \end{bmatrix}$$

The multiplication of matrices is not commutative: i.e., $AB \neq BA$. The multiplicative identity matrix I has the property that $A \cdot I = I \cdot A = A$, and it is given by:

$$I = \begin{bmatrix} 1 & 0 \\ 0 & 1 \end{bmatrix}$$

A matrix A may be multiplied by a scalar λ, and this yields the matrix λA where each entry in A is multiplied by the scalar λ. That is the entries in the matrix λA are λa_{ij}.

27.3   Matrix Operations

More general sets of linear equations may be solved with $m \times n$ matrices (i.e., a matrix with m rows and n columns) or square $n \times n$ matrices. In this section, we consider several matrix operations including addition, subtraction, multiplication of matrices, scalar multiplication, and the transpose of a matrix.

The addition and subtraction of two matrices A and B are meaningful if and only if A and B have the same dimensions: i.e., they are both $m \times n$ matrices. In this case, $A + B$ is defined by adding the corresponding entries:

$$(A + B)_{ij} = A_{ij} + B_{ij}$$

The additive identity matrix for the square $n \times n$ matrices is denoted by 0, where 0 is a $n \times n$ matrix whose entries are zero: i.e., $r_{ij} = 0$ for all i, j where $1 \leq i \leq n$ and $1 \leq j \leq n$.

The scalar multiplication of a matrix A by a scalar k is meaningful and the resulting matrix kA is given by:

$$(kA)_{ij} = kA_{ij}$$

The multiplication of two matrices A and B is meaningful if and only if the number of columns of A is equal to the number of rows of B (Fig. 27.2): i.e., A

$$\begin{pmatrix} a_{11} & a_{12} & a_{13} & \cdots & a_{1n} \\ a_{21} & a_{22} & a_{23} & \cdots & a_{2n} \\ a_{31} & a_{32} & a_{33} & \cdots & a_{3n} \\ \cdots & \cdots & \cdots & \cdots & \cdots \\ \cdots & \cdots & \cdots & \cdots & \cdots \\ a_{m1} & a_{m2} & a_{m3} & \cdots & a_{mn} \end{pmatrix} \begin{pmatrix} b_{11} & b_{12} & \cdots & b_{1p} \\ b_{21} & b_{22} & \cdots & b_{2p} \\ b_{31} & b_{32} & \cdots & b_{3p} \\ \cdots & \cdots & \cdots & \cdots \\ \cdots & \cdots & \cdots & \cdots \\ \cdots & \cdots & \cdots & \cdots \\ b_{n1} & b_{n2} & \cdots & b_{np} \end{pmatrix} = \begin{pmatrix} c_{11} & c_{12} & \cdots & c_{1p} \\ c_{21} & c_{22} & \cdots & c_{2p} \\ c_{31} & c_{32} & \cdots & c_{3p} \\ \cdots & \cdots & \cdots & \cdots \\ \cdots & \cdots & \cdots & \cdots \\ c_{m1} & c_{m2} & \cdots & c_{mp} \end{pmatrix}$$

$$\text{\textit{m} rows, \textit{n} columns} \qquad\qquad \text{\textit{n} rows, \textit{p} columns} \qquad\qquad \text{\textit{m} rows, \textit{p} columns}$$

Fig. 27.2 Multiplication of two matrices

is an $m \times n$ matrix and B is a $n \times p$ matrix and the resulting matrix AB is a $m \times p$ matrix.

Let $A = (a_{ij})$ where i ranges from 1 to m and j ranges from 1 to n, and let $B = (b_{jl})$ where j ranges from 1 to n and l ranges from 1 to p. Then AB is given by (c_{il}) where i ranges from 1 to m and l ranges from 1 to$p$ with c_{il} given by:

$$c_{il} = \sum_{k=1}^{n} a_{ik} b_{kl}.$$

That is, the entry (c_{il}) is given by multiplying the ith row in A by the lth column in B followed by a summation. Matrix multiplication is not commutative: i.e., $AB \neq BA$.

The identity matrix I is a $n \times n$ matrix and the entries are given by r_{ij} where $r_{ii} = 1$ and $r_{ij} = 0$ where $i \neq j$ (Fig. 27.3). A matrix that has nonzero entries only on the diagonal is termed a *diagonal matrix*. A triangular matrix is a square matrix in which all the entries above or below the main diagonal are zero. A matrix is an *upper triangular* matrix if all entries below the main diagonal are zero and *lower triangular* if all of the entries above the main diagonal are zero. Upper triangular and lower triangular matrices form a sub-algebra of the algebra of square matrices.

Fig. 27.3 Identity matrix I_n

$$\begin{pmatrix} 1 & 0 & 0 & \cdots & 0 \\ 0 & 1 & 0 & \cdots & 0 \\ 0 & 0 & 1 & \cdots & 0 \\ \cdots & \cdots & \cdots & \cdots & \cdots \\ \cdots & \cdots & \cdots & \cdots & \cdots \\ \cdots & \cdots & \cdots & \cdots & \cdots \\ 0 & 0 & 0 & \cdots & 1 \end{pmatrix}$$

$$
\begin{pmatrix}
a_{11} & a_{12} & a_{13} & \cdots & a_{1n} \\
a_{21} & a_{22} & a_{23} & \cdots & a_{2n} \\
a_{31} & a_{32} & a_{33} & \cdots & a_{3n} \\
\cdots & \cdots & \cdots & \cdots & \cdots \\
\cdots & \cdots & \cdots & \cdots & \cdots \\
a_{m1} & a_{m2} & a_{m3} & \cdots & a_{mn}
\end{pmatrix}^{T}
=
\begin{pmatrix}
a_{11} & a_{21} & a_{31} & \cdots & a_{m1} \\
a_{12} & a_{22} & a_{32} & \cdots & a_{m2} \\
a_{13} & a_{23} & a_{33} & \cdots & a_{m3} \\
\cdots & \cdots & \cdots & \cdots & \cdots \\
\cdots & \cdots & \cdots & \cdots & \cdots \\
a_{1n} & a_{2n} & a_{3n} & \cdots & a_{mn}
\end{pmatrix}
$$

m rows, n columns   n rows, m columns

Fig. 27.4 Transpose of a matrix

A key property of the identity matrix is that for all $n \times n$ matrices A we have:

$$AI = IA = A$$

The inverse of a $n \times n$ matrix A is a matrix A^{-1} such that:

$$AA^{-1} = A^{-1}A = I$$

The inverse A^{-1} exists if and only if the determinant of A is nonzero.

The *transpose* of a matrix $A = (a_{ij})$ involves changing the rows to columns and vice versa to form the transpose matrix A^{T}. The result of the operation is that the $m \times n$ matrix A is converted to the $n \times m$ matrix A^{T} (Fig. 27.4). It is defined by:

$$\left(A^{T}\right)_{ij} = \left(A_{ji}\right) \quad 1 \le j \le n \text{ and } 1 \le i \le m$$

A matrix is *symmetric* if it is equal to its transpose: i.e., $A = A^{T}$.

27.4   Determinants

The determinant is a function defined on square matrices, and its value is a scalar. A key property of determinants is that a matrix is invertible if and only if its determinant is nonzero. The determinant of a 2×2 matrix is given by:

$$\begin{vmatrix} a & b \\ c & d \end{vmatrix} = ad - bc$$

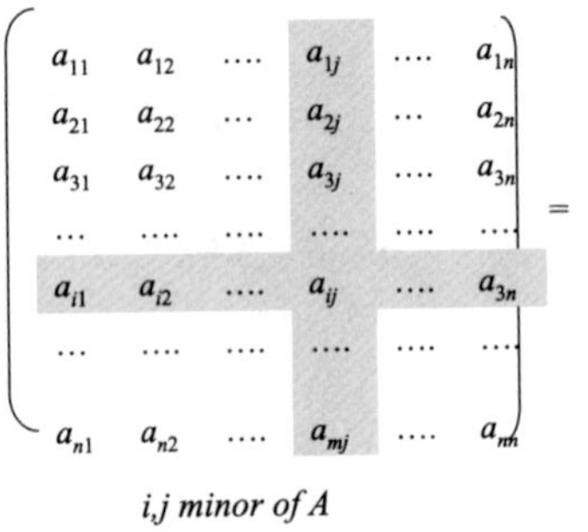

Fig. 27.5 Determining the
(i, j) minor of A

$$
\begin{pmatrix}
a_{11} & a_{12} & \cdots\cdots & a_{1j} & \cdots\cdots & a_{1n} \\
a_{21} & a_{22} & \cdots & a_{2j} & \cdots & a_{2n} \\
a_{31} & a_{32} & \cdots\cdots & a_{3j} & \cdots\cdots & a_{3n} \\
\cdots & \cdots\cdots & \cdots\cdots & \cdots\cdots & \cdots\cdots & \cdots\cdots \\
a_{i1} & a_{i2} & \cdots\cdots & a_{ij} & \cdots\cdots & a_{3n} \\
\cdots & \cdots\cdots & \cdots\cdots & \cdots\cdots & \cdots\cdots & \cdots\cdots \\
a_{n1} & a_{n2} & \cdots\cdots & a_{mj} & \cdots\cdots & a_{nh}
\end{pmatrix} =
$$

i,j minor of A

The determinant of a 3×3 matrix is given by:

$$
\begin{vmatrix}
a & b & c \\
d & e & f \\
g & h & i
\end{vmatrix} = aei + bfg + cdh - afh - bdi - ceg
$$

Cofactors

Let A be an $n \times n$ matrix. For $1 \le i, j \le n$, the (i, j) *minor* of A is defined to be the $(n - 1) \times (n - 1)$ matrix obtained by deleting the ith row and jth column of A (Fig. 27.5).

The shaded row is the ith row, and the shaded column is the jth column. These are both deleted from A to form the (i, j) minor of A, and this is a $(n - 1) \times (n - 1)$ matrix.

The (i, j) *cofactor* of A is defined to be $(-1)^{i+j}$ times the determinant of the (i, j) minor. The (i, j) cofactor of A is denoted by $K_{ij}(A)$.

The cofactor matrix $Cof\ A$ is formed in this way where the (i, j)th element in the cofactor matrix is the (i, j) cofactor of A.

Definition of Determinant

The determinant of a matrix is defined as:

$$
\det A = \sum_{j=1}^{n} A_{ij} K_{ij}
$$

In other words the determinant of A is determined by taking any row of A and multiplying each element by the corresponding cofactor and adding the results. The determinant of the product of two matrices is the product of their determinants.

$$
\det(AB) = \det A \times \det B
$$

Definition The *adjugate* of A is the $n \times n$ matrix $\text{Adj}(A)$ whose (i, j) entry is the (j, i) cofactor $K_{ji}(A)$ of A. That is, the adjugate of A is the transpose of the cofactor matrix of A.

Inverse of A

The inverse of A is determined from the determinant of A and the adjugate of A. That is,

$$A^{-1} = \frac{1}{\det A} \text{Adj } A = \frac{1}{\det A} (\text{Cof } A)^{\text{T}}$$

A matrix is invertible if and only if its determinant is nonzero: i.e., A is invertible if and only if $\det(A) \neq 0$.

Cramer's Rule

Cramer's rule is a theorem that expresses the solution to a system of linear equations with several unknowns using the determinant of a matrix. There is a unique solution if the determinant of the matrix is nonzero.

For a system of linear equations of the $A\underline{x} = \underline{v}$ where $\underline{x}$ and $\underline{v}$ are n-dimensional column vectors, then if $\det A \neq 0$ then the unique solution for each x_i is

$$x_i = \frac{\det U_i}{\det A}$$

where U_i is the matrix obtained from A by replacing the ith column in A by the v-column.

Characteristic Equation

For every $n \times n$ matrix A there is a polynomial equation of degree n satisfied by A. The *characteristic polynomial* of A is a polynomial in x of degree n. It is given by:

$$cA(x) = \det(xI - A).$$

Cayley-Hamilton Theorem

Every matrix A satisfies its characteristic polynomial: i.e., $p(A) = 0$ where $p(x)$ is the characteristic polynomial of A.

27.5 Eigen Vectors and Values

A number λ is an eigenvalue of a $n \times n$ matrix A if there is a nonzero vector v such that the following equation holds:

$$Av = \lambda v$$

The vector v is termed an eigenvector and the equation is equivalent to:

$$(A - \lambda I)v = 0$$

This means that $(A - \lambda I)$ is a zero divisor, and hence, it is not an invertible matrix. Therefore,

$$\det(A - \lambda I) = 0$$

The polynomial function $p(\lambda) = \det(A - \lambda I)$ is called the characteristic polynomial of A, and it is of degree n. The characteristic equation is $p(\lambda) = 0$ and as the polynomial is of degree n there are at most n roots of the characteristic equation, and so there at most n eigenvalues.

The *Cayley-Hamilton theorem* states that every matrix satisfies its characteristic equation: i.e., the application of the characteristic polynomial to the matrix A yields the zero matrix.

$$p(A) = 0$$

27.6   Gaussian Elimination

Gaussian elimination with backward substitution is an important method used in solving a set of linear equations. A matrix is used to represent the set of linear equations, and Gaussian elimination reduces the matrix to a *triangular* or *reduced form*, which may then be solved by backward substitution.

This allows the set of n linear equations $(E_1–E_n)$ defined below to be solved by applying operations to the equations to reduce the matrix to triangular form. This reduced form is easier to solve and it provides exactly the same solution as the original set of equations. The set of equations is defined as:

$$\begin{aligned}
E_1 &: a_{11}x_1 + a_{12}x_2 + \cdots + a_{1n}x_n = b_1 \\
E_2 &: a_{21}x_1 + a_{22}x_2 + \cdots + a_{2n}x_n = b_2 \\
&\quad \vdots \qquad \vdots \quad \vdots \qquad\qquad \vdots \quad \vdots \\
E_n &: a_{n1}x_1 + a_{n2}x_2 + \cdots + a_{nn}x_n = b_n
\end{aligned}$$

Three operations are permitted on the equations, and these operations transform the linear system into a reduced form. They are:

(a) Any equation may be multiplied by a nonzero constant.
(b) An equation E_i may be multiplied by a constant and added to another equation E_j, with the resulting equation replacing E_j
(c) Equations E_i and E_j may be transposed with E_j replacing E_i and vice versa.

This method for solving a set of linear equations is best illustrated by an example, and we consider an example taken from [1]. Then the solution to a set of linear equations with four unknowns may be determined as follows:

$$\begin{aligned}
E_1 &: \quad x_1 + x_2 \quad + 3x_4 = 4 \\
E_2 &: \quad 2x_1 + x_2 - x_3 + x_4 = 1
\end{aligned}$$

$$E_3: \quad 3x_1 - x_2 - x_3 + 2x_4 = -3$$
$$E_4: \quad -x_1 + 2x_2 + 3x_3 - x_4 = 4$$

First, the unknown x_1 is eliminated from E_2, E_3, and E_4 and this is done by replacing E_2 with $E_2 - 2E_1$; replacing E_3 with $E_3 - 3E_1$; and replacing E_4 with $E_4 + E_1$. The resulting system is

$$E_1: x_1 + x_2 \quad + 3x_4 = 4$$
$$E_2: \quad -x_2 - x_3 - -5x_4 = -7$$
$$E_3: \quad -4x_2 - x_3 - 7x_4 = -15$$
$$E_4: \quad 3x_2 + 3x_3 + 2x_4 = 8$$

The next step is then to eliminate x_2 from E_3 and E_4. This is done by replacing E_3 with $E_3 - 4E_2$ and replacing E_4 with $E_4 + 3E_2$. The resulting system is now in triangular form, and the unknown variable may be solved easily by backward substitution. That is, we first use equation E_4 to find the solution to x_4 and then we use equation E_3 to find the solution to x_3. We then use equations E_2 and E_1 to find the solutions to x_2 and x_1.

$$E_1: \quad x_1 + x_2 \quad + 3x_4 = 4$$
$$E_2: \quad -x_2 - x_3 - 5x_4 = -7$$
$$E_3: \quad 3x_3 + 13x_4 = 13$$
$$E_4: \quad -13x_4 = -13$$

The usual approach to Gaussian elimination is to do it with an augmented matrix. That is, the set of equations is a $n \times n$ matrix and it is augmented by the column vector to form the augmented $n \times n + 1$ matrix. Gaussian elimination is then applied to the matrix to put it into triangular form, and it is then easy to solve the unknowns.

The other common approach to solving a set of linear equation is to employ Cramer's rule, which was discussed in Sect. 27.4. Finally, another possible (but computationally expensive) approach is to compute the determinant and inverse of A and to then compute $\underline{x} = A^{-1}\underline{v}$.

27.7 Review Questions

1. Show how 2×2 matrices may be added and multiplied.
2. What is the additive identity for 2×2 matrices? The multiplicative identity?
3. What is the determinant of a 2×2 matrix?
4. Show that a 2×2 matrix is invertible if its determinant is nonzero.
5. Describe general matrix algebra including addition and multiplication, determining the determinant and inverse of a matrix.

6. What is Cramer's rule?
7. Show how Gaussian elimination may be used to solve a set of linear equations.
8. Write a program to find the inverse of a 3×3 and then a $(n \times n)$ matrix.

27.8 Summary

A matrix is a rectangular array of numbers that consists of horizontal rows and vertical columns. A matrix with m rows and n columns is termed an $m \times n$ matrix, where m and n are its dimensions. A matrix with an equal number of rows and columns (e.g., n rows and n columns) is termed a square matrix.

Matrices arose in practice as a means of solving a set of linear equations, and matrices of the same dimensions may be added, subtracted, and multiplied by a scalar. Two matrices A and B may be multiplied provided that the number of columns of A equals the number of rows in B.

Matrices have an identity matrix under addition and multiplication, and a square matrix has an inverse provided that its determinant is nonzero. The inverse of a matrix involves determining its determinant, constructing the cofactor matrix, and transposing the cofactor matrix.

The solution to a set of linear equations may be determined by Gaussian elimination to convert the matrix to upper triangular form and then employing backward substitution. Another approach is to use Cramer's rule. Eigenvalues and eigenvectors lead to the characteristic polynomial and every matrix satisfies its characteristic polynomial.

Reference

1. Burden RL, Faires JD (1989) Numerical analysis, 4th edn. PWS Kent

Complex Numbers and Quaternions

28

Key Topics

Complex numbers

Argand diagram

Polar representation

De Moivre's theorem

Complex conjugate

Quaternions

28.1 Introduction

A complex number z is a number of the form $a + bi$ where a and b are real numbers and $i^2 = -1$. Cardona, who was a sixteenth century Italian mathematician, introduced complex numbers, and he used them to solve cubic equations. The set of complex numbers is denoted by $\mathbb{C}$, and each complex number has two parts namely the real part $\mathrm{Re}(z) = a$, and the imaginary part $\mathrm{Im}(z) = b$. The set of complex numbers is an extension of the set of real numbers, and this is clear since every real number is a complex number with an imaginary part of zero. A complex number with a real part of zero (i.e., $a = 0$) is termed an imaginary number. Complex numbers have many applications in physics, engineering, and applied mathematics.

© The Author(s), under exclusive license to Springer Nature Switzerland AG 2023
G. O'Regan, *Mathematical Foundations of Software Engineering*,
Texts in Computer Science, https://doi.org/10.1007/978-3-031-26212-8_28

Fig. 28.1 Argand diagram

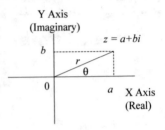

A complex number may be viewed as a point in a two-dimensional Cartesian coordinate system (called the complex plane or Argand diagram), where the complex number $a + bi$ is represented by the point (a, b) on the complex plane (Fig. 28.1). The real part of the complex number is the horizontal component, and the imaginary part is the vertical component.

Quaternions are an extension of complex numbers. A quaternion number is a quadruple of the form $(a + bi + cj + dk)$ where $i^2 = j^2 = k^2 = ijk = -1$. The set of quaternions is denoted by $\mathbb{H}$, and the quaternions form an algebraic system known as a division ring. The multiplication of two quaternions is not commutative: i.e., given $q_1, q_2 \in \mathbb{H}$ then $q_1, q_2 \neq q_2, q_1$. Quaternions were one the first non-commutative algebraic structures to be discovered (as matrix algebra came later).

The Irish mathematician, Sir William Rowan Hamilton,[1] discovered quaternions. Hamilton was trying to generalize complex numbers to triples without success. He had a moment of inspiration along the banks of the Royal Canal in Dublin, and he realized that if he used quadruples instead of triples that a generalization from the complex numbers was possible. He was so overcome with emotion at his discovery that he traced the famous quaternion formula[2] on Brooms Bridge in Dublin. This formula is given by:

$$i^2 = j^2 = k^2 = ijk = -1$$

---

[1] There is a possibility that the German mathematician, Gauss, discovered quaternions earlier, but he did not publish his results.

[2] Eamonn DeValera (a former taoiseach and president of Ireland) was previously a mathematics teacher, and his interests included maths physics and quaternions. He carved the quaternion formula on the door of his prison cell in Lincoln Jail, England, during the Irish struggle for independence. He escaped from Lincoln Jail in February 1919.

Quaternions have many applications in physics and quantum mechanics and are applicable to the computing field. They are useful and efficient in describing rotations and are therefore applicable to computer graphics, computer vision, and robotics.

28.2 Complex Numbers

There are several operations on complex numbers such as addition, subtraction, multiplication, division, and so on (Table 28.1). Consider two complex numbers $z_1 = a + bi$ and $z_2 = c + di$. Then,

Properties of Complex Numbers
The absolute value of a complex number z is denoted by $|z| = \sqrt{(a^2 + b^2)}$ and is just its distance from the origin. It has the following properties:

(i) $|z| \geq 0$ and $|z| = 0$ if and only if $z = 0$.
(ii) $|z| = |z^*|$
(iii) $|z_1 + z_2| \leq |z_1| + |z_2|$ (This is known as the triangle inequality)
(iv) $|z_1 z_2| = |z_1| \, |z_2|$
(v) $|1/z| = 1/|z|$
(vi) $|z_1/z|_2 = |z_1|/|z_2|$

Table 28.1 Operations on Complex Numbers

Operation	Definition
Addition	$z_1 + z_2 = (a + bi) + (c + di) = (a + c) + (b + d)i$ The addition of two complex numbers may be interpreted as the addition of two vectors
Subtraction	$z_1 - z_2 = (a + bi) - (c + di) = (a - c) + (b - d)i$
Multiplication	$z_1 z_2 = (a + bi) \cdot (c + di) = (ac - bd) + (ad + cb)i$
Division	This operation is defined for $z_2 \neq 0$ $\frac{z_1}{z_2} = \frac{a+bi}{c+di} = \frac{ac+bd}{c^2+d^2} + \frac{bc-ad}{c^2+d^2} i$
Conjugate	The conjugate of a complex number $z = a + bi$ is given by $z^* = a - bi$ Clearly, $z^{**} = z$ and $(z_1 + z_2)^* = z_1{}^* + z_2{}^*$ Further, $\mathrm{Re}(z) = z + z^*/2$ and $\mathrm{Im}(z) = z - z^*/2i$
Absolute value	The absolute value or modulus of a complex number $z = a + bi$ is given by $|z| = \sqrt{(a^2 + b^2)}$. Clearly, $z \cdot z^* = |z|^2$
Reciprocal	The reciprocal of a complex number z is defined for $z \neq 0$ and is given by: $\frac{1}{z} = \frac{1}{a+bi} = \frac{a-bi}{a^2+b^2} = \frac{z^*}{|z|^2}$

Fig. 28.2 Interpretation of complex conjugate

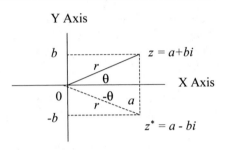

Proof (iii)

$$|z_1 + z_2|^2 = (z_1 + z_2)(z_1 + z_2)^*$$
$$= (z_1 + z_2)(z_1^* + z_2^*)$$
$$= z_1 z_1^* + z_1 z_2^* + z_2 z_1^* + z_2 z_2^*$$
$$= |z_1|^2 + z_1 z_2^* + z_2 z_1^* + |z_2|^2$$
$$= |z_1|^2 + z_1 z_2^* + (z_1 z_2^*)^* + |z_2|^2$$
$$= |z_1|^2 + 2\mathrm{Re}(z_1 z_2^*) + |z_2|^2$$
$$\leq |z_1|^2 + 2|z_1 z_2^*| + |z_2|^2$$
$$= |z_1|^2 + 2|z_1||z_2^*| + |z_2|^2$$
$$= |z_1|^2 + 2|z_1||z_2| + |z_2|^2$$
$$= (|z_1| + |z_2|)^2$$

Therefore, $|z_1 + z_2| \leq |z_1| + |z_2|$ and so the triangle inequality is proved.

The modulus of z is used to define a distance function between two complex numbers, and $d(z_1, z_2) = |z_1 - z_2|$. This turns the complex numbers into a metric space.[3]

Interpretation of Complex Conjugate

The complex conjugate of the complex number $z = a + bi$ is defined as $z^* = a - bi$, and this is the reflection of z about the real axis in Fig. 28.2.

The modulus $|z|$ of the complex number z is the distance of the point z from the origin.

---

[3] A non-empty set X with a distance function d is a metric space if

(i) $d(x, y) \geq 0$ and $d(x, y) = 0 \Leftrightarrow x = y$
(ii) $d(z,y) = d(y,x)$
(iii) $d(x, y) \leq d(x, z) + d(z, y)$

Polar Representation of Complex Numbers

The complex number $z = a + bi$ may also be represented in polar form (r, θ) in terms of its modulus $|z|$ and the argument θ.

$$\cos \theta = \frac{a}{\sqrt{a^2 + b^2}} = \frac{a}{|z|}$$

$$\sin \theta = \frac{b}{\sqrt{a^2 + b^2}} = \frac{b}{|z|}$$

Let r denote the modulus of z: i.e., $r = |z|$. Then, z may be represented by $z = (r \cos \theta + ir \sin \theta) = r (\cos \theta + i \sin \theta)$. Clearly, $\text{Re}(z) = r \cos \theta$ and $\text{Im}(z) = r \sin \theta$. Euler's formula (discussed below) states that $re^{i\theta} = r (\cos \theta + i \sin \theta)$.

Each real number θ for which $z = |z| (\cos \theta + i \sin \theta)$ is said to be an argument of z (denoted by arg z). There is, of course, more than one argument θ that will satisfy $z = r (\cos \theta + i \sin \theta)$, and the full set of arguments is given by arg $z = \theta + 2k\pi$, where $k \in \mathbb{Z}$ and satisfies $z = re^{i(\theta + 2k\pi)}$.

The *principle argument* of z (denoted by Arg $z = \theta$) is the unique real number chosen so that $\theta \in (- \pi, \pi]$. That is, arg z denotes a set of arguments, whereas Arg z denotes a unique argument. The following are properties of arg z:

$$\arg\left(z^{-1}\right) = - \arg z$$
$$\arg(zw) = \arg z + \arg w$$

Euler's Formula

Euler's remarkable formula expresses the relationship between the exponential function for complex numbers and trigonometric functions. It is named after the eighteenth century Swiss mathematician, Euler.

It may be interpreted as the function $e^{i\theta}$ traces out the unit circle in the complex plane as the angle θ ranges through the real numbers (Fig. 28.3). Euler's formula provides a way to convert between Cartesian coordinates and polar coordinates (r, θ). It states that:

$$e^{i\theta} = \cos \theta + i \sin \theta.$$

Fig. 28.3 Interpretation of Eulers' formula

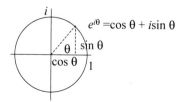

Further, the complex number $z = a + bi$ may be represented in polar coordinates as $z = r(\cos\theta + i\sin\theta) = re^{i\theta}$.

Next, we prove Euler's formula: i.e., $e^{i\theta} = \cos\theta + i\sin\theta$.

Proof Recall the exponential expansion for e^x:

$$e^x = 1 + x + x^2/2! + x^3/3! + \cdots + x^r/r! + \ldots.$$

The expansion of $e^{i\theta}$ is then given by:

$$e^{i\theta} = 1 + i\theta + \frac{(i\theta)^2}{2!} + \frac{(i\theta)^3}{3!} + \cdots + \frac{(i\theta)^r}{r!} + \ldots.$$

$$= 1 + i\theta - \frac{\theta^2}{2!} - \frac{i\theta^3}{3!} + \frac{\theta^4}{4!} + \frac{i\theta^5}{5!} + \ldots \frac{(i\theta)^r}{r!} + \ldots$$

$$= \left[1 - \frac{\theta^2}{2!} + \frac{\theta^4}{4!} - \frac{\theta^6}{6!} + \ldots \right] + i\left[\theta - \frac{\theta^3}{3!} + \frac{\theta^5}{5!} - \frac{\theta^7}{7!} + \ldots \right]$$

$$= \cos\theta + i\sin\theta$$

(This follows from the Taylor Series expansion of $\sin\theta$ and $\cos\theta$).

Euler's Identity

This remarkable identity follows immediately and is stated as:

$$e^{i\pi} = -1 \quad \text{(it is also written as } e^{i\pi} + 1 = 0\text{)}$$

De Moivre's Theorem

$$(\cos\theta + i\sin\theta)^n = (\cos n\theta + i\sin n\theta) \quad \text{(where } n \in \mathbb{Z}\text{)}$$

Proof This result is proved by mathematical induction, and the result is clearly true for the base case n = 1.

Inductive Step:

The inductive step is to assume that the theorem is true for $n = k$ and to then show that it is true for $n = k + 1$. That is, we assume that

$$(\cos\theta + i\sin\theta)^k = (\cos k\theta + i\sin k\theta) \quad \text{(for some } k > 1\text{)}$$

We next show that the result is true for $n = k + 1$:

$$(\cos\theta + i\sin\theta)^{k+1} =$$

$$= (\cos\theta + i\sin\theta)^k(\cos\theta + i\sin\theta)$$
$$= (\cos k\theta + i\sin k\theta)(\cos\theta + i\sin\theta) \quad \text{(from inductive step)}$$
$$= (\cos k\theta\cos\theta - \sin k\theta\sin\theta) + i(\cos k\theta\sin\theta + \sin k\theta\cos\theta)$$
$$= \cos(k\theta + \theta) + i\sin(k\theta + \theta)$$
$$= \cos(k+1)\theta + i\sin(k+1)\theta$$

Therefore, we have shown that if the result is true for some value of n say $n = k$, then the result is true for $n = k + 1$. We have shown that the base case of $n = 1$ is true, and it therefore follows that the result is true for $n = 2, 3, \ldots$ and for all natural numbers. The result may also be shown to be true for the integers.

Complex Roots

Suppose that z is a nonzero complex number and that n is a positive integer. Then z has exactly n distinct complex nth roots and these roots are given in polar form by:

$$\sqrt[n]{|z|}\left[\cos\left\{\frac{\text{Arg } z + 2k\pi}{n}\right\} + i\sin\left\{\frac{\text{Arg } z + 2k\pi}{n}\right\}\right]$$

for $k = 0, 1, 2, \ldots, n - 1$

Proof The objective is to find all complex numbers w such that $w^n = z$ where $w = |w|(\cos\Phi + i\sin\Phi)$. Using De Moivre's Theorem this results in:

$$|w|^n(\cos n\Phi + i\sin n\Phi) = |z|(\cos\theta + i\sin\theta)$$

Therefore, $|w| = \sqrt[n]{|z|}$ and $n\Phi = \theta + 2k\pi$ for some k. That is,

$$\Phi = (\theta + 2k\pi)/n = (\text{Arg } z + 2k\pi)/n$$

The choices $k = 0, 1, \ldots, n - 1$ produce the distinct nth roots of z.
    The *principle* nth root of z (denoted by $\sqrt[n]{z}$) is obtained by taking $k = 0$.

Fundamental Theorem of Algebra

Every polynomial equation with complex coefficients has complex solutions, and the roots of a complex polynomial of degree n exist, and the n roots are all complex numbers.

Example Describe the set S of complex numbers that satisfy $|z - 1| = 2|z + 1|$.

Solution

Let $z = x + iy$ then we note that $z \in S$.

$$\Rightarrow |z - 1| = 2|z + 1|$$

$\Rightarrow |z - 1|^2 = 4|z + 1|^2$

$\Rightarrow |z|^2 - 2\text{Re}(z) + 1 = 4(|z|^2 + 2\text{Re}(z) + 1)$

$\Rightarrow |z|^2 - 2\text{Re}(z) + 1 = 4|z|^2 + 8\text{Re}(z) + 4$

$\Rightarrow 3|z|^2 + 10\text{Re}(z) + 3 = 0$

$\Rightarrow 3(x^2 + y^2) + 10x = -3$

$\Rightarrow x^2 + y^2 + {}^{10}/_3 x = -1$

$\Rightarrow x^2 + {}^{10}/_3 x + ({}^5/_3)^2 + y^2 = -1 + ({}^5/_3)^2$............(completing the square)

$\Rightarrow (x + {}^5/_3)^2 + y^2 = ({}^{16}/_9)$

$\Rightarrow (x + {}^5/_3)^2 + y^2 = ({}^4/_3)^2$

This is the formula for a circle with radius ${}^4/_3$ and centre $(-{}^5/_3, 0)$.

Exponentials and Logarithms of Complex Numbers

We discussed Euler's formula and noted that any complex number $z = x + iy$ may be written as $z = r(\cos\theta + i\sin\theta) = re^{i\theta}$. This leads naturally

$$\exp(z) = e^z = e^{x+iy} = e^x e^{iy} = e^x(\cos y + i\sin y) \qquad \text{where } z = x + iy.$$

There are several properties of exponentials:

(i) $e^0 = 1$

(ii) $e^w = 1 \Leftrightarrow w = 2k\pi i \ k \quad k = 0, \pm 1, \pm 2, \ldots..$

(iii) $|e^z| = e^{\text{Re } z}$

(iv) $\arg(e^z) = \text{Im}(z)$

(v) $(e^z)^n = e^{zn}$

(vi) $(e^z)^{-1} = e^{-z}$

(vii) $e^z e^w = e^{z+w}$

(viii) $e^z = e^w \Leftrightarrow w = z + 2k\pi i$

We say that w is a logarithm of z (there are an infinite number of them) if

$$\ln z = w \Leftrightarrow e^w = z$$

The *principal logarithm* of z (denoted by Log z) is given by:

$$\text{Log } z = w = \ln|z| + i \text{ Arg } z$$

$$e^w = e^{\ln|z| + i \text{ Arg } z}$$

$$= e^{\ln|z|} e^{i \text{ Arg } z}$$

$$= |z|(\text{Cos}\theta + i\text{Sin}\theta)$$

$$= z$$

Further, the complete set of logarithms of z may be determined since any logarithm w of z (log z) satisfies:

$$e^w = e^{\text{Log } z}$$

$\Rightarrow w = \text{Log } z + 2k\pi i$
$\Rightarrow w = \ln |z| + i\text{Arg } z + 2k\pi i$
$\Rightarrow w = \ln |z| + i(\text{Arg } z + 2k\pi)$

Raising Complex Numbers to Complex Powers

The value of x^t may be determined by the exponential and logarithm functions for real numbers

$$\ln x^t = t \ln x$$
$$x^t = e^{t \ln x}$$

The *principal λ power* of z is defined in terms of the principal logarithm as follows:

$$z^\lambda = e^{\lambda \text{ Log } z}$$

The general λ power of z is of the form:

$$z^\lambda e^{2k\pi\lambda i}$$

Complex Derivatives

A function $f: A \to \mathbb{C}$ is said to be differentiable at a point z_0 if f is continuous at z_0

$$f'(z_0) = \lim_{z \to z_0} \frac{f(z) - f(z_0)}{z - z_0}$$

and if the limit below exists. The derivative at z_0 is denoted by $f'(z_0)$.
It is often written as

$$f'(z_0) = \lim_{h \to 0} \frac{f(z_0 + h) - f(z_0)}{h}$$

28.3 Quaternions

The Irish mathematician, Sir William Rowan Hamilton, discovered quaternions in the nineteenth century (Fig. 28.4). Hamilton was born in Dublin in 1805 and attended Trinity College, Dublin. He was appointed professor of astronomy in 1827 while still an undergraduate. He made important contributions to optics, classical mechanics, and mathematics.

Hamilton had been trying to generalize complex numbers to triples without success. However, he had a sudden flash of inspiration on the idea of quaternion algebra at Broom's Bridge in 1843 while he was walking with his wife from his home at Dunsink Observatory to the Royal Irish Academy in Dublin. This route followed the towpath of the Royal Canal, and Hamilton was so overcome with emotion at his discovery of quaternions that he carved the quaternion formula into the stone on the bridge. Today, there is a plaque at Broom's Bridge that commemorates Hamilton's discovery (Fig. 28.5).

$$i^2 = j^2 = k^2 = ijk = -1$$

Quaternions are an extension of complex numbers, and Hamilton had been trying to extend complex numbers to three-dimensional space without success from

Fig. 28.4 William Rowan Hamilton

Fig. 28.5 Plaque at Broom's Bridge

the 1830s. Complex numbers are numbers of the form $(a + bi)$ where $i^2 = -1$ and may be regarded as points on a 2-dimensional plane. A quaternion number is of the form $(a + bi + cj + dk)$ where $i^2 = j^2 = k^2 = ijk = -1$ and may be regarded as points in 4-dimensional space.

The set of quaternions is denoted $\mathbb{H}$ by and the quaternions form an algebraic system known as a division ring. The multiplication of two quaternions is not commutative: i.e., given $q_1, q_2 \in \mathbb{H}$ then $q_1 q_2 \neq q_2 q_1$. Quaternions were the first non-commutative algebraic structure to be discovered, and other non-commutative algebras (e.g., matrix algebra) were discovered in later years.

Quaternions have applications in physics, quantum mechanics, and theoretical and applied mathematics. Gibbs and Heaviside later developed the vector analysis field from quaternions (see Chap. 29), and vector analysis replaced quaternions from the 1880s. Quaternions have become important in computing in recent years, as they are useful and efficient in describing rotations. They are applicable to computer graphics, computer vision, and robotics.

28.4   Quaternion Algebra

Hamilton had been trying to extend the 2-dimensional space of the complex numbers to a 3-dimensional space of triples. He wanted to be able to add, multiply, and divide triples of numbers but he was unable to make progress on the problem of the division of two triples.

The generalization of complex numbers to the 4-dimensional quaternions rather than triples allows the division of two quaternions to take place. The quaternion is a number of the form $(a + bi + cj + dk)$ where $1, i, j, k$ are the basis elements (where 1 is the identity) and satisfy the following properties:

$$i^2 = j^2 = k^2 = ijk = -1 \quad \text{(Quaternion Formula)}$$

This formula leads to the following properties:

$$ij = k = -ji$$
$$jk = i = -kj$$
$$ki = j = -ik$$

These properties can be easily derived from the quaternion formula. For example:

$$ijk = -1$$
$$\Rightarrow ijkk = -k \quad \text{(Right multiplying by } k)$$
$$\Rightarrow ij(-1) = -k \quad \text{(since } k^2 = -1)$$
$$\Rightarrow ij = k$$

Table 28.2 Basic quaternion multiplication

×	1	i	j	k
1	1	i	j	k
i	i	-1	k	$-j$
j	j	$-k$	-1	i
k	k	j	$-i$	-1

Similarly, from

$$ij = k$$
$$\Rightarrow i^2 j = ik \quad \text{(Left multiplying by } i)$$
$$\Rightarrow -j = ik \quad \text{(since } i^2 = -1)$$
$$\Rightarrow j = -ik$$

Table 28.2 represents the properties of quaternions under multiplication.

Hamilton saw that the i, j, k terms could represent the three Cartesian unit vectors **i**, **j**, **k**. The quaternions ($\mathbb{H}$) are a 4-dimensional vector space over the real numbers with three operations: addition, scalar multiplication, and quaternion multiplication.

Addition and Subtraction of Quaternions

The addition of two quaternions $q_1 = (a_1 + b_1 i + c_1 j + d_1 k)$ and $q_2 = (a_2 + b_2 i + c_2 j + d_2 k)$ is given by

$$q_1 + q_2 = (a_1 + a_2) + (b_1 + b_2)i + (c_1 + c_2)j + (d_1 + d_2)k$$
$$q_1 - q_2 = (a_1 - a_2) + (b_1 - b_2)i + (c_1 - c_2)j + (d_1 - d_2)k$$

Identity Element

The addition identity is given by the quaternion $(0 + 0i + 0j + 0k)$, and the multiplicative identity is given by $(1 + 0i + 0j + 0k)$.

Multiplication of Quaternions

The multiplication of two quaternions q_1 and q_2 is determined by the product of the basis elements and the distributive law. It yields (after a long calculation that may be simplified with a representation of 2×2 matrices over Complex Numbers as described later in the chapter):

$$\begin{aligned} q_1 . q_2 = {} & a_1 a_2 + a_1 b_2 i + a_1 c_2 j + a_1 d_2 k \\ & + b_1 a_2 i + b_1 b_2 ii + b_1 c_2 ij + b_1 d_2 ik \\ & + c_1 a_2 j + c_1 b_2 ji + c_1 c_2 jj + c_1 d_2 jk \\ & + d_1 a_2 k + d_1 b_2 ki + d_1 c_2 kj + d_1 d_2 kk \end{aligned}$$

This may then be simplified to:

$$q_1 \cdot q_2 = a_1 a_2 - b_1 b_2 - c_1 c_2 - d_1 d_2$$
$$+ (a_1 b_2 + b_1 a_2 + c_1 d_2 - d_1 c_2) i$$
$$+ (a_1 c_2 - b_1 d_2 + c_1 a_2 + d_1 b_2) j$$
$$+ (a_1 d_2 + b_1 c_2 - c_1 b_2 + d_1 a_2) k$$

The multiplication of two quaternions may be defined in terms of matrix multiplication. It is easy to see that the product of the two quaternions above is equivalent to:

$$q_1 q_2 = \begin{pmatrix} a_1 & -b_1 & -c_1 & -d_1 \\ b_1 & a_1 & -d_1 & c_1 \\ c_1 & d_1 & a_1 & -b_1 \\ d_1 & -c_1 & b_1 & a_1 \end{pmatrix} \begin{pmatrix} a_2 \\ b_2 \\ c_2 \\ d_2 \end{pmatrix}$$

This may also be written as:

$$(a_2 b_2 c_2 d_2) \begin{pmatrix} a_1 & b_1 & c_1 & d_1 \\ -b_1 & a_1 & d_1 & -c_1 \\ -c_1 & -d_1 & a_1 & b_1 \\ -d_1 & c_1 & -b_1 & a_1 \end{pmatrix} = q_1 q_2$$

Property of Quaternions Under Multiplication

The quaternions are *not commutative* under multiplication. That is,

$$q_1 q_2 \neq q_2 q_1$$

The quaternions are associative under multiplication. That is,

$$q_1 (q_2 q_3) = (q_1 q_2) q_3$$

Conjugation

The conjugation of a quaternion is analogous to the conjugation of a complex number. The conjugate of a complex number $z = (a + bi)$ is given by $z^* = (a - bi)$. Similarly, the conjugate of a quaternion is determined by reversing the sign of the vector part of the quaternion. That is, the conjugate of $q = (a + bi + cj + dk)$ (denoted by q^*) is given by $q^* = (a - bi - cj - dk)$. Similarly, $q^{**} = q$.

Scalar and Vector Parts of Quaternion

A quaternion $(a + bi + cj + dk)$ consists of a *scalar* part a, and a *vector* part $bi + cj + dk$. The scalar part is always real, and the vector part is imaginary. That is, the quaternion q may be represented $q = (s, v)$ where s is the scalar part and v is the vector part. The scaler part of a quaternion is given by $s = (q + q^*)/2$, and the vector part is given by $v = (q - q^*)/2$.

The vector part of a quaternion may be regarded as a coordinate vector in three-dimensional space $\mathbb{R}^3$, and the algebraic operations of quaternions reflect the geometry of three-dimensional space. The elements i, j, k represents imaginary basis vectors of the quaternions and the basis elements in $\mathbb{R}^3$.

The *norm* of a quaternion q (denoted by $\|q\|$) is given by:

$$\|q\| = \sqrt{qq^*} = \sqrt{q^*q} = \sqrt{a^2 + b^2 + c^2 + d^2}$$

A quaternion of norm one is termed a unit quaternion (i.e., $\|u\| = 1$). Any quaternion u where u is defined by $u = q/\|q\|$ is a unit quaternion. Unit quaternions may be identified with rotations in $\mathbb{R}^3$. Given $\alpha \in \mathbb{R}$ then $\|\alpha q\| = |\alpha| \, \|q\|$.

The product of two quaternions may also be given in terms of the scalar and vector parts. That is, if $q_1 = (s_1, v_1)$ and $q_2 = (s_2, v_2)$ then the product of q_1 and q_2 is given by:

$$q_1 q_2 = (s_1 s_2 - v_1.v_2, s_1 v_2 + s_2.v_1 + v_1 \times v_2)$$

where "." is the dot product and "$\times$" is the cross product as defined in Chap. 29.

Inverse of a Quaternion

The inverse of a quaternion q is given by q^{-1} where

$$q^{-1} = q^*/\|q\|^2$$

and $qq^{-1} = q^{-1}q = 1$

Given two quaternions p and q we have:

$$\|pq\| = \|p\| \|q\|$$

The norm is used to define the distance between two quaternions, and the distance between two quaternions p and q (denoted by $d(p, q)$) is given by:

$$\|p - q\|$$

Representing Quaternions with 2×2 Matrices Over Complex Numbers

The quaternions have an interpretation under the 2×2 matrices where the basis elements i, j, k may be interpreted as matrices. Recall, that the multiplicative identity for 2×2 matrices is

$$1 = \begin{bmatrix} 1 & 0 \\ 0 & 1 \end{bmatrix} \quad -1 = \begin{bmatrix} -1 & 0 \\ 0 & -1 \end{bmatrix}$$

Consider then the quaternion basis elements defined as follows:

$$i = \begin{bmatrix} 0 & 1 \\ -1 & 0 \end{bmatrix} \quad j = \begin{bmatrix} 0 & i \\ i & 0 \end{bmatrix} \quad k = \begin{bmatrix} i & 0 \\ 0 & -i \end{bmatrix}$$

Then a simple calculation shows that:

$$i^2 = j^2 = k^2 = ijk = \begin{bmatrix} -1 & 0 \\ 0 & -1 \end{bmatrix} = -1$$

Then the quaternion $q = (a + bi + cj + dk)$ may also be defined as

$$a \begin{bmatrix} 1 & 0 \\ 0 & 1 \end{bmatrix} + b \begin{bmatrix} 0 & 1 \\ -1 & 0 \end{bmatrix} + c \begin{bmatrix} 0 & i \\ i & 0 \end{bmatrix} + d \begin{bmatrix} i & 0 \\ 0 & -i \end{bmatrix}$$

This may be simplified to the complex matrix

$$q = \begin{bmatrix} a + di & b + ci \\ -b + ci & a - di \end{bmatrix}$$

and this is equivalent to:

$$q = \begin{bmatrix} u & v \\ -v^* & u^* \end{bmatrix}$$

where $u = a + di$ and $v = b + ci$.

The addition and multiplication of quaternions are then just the usual matrix addition and multiplication. Quaternions may also be represented by 4×4 real matrices.

28.4.1 Quaternions and Rotations

Quaternions may be applied to computer graphics; computer vision, and robotics, and unit quaternions provide an efficient mathematical way to represent rotations in three dimensions (Fig. 28.6). They offer an alternative to Euler angles and matrices.

The unit quaternion $q = (s, v)$ that computes the rotation about the unit vector u by an angle θ is given by:

$$(\text{Cos}(\theta/2), u \, \text{Sin}(\theta/2))$$

The scalar part is given by $s = \text{Cos}(^\theta/_2)$, and the vector part is given by $v = u \, \text{Sin}(^\theta/_2)$.

Fig. 28.6 Quaternions and
Rotations

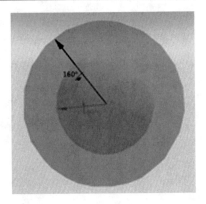

A point p in space is represented by the quaternion $P = (0, p)$. The result of
the rotation of p is given by:

$$P_q = qPq^{-1}$$

The norm of q is 1, and so q^{-1} is given by $(\text{Cos}(^\theta/_2), -u\,\text{Sin}(^\theta/_2))$.

Suppose we have two rotations represented by the unit quaternions q_1 and q_2,
and that we first wish to perform q_1 followed by the rotation q_2. Then, the composition of the two relations is given by applying q_2 to the result of applying q_1.
This is given by the following:

$$P_{(q_2^\circ q_1)} = q_2\left(q_1 P q_1^{-1}\right) q_2^{-1}$$
$$= q_2\, q_1 P q_1^{-1} q_2^{-1}$$
$$= (q_2\, q_1) P (q_2\, q_1)^{-1}$$

28.5   Review Questions

1. What is a complex number?
2. Show how a complex number may be represented as a point in the
   complex plane.
3. Show that $|z_1 z_2| = |z_1|\,|z_2|$.
4. Evaluate the following
   (a) $(-1)^{1/4}$
   (b) $1^{1/5}$
5. What is the fundamental theorem of algebra?
6. Show that $^d/_{dz}\, z^n = nz^{n-1}$.
7. What is a quaternion?

8. Investigate the application of quaternions to computer graphics, computer vision, and robotics.

28.6 Summary

A complex number z is a number of the form $a + bi$ where a and b are real numbers and $i^2 = -1$. The set of complex numbers is denoted by $\mathbb{C}$, and a complex number has two parts namely its real part a and imaginary part b. The complex numbers are an extension of the set of real numbers, and complex numbers with a real part $a = 0$ are termed imaginary numbers. Complex numbers have many applications in physics, engineering, and applied mathematics.

The Irish mathematician, Sir William Rowan Hamilton, discovered quaternions. Hamilton had been trying to extend the 2-dimensional space of the complex numbers to a 3-dimensional space of triples. He wanted to be able to add, multiply, and divide triples of numbers, but he was unable to make progress on the problem of the division of two triples. His insight was that if he considered quadruples rather than triples that this structure would give him the desired mathematical properties. Hamilton also made important contributions to optics, classical mechanics, and mathematics.

The generalization of complex numbers to the 4-dimensional quaternions rather than triples allows the division of two quaternions to take place. The quaternion is a number of the form $(a + bi + cj + dk)$ where $1, i, j, k$ are the basis elements (where 1 is the identity) and satisfy the quaternion formula:

$$i^2 = j^2 = k^2 = ijk = -1$$

Vectors

29

Key Topics

Addition and subtraction of vectors

Scalar multiplication

Dot product

Cross product

29.1   Introduction

A *vector* is an object that has magnitude and direction, and the origin of the term is from the Latin word '*vehere*' meaning to carry or to convey. Vectors arose out of Hamilton's discovery of quaternions in 1843, and quaternions arose from Hamilton's attempts to generalize complex numbers to triples, where a quaternion consists of a scalar part and a vector part (see Chap. 28). They were subsequently refined by Gibbs who treated the vector part of a quaternion as an independent object in its own right. Gibbs defined the scalar and dot product of vectors using the relevant part of quaternion algebra, and his approach led to the new discipline of vector analysis. Gibbs's vector notation became dominant with quaternion algebra fading into obscurity until more recent times.

A vector is represented as a directed line segment such that the length represents the magnitude of the vector, and the arrow indicates the direction of the vector. It is clear that velocity is a vector in the example of a jogger running at 9 km/h in a westerly direction, whereas speed is not a vector since the direction is not specified when we state that a runner travelled 6 km at an average speed of 9 km/h. The magnitude of the velocity vector (9 km/h) indicates how large it is and the direction is 270° (the direction is usually given by a unit vector).

© The Author(s), under exclusive license to Springer Nature Switzerland AG 2023
G. O'Regan, *Mathematical Foundations of Software Engineering*,
Texts in Computer Science, https://doi.org/10.1007/978-3-031-26212-8_29

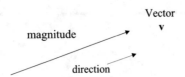

The magnitude of a vector **v** is denoted by ‖v‖, and all vectors (apart from the zero vector **0** which has magnitude 0 and does not point in any particular direction) have a positive magnitude. The vector - **v** represents the same vector as **v** except that it points in the opposite direction. The unit vector **u** has a magnitude of 1. Two vectors are equal if they have the same magnitude and direction.

The addition of two vectors **a** and **b** may be seen visually below (triangle law), and the addition of two vectors is commutative with **a** + **b** = **b** + **a** (parallelogram law). Similarly, vectors may be subtracted with **a** − **b** = **a** + (-**b**).

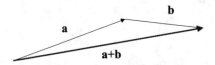

The multiplication of a vector by a scalar changes its magnitude (length) but not its direction, except that the multiplication by a negative scalar reverses the direction of the vector. For example, the multiplication of a vector by 0.5 results in a vector half as large in the same direction, whereas multiplication by –2 results in a vector twice as large pointing in the opposite direction.

There are two ways of multiplying two vectors **v** and **w** together namely the *cross product* (**v** × **w**) and the *dot product* (**v** · **w**). The cross product (or vector product) is given by the formula **v** × **w** = ‖v‖ ‖w‖ Sin θ **n**, where θ is the angle between **v** and **w**, and **n** is the unit vector perpendicular to the plane containing **v** and **w**. The cross product is anti-commutative: i.e., **v** × **w** = - **w** × **v**.

The dot product (or scalar product) of two vectors is given by the formula **v.w** = ‖v‖ ‖w‖ Cos θ, where θ is the angle between **v** and **w**. The dot product is often employed to find the angle between the two vectors, and the dot product is commutative with **v.w** = **w.v**.

Example 29.1 An aircraft can fly at 300 and a 40 kph wind is blowing in a south easterly direction. The aircraft sets off due north. What is its velocity with respect to the ground?

Solution
The aircraft is flying at 300 kph N and the wind is blowing at 40 kph SE and so the resulting velocity is the vector **v**.

The magnitude of **v** is given from the cosine rule:

$$\|\mathbf{v}\|^2 = 300^2 + 40^2 - 2 * 300 * 40\text{Cos}45°$$
$$\|\mathbf{v}\| = 213.2\,\text{kph}$$

The direction of **v** is given by the angle φ which is determined from the sine rule.

$$\frac{\text{Sin}45}{273.2} = \frac{\text{Sin}\varphi}{40}$$
$$0.1035 = \text{Sin}\varphi$$
$$\varphi = 5.96°$$

Thus the velocity of the aircraft is 273.2 kph in the direction of $(90 - 5.96) = 84.04°$.

29.2 Vectors in Euclidean Space

A vector in Euclidean space is represented as an ordered pair (tuple) in $\mathbb{R}^2$, an ordered triplet (3-tuple) of numbers in $\mathbb{R}^3$, and an ordered n-tuple in $\mathbb{R}^n$. For example, the point (a, b) is a vector in two-dimensional space and the triple (a, b, c) is a vector in three-dimensional space. The vector is drawn by joining an arrow from the origin to the point (in 2-dimensional or 3-dimensional space) (Fig. 29.1)

The addition and subtraction of two vectors $\mathbf{v} = (a, b, c)$ and $\mathbf{w} = (x, y, z)$ are given by:

$$\mathbf{v} + \mathbf{w} = (a + x, b + y, c + z)$$

Fig. 29.1 The Vector (a,b) in Euclidean Plane

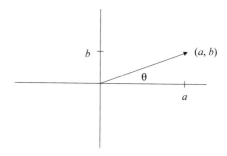

$$\mathbf{v} - \mathbf{w} = (a - x, b - y, c - z)$$

The multiplication of a vector $\mathbf{v}$ by a scalar k is given by:

$$k\mathbf{v} = (ka, kb, kc)$$

For any vector $\mathbf{v} = (a, b)$ in 2-dimensional space then from Pythagoras's Theorem we have that the magnitude of $\mathbf{v}$ is given by:

$$\|\mathbf{v}\| = \sqrt{(a^2 + b^2)}.$$

Similarly, for any vector $\mathbf{v} = (a, b, c)$ in 3-dimensional space then from Pythagoras's Theorem we have:

$$\|\mathbf{v}\| = \sqrt{(a^2 + b^2 + c^2)}.$$

We can create the unit vector along $\mathbf{v}$ in 2-dimensional space by:

$$\frac{\mathbf{v}}{\|\mathbf{v}\|} = \left(\frac{a}{\|\mathbf{v}\|}, \frac{b}{\|\mathbf{v}\|}\right)$$

The zero vector $\mathbf{0}$ is $(0, 0)$ in 2-dimensional space and $(0, 0, 0)$ in 3-dimensional space.

The standard basis for three-dimensional space is $\mathbf{i} = (1, 0, 0)$, $\mathbf{j} = (0, 1, 0)$, and $\mathbf{k} = (0, 0, 1)$, and any vector $\mathbf{v} = (a, b, c)$ may be expressed as a linear combination of the basis elements:

$$\mathbf{v} = a\mathbf{i} + b\mathbf{j} + c\mathbf{k}$$

Example 29.2 Calculate the magnitude of $\mathbf{v}$ where $\mathbf{v}$ is $(2,3,1)$. Determine the unit vector along $\mathbf{v}$.

Solution

The magnitude of $\mathbf{v}$ (or $\|\mathbf{v}\|$) is given by:

$$\begin{aligned}
\|\mathbf{v}\| &= \sqrt{(a^2 + b^2 + c^2)} \\
&= \sqrt{(2^2 + 3^2 + 1^2)} \\
&= \sqrt{14} \\
&= 3.74
\end{aligned}$$

The unit vector along $\mathbf{v}$ is given by:

$$\begin{aligned}
\frac{\mathbf{v}}{\|\mathbf{v}\|} &= \left(\frac{a}{\|\mathbf{v}\|}, \frac{b}{\|\mathbf{v}\|}, \frac{c}{\|\mathbf{v}\|}\right) \\
&= \left(\frac{2}{3.74}, \frac{3}{3.74}, \frac{1}{3.74}\right) \\
&= (0.53, 0.8, 0.27)
\end{aligned}$$

29.2.1 Dot Product

The result of the dot product (also called scalar product) of two vectors is a scalar, and the dot product of two vectors $\mathbf{v} = (a, b, c)$ and $\mathbf{w} = (x, y, z)$ is given by:

$$\mathbf{v} \cdot \mathbf{w} = ax + by + cz$$

We previously defined the dot product as.

$$\mathbf{v} \cdot \mathbf{w} = \|\mathbf{v}\| \|\mathbf{w}\| \cos \theta$$

where θ is the angle between $\mathbf{v}$ and $\mathbf{w}$

The dot product of the basis elements is given by

$$\mathbf{i} \cdot \mathbf{j} = 1.1.\cos \pi/2 = 0$$
$$\mathbf{j} \cdot \mathbf{k} = 1.1.\cos \pi/2 = 0$$
$$\mathbf{k} \cdot \mathbf{i} = 1.1.\cos \pi/2 = 0$$

The two expressions for the calculation of the dot product allow the angle between $\mathbf{v}$ and $\mathbf{w}$ to be determined:

$$\|\mathbf{v}\| \|\mathbf{w}\| \cos\theta = ax + by + cz$$
$$\cos\theta = \frac{ax + by + cz}{\|\mathbf{v}\| \|\mathbf{w}\|}$$
$$\theta = \cos^{-1}\left(\frac{ax + by + cz}{\|\mathbf{v}\| \|\mathbf{w}\|}\right)$$

Example 29.3 Calculate the dot product of $\mathbf{v}$ and $\mathbf{w}$ where $\mathbf{v}$ is (2,3,1) and $\mathbf{w}$ is (3,4,1). Determine the angle between $\mathbf{v}$ and $\mathbf{w}$.

Solution

The dot products of $\mathbf{v}$ and $\mathbf{w}$ are given by:

$$\mathbf{v} \cdot \mathbf{w} = ax + by + cz$$
$$= 2.3 + 3.4 + 1.1$$
$$= 19$$

The magnitudes of $\mathbf{v}$ and $\mathbf{w}$ are given by:

$$\|\mathbf{v}\| = \sqrt{(a^2 + b^2 + c^2)} == \sqrt{(2^2 + 3^2 + 1^2)} = 3.74$$
$$\|\mathbf{w}\| = \sqrt{(x^2 + y^2 + z^2)} == \sqrt{(3^2 + 4^2 + 1^2)} = \sqrt{26} = 5.1$$

$\|\mathbf{v}\| \|\mathbf{w}\| \cos\theta = ax + by + cz$

$\Rightarrow 3.74 * 5.1 \cos\theta = 19$

⇨ $19.07 \, Cos\theta = 19$

⇨ $Cos\theta = 0.9963$

⇨ $\theta = 4.92°$

29.2.2 Cross Product

The result of the cross product (also called vector product) of two vectors **v** and **w** is another vector that is perpendicular to both **v** and **w**. The cross product of the vectors $\mathbf{v} = (a, b, c)$ and $\mathbf{w} = (x, y, z)$ is given by:

$$\mathbf{v} \times \mathbf{w} = (bz - cy, cx - az, ay - bx)$$

The cross product is also defined by:

$$\mathbf{v} \times \mathbf{w} = \|\mathbf{v}\| \|\mathbf{w}\| \, Sin \, \theta \, \mathbf{n}$$

θ is the angle between **v** and w

n   is the unit vector perpendicular to the plane containing **v** and **w**.

The cross product is anti-commutative

$$\mathbf{v} \times \mathbf{w} = -\mathbf{w} \times \mathbf{v}$$

The cross product of the basis elements is given by

$$\mathbf{i} \times \mathbf{j} = \mathbf{k}$$
$$\mathbf{j} \times \mathbf{k} = \mathbf{i}$$
$$\mathbf{k} \times \mathbf{i} = \mathbf{j}$$

The cross product may also be expressed as the determinant of a 3-dimensional matrix which has the basis elements on the first row, and the vectors in the remaining rows. This is seen as follows:

$$\mathbf{v} \times \mathbf{w} = \begin{vmatrix} i & j & k \\ a & b & c \\ x & y & z \end{vmatrix}$$

The magnitude of $\mathbf{v} \times \mathbf{w}$ may be interpreted as the area of a parallelogram having **v** and **w** as sides.

Example 29.4 Calculate the cross product of **v** and **w** where **v** is (2,3,1) and **w** is (3,4,1). Show that its cross product is anti-commutative. Show that **v** × **w** is perpendicular to **v** and **w**.

Solution
The cross product of **v** and **w** is given by:

$$\mathbf{v} \times \mathbf{w} = \begin{vmatrix} \mathbf{i} & \mathbf{j} & \mathbf{k} \\ 2 & 3 & 1 \\ 3 & 4 & 1 \end{vmatrix}$$

$$= (3 - 4)\mathbf{i} - (2 - 3)\mathbf{j} + (8 - 9)\mathbf{k}$$

$$= -\mathbf{i} + \mathbf{j} - \mathbf{k}$$

$$= (-1, 1, -1)$$

The cross product of **w** and **v** is given by:

$$\mathbf{w} \times \mathbf{v} = \begin{vmatrix} \mathbf{i} & \mathbf{j} & \mathbf{k} \\ 3 & 4 & 1 \\ 2 & 3 & 1 \end{vmatrix}$$

$$= (4 - 3)\mathbf{i} - (3 - 2)\mathbf{j} + (9 - 8)\mathbf{k}$$

$$= \mathbf{i} - \mathbf{j} + \mathbf{k}$$

$$= (1, -1, 1)$$

$$= -\mathbf{v} \times \mathbf{w}$$

Finally, to show that **v** × **w** is perpendicular to both **v** and **w** we compute the dot product of **v** × **w** with **v** and **w** (if they are perpendicular the dot product will be zero) and we get:

$$(\mathbf{v} \times \mathbf{w}) \cdot \mathbf{v} = (-1, 1, -1) \cdot (2, 3, 1) = -2 + 3 - 1 = 0$$
$$(\mathbf{v} \times \mathbf{w}) \cdot \mathbf{w} = (-1, 1, -1) \cdot (3, 4, 1) = -3 + 4 - 1 = 0$$

29.2.3 Linear Dependent and Independent Vectors

A set of vectors $\mathbf{v}_1, \mathbf{v}_2, \dots \mathbf{v}_n$ in a vector space V are said to be linearly dependent if there is a set of values $\lambda_1, \lambda_2, \dots \lambda_n$ (where not all of the λ_i are zero) and

$$\lambda_1 \mathbf{v}_1 + \lambda_2 \mathbf{v}_2 + \dots + \lambda_n \mathbf{v}_n \lambda = \mathbf{0}$$

A set of vectors $\mathbf{v}_1, \mathbf{v}_2, \dots \mathbf{v}_n$ in a vector space V are said to be linearly independent if whenever $\lambda_1 \mathbf{v}_1 + \lambda_2 \mathbf{v}_2 + \dots + \lambda_n \mathbf{v}_n \lambda = \mathbf{0}$ then $\lambda_1 = \lambda_2 = \dots = \lambda_n = 0$.

Example 29.5 Determine whether the following pairs of vectors are linearly independent or linearly dependent.

(i) $\mathbf{v}$ is $(1, 2, 3)$ and $\mathbf{w}$ is $(2,4,6)$.
(ii) $\mathbf{v}$ is $(2,3,1)$ and $\mathbf{w}$ is $(3,4,1)$.

Solution

(i) We are seeking λ_1 and λ_2 such that $\lambda_1\mathbf{v} + \lambda_2\mathbf{w} = \mathbf{0}$.

$$\lambda_1\mathbf{v} + \lambda_2\mathbf{w}$$
$$= \lambda_1(1, 2, 3) + \lambda_2(2, 4, 6)$$
$$= (\lambda_1 + 2\lambda_2, 2\lambda_1 + 4\lambda_2, 3\lambda_1 + 6\lambda_2)$$

Thus when $\lambda_1\mathbf{v} + \lambda_2\mathbf{w} = \mathbf{0}$ we have:
$\lambda_1 + 2\lambda_2 = 0$, $2\lambda_1 + 4\lambda_2 = 0$ and $3\lambda_1 + 6\lambda_2 = 0$.
That is, $\lambda_1 + 2\lambda_2 = 0$ or $\lambda_1 = -2\lambda_2$ and so we may take $\lambda_1 = 2$, $\lambda_2 = -1$.
That is, $\mathbf{v}$ is $(1, 2, 3)$ and $\mathbf{w}$ is $(2,4,6)$ are linearly dependent.

(ii) $\lambda_1\mathbf{v} + \lambda_2\mathbf{w}$
$$= \lambda_1(2, 3, 1) + \lambda_2(3, 4, 1).$$
$$= (2\lambda_1 + 3\lambda_2, 3\lambda_1 + 4\lambda_2, \lambda_1 + \lambda_2).$$

Thus when $\lambda_1\mathbf{v} + \lambda_2\mathbf{w} = \mathbf{0}$ we have three equations with two unknowns and so we solve for λ_1 and λ_2:

a. $2\lambda_1 + 3\lambda_2 = 0$
b. $3\lambda_1 + 4\lambda_2 = 0$
c. $\lambda_1 + \lambda_2 = 0$

We deduce $\lambda_1 = -\lambda_2$ from equation (c) and we substitute for λ_1 in equation (a) to get:
$2(-\lambda_2) + 3\lambda_2 = 0.$
$\Rightarrow \lambda_2 = 0$
$\Rightarrow \lambda_1 = 0$

That is, $\mathbf{v} = (2, 3, 1)$ and $\mathbf{w} = (3, 4, 1)$ are linearly independent.

29.3 Review Questions

1. What is a vector?
2. Explain the triangle and parallelogram laws.
3. An aircraft is capable of flying at 500 kph and a 60 kph wind is blowing in a south westerly direction. The aircraft sets off due north. What is its velocity with respect to the ground?
4. Determine the dot product of u and v where $u = (3, 1, 5)$ and $v = (2, 4, 3)$.
5. Determine the angle between u and v where $u = (3, 1, 5)$ and $v = (2, 4, 3)$.
6. Determine the cross product of u and v where $u = (2, 1, 4)$ and $v = (1, 3, 3)$.
7. Determine if the vectors u and v are linearly independent.

29.4 Summary

A *vector* is an object that has magnitude and direction, and it is represented as a directed line segment such that the length represents the magnitude of the vector and the arrow indicates the direction of the vector. A vector in Euclidean space is represented as an ordered pair in $\mathbb{R}^2$, an ordered triplet of numbers in $\mathbb{R}^3$, and an ordered n-tuple in $\mathbb{R}^n$.

The dot product of two vectors is a scalar, whereas the cross product of two vectors v and w results in another vector that is perpendicular to both v and w.

A set of vectors $v_1, v_2, \ldots v_n$ in a vector space V are said to be linearly dependent if there is a set of values $\lambda_1, \lambda_2, \ldots \lambda_n$ (where not all of the λ_i are zero) and $\lambda_1 v_1 + \lambda_2 v_2 + \cdots + \lambda_n v_n \lambda = 0$. Otherwise, they are said to be linearly independent.

Basic Financial Mathematics

30

Key Topics

Simple interest

Compound interest

Treasury bills

Promissory notes

Annuities

Present and future values

Equivalent values

30.1   Introduction

Banks are at the heart of the financial system, and they play a key role in facilitating the flow of money throughout the economy. Banks provide a service where those with excess funds (i.e., the savers) may lend to those who need funds (i.e., the borrowers). Banks pay interest to its savers[1] and earn interest from its borrowers, and the spread between the interest rate given to savers and the interest rate charged to borrowers provides banks with the revenue to provide service to their clients, and to earn a profit from the provision of the services.

---

[1] We are assuming that the country is not in a cycle of negative interest rates where investors are essentially charged to place their funds on deposit with the bank.

We distinguish between simple and compound interest, with simple interest calculated on the principal only, whereas compound interest is calculated on both the principal and the accumulated interest of previous compounding periods. That is, simple interest is always calculated on the original principal, whereas for compound interest, the interest is added to the principal sum, so that interest is also earned on the added interest for the next compounding period.

The future value is what the principal will amount to in future at a given rate of interest, whereas the present value of an amount to be received in future is the principal that would grow to that amount at a given rate of interest.

An annuity is a sequence of fixed equal payments made over a period of time, and it is usually paid at the end of the payment interval. For example, for a hire purchase contract the buyer makes an initial deposit for an item and then pays an equal amount per month (the payment is generally at the end of the month) up to a fixed end date. Personal loans from banks are paid back in a similar manner but without an initial deposit.

An interest-bearing debt is amortized if both the principal and interest are repaid by a series of equal payments (except for possibly the last payment) made at equal intervals of time. The debt is repaid by an amortization annuity, where each payment consists of both the repayment of the capital borrowed and the interest due on the capital for that time interval.

30.2   Simple Interest

Savers receive interest for placing deposits at the bank for a period of time, whereas lenders pay interest on their loans to the bank. Simple interest is generally paid on *term deposits* (these are usually short-term fixed-term deposits for 3, 6, or 12 months) or short-term investments or loans. The interest earned on a savings account depends on the principal amount placed on deposit at the bank, the period of time that it will remain on deposit, and the specified rate of interest for the period.

For example, if Euro 1000 is placed on deposit at a bank with an interest rate of 10% per annum for two years, then it will earn a total of Euro 200 in simple interest for the period. The interest amount is calculated by

$$\frac{1000 * 10 * 2}{100} = \text{EURO } 200$$

The general formula for calculating the amount of simple interest A due for placing principal P on deposit at a rate r of interest (where r is expressed as a percentage) for a period of time T (in years) is given by:

$$I = \frac{P \times r \times t}{100}$$

If the rate of interest r is expressed as a decimal then the formula for the interest earned is simply:

$$I = P \times r \times t$$

It is essential in using the interest rate formula that the units for time and rate of interest are the same:

Example 30.1 (Simple Interest) Calculate the simple interest payable for the following short-term investments.

1. £5000 placed on deposit for six months ($^1/_2$ year) at an interest rate of 4%.
2. £3000 placed on deposit for one month ($^1/_{12}$ year) at an interest rate of 5%.
3. £10,000 placed on deposit for one day ($^1/_{365}$ year) at an interest rate of 7%.

Solution (Simple Interest)

1. $A = 5000 * 0.04 * 0.5 = £100$
2. $A = 3000 * 0.05 * 0.08333 = £12.50$
3. $A = 10000 * 0.07 * 0.00274 = £1.92$

We may derive various formulae from the simple interest formula $A = P \times r \times T$.

$$P = \frac{I}{rt} \quad r = \frac{I}{Pt} \quad t = \frac{I}{Pr}$$

Example 30.2 (Finding the Principal, Rate or Time) Find the value of the principal or rate or time in the following.

1. What principal will earn interest of €24.00 at 4.00% in eight months?
2. Find the annual rate of interest for a principal of €800 to earn €50 in interest in nine months.
3. Determine the number of months required for a principal of €2000 to earn €22 in interest at a rate of 5%.

Solution

We use the formulae derived from the simple interest formula to determine these.

1. $P = \frac{I}{rt} = \frac{24}{0.04 \times 0.6666} = €900$
2. $r = \frac{I}{Pt} = \frac{50}{800 \times 0.75} = 0.0833 = 8.33\%$
3. $t = \frac{I}{Pr} = \frac{22}{2000 \times 0.05} = 0.22$ years $= 2.64$ months

30.2.1 Computing Future and Present Values

The future value is what the principal will amount to in future at a given rate of interest, whereas the present value of an amount to be received in future is the principal that would grow to that amount at a given rate of interest.

30.2.2 Computing Future Value

A fixed-term account is an account that is opened for a fixed period of time (typically 3, 6, or 12 months). The interest rate is fixed during the term, and thus, the interest due at the maturity date is known in advance. That is, the customer knows what the *future value* (FV) of the investment will be and knows what is due on the maturity date of the account (this is termed the *maturity value*).

On the *maturity date* both the interest due and the principal are paid to the customer, and the account is generally closed on this date. In some cases, the customer may agree to roll over the principal, or the principal and interest for a further fixed period of time, but there is no obligation on the customer to do so. The account is said to mature on the maturity date, and the maturity value (MV) or future value (FV) is given by the sum of the principal and interest:

$$MV = FV = P + I$$

Further, since $I = Prt$ we can write this as $MV = P + Prt$ or

$$FV = MV = P(1 + rt)$$

Example 30.3 (Computing Maturity Value) Jo invests €10,000 in a short-term investment for three months at an interest rate of 9%. What is the maturity value of her investment?

Solution (Computing Maturity Value)

$$MV = P(1 + rt)$$
$$= 10{,}000(1 + 0.09 * 0.25)$$
$$= €10{,}225$$

30.2.3 Computing Present Values

The present value of an amount to be received at a given date in future is the principal that will grow to that amount at a given rate of interest over that period of time. We computed the maturity value of a given principal at a given rate of interest r over a period of time t as:

$$MV = P(1 + rt)$$

Therefore, the present value (PV $= P$) of an amount V to be received t years in future at an interest rate r (simple interest) is given by:

$$PV = P = \frac{V}{(1 + rt)}$$

Example 30.4 (Computing Present Value) Compute the present value of an investment eight months prior to the maturity date, where the investment earns interest of 7% per annum and has a maturity value of €920.

Solution (Computing Present Value)

$$V = 920, r = 0.07, t = 8/12 = 0.66$$

$$PV = P = \frac{V}{(1 + rt)} = \frac{920}{(1 + 0.07 * 0.66)} = €879.37$$

Example 30.5 (Equivalent Values) Compute the payment due for the following:

1. A payment of $2000 is due in one year. It has been decided to repay early and payment is to be made today. What is the equivalent payment that should be made given that the interest rate is 10%?
2. A payment of $2000 is due today. It has been agreed that payment will instead be made six months later. What is the equivalent payment that will be made at that time given that the interest rate is 10%?

Solution (Equivalent Values)

1. The original payment date is 12 months from today and is now being made 12 months earlier than the original date. Therefore, we compute the present value of $2000 for 12 months at an interest rate of 10%.

$$PV = P = \frac{V}{(1 + rt)} = \frac{2000}{(1 + 0.1 * 1)} = €1818.18$$

2. The original payment date is today but has been changed to six months later, and so we compute the future value of $2000 for six months at an interest rate of 10%.

$$FV = P(1 + rt) = 2000(1 + 0.1 * 0.5) = 2000(1.05) = \$2100$$

Example 30.6 (Equivalent Values) A payment of €5000 that is due today is to be replaced by two equal payments (we call the unknown payments value x) due in four and eight months, where the interest rate is 10%. Find the value of the replacement payments.

Solution (Equivalent Values)

The sum of the present value of the two (equal but unknown) payments is €5000.

The present value of x (received in four months) is

$$\text{PV} = P = \frac{V}{(1+rt)} = \frac{x}{(1+0.1*0.33)} = \frac{x}{1.033} = 0.9678x$$

The present value of x (received in eight months) is

$$\text{PV} = P = \frac{V}{(1+rt)} = \frac{x}{(1+0.1*0.66)} = \frac{x}{1.066} = 0.9375x$$

Therefore,

$$0.9678x + 0.9375x = €5000$$
$$\Rightarrow 1.9053x = €5000$$
$$\Rightarrow x = €2624.26$$

30.3   Compound Interest

The calculation of compound interest is more complicated as may be seen from the following example:

Example 30.7 (Compound Interest) Calculate the interest earned and what the new principal will be on Euro 1000, which is placed on deposit at a bank, with an interest rate of 10% per annum (compound) for three years.

Solution

At the end of year 1, Euro 100 of interest is earned, and this is added to the existing principal making the new principal (at the start of year 2) €1000 + €100 = Euro 1100. At the end of year 2, Euro 110 is earned in interest, and this is added to the principal making the new principal (at the start of year 3) €1100 + €110 = Euro 1210. Finally, at the end of year 3 a further Euro 121 is earned in interest, and so the new principal is Euro 1331 and the total interest earned for the three years is the sum of the interest earned for each year (i.e., Euro 331). This may be seen from Table 30.1.

The new principal each year is given by a geometric sequence (recall a geometric sequence is a sequence in the form $a, ar, ar^2, \ldots ar^n$). For this example, we have $a = 1000$, and as the interest rate is $10\% = {}^1/_{10} = 0.1$ we have $r = (1+0.1)$, and so the sequence is:

$$1000, 1000(1.1), 1000(1.1)^2, 1000(1.1)^3 \ldots$$

Table 30.1 Calculation of compound interest

Year	Principal	Interest earned	New principal
1	€1000	€100	€1100
2	€1100	€110	€1210
3	€1210	€121	€1331

That is, if a principal amount P is invested for n years at a rate r of interest (r is expressed as a decimal) then it will amount to:

$$A = \text{FV} = P(1 + r)^n$$

For our example above, $A = 1000$, $t = 3$ and $r = 0.1$ Therefore,

$$A = 1000(1.1)^3$$
$$= €1331 \text{ (as before)}$$

A principal amount P invested for n years at a rate r of simple interest (r is expressed as a decimal) will amount to:

$$A = \text{FV} = P(1 + rt)$$

The principal €1000 invested for three years at a rate of interest of 10% (simple interest) will amount to:

$$A = 1000(1 + 0.1^*3) = 1000(1.3) = €1300$$

There are variants of the compound interest formula to cover situations where there are m-compounding periods per year. For example, interest may be compounded annually, semi-annually (with two compounding periods per year), quarterly (with four compounding periods per year), monthly (with 12 compounding periods per year), or daily (with 365 compounding periods per year).

The periodic rate of interest (i) per compound period is given by the nominal annual rate of interest (r) divided by the number of compounding periods (m) per year:

$$i = \frac{\text{Nominal Rate}}{\text{\# compounding periods}} = \frac{r}{m}$$

For example, if the nominal annual rate is 10% and interest is compounded quarterly then the period rate of interest per quarter is $^{10}/_4 = 2.5\%$. That is, a compound interest of 2.5% is calculated at the end of each quarter and applied to the account.

The number of compounding periods for the total term of a loan or investment is given by the number of compounding periods per year (m) multiplied by the number of years of the investment or loan.

$$n = \#\text{years} \times m$$

Example 30.8 (Compound Interest—Multiple Compounding Periods) An investor places £10,000 on a term deposit that earns interest at 8% per annum compounded quarterly for three years and nine months. At the end of the term the interest rate changes to 6% compounded monthly and it is invested for a further term of two years and three months.

1. How many compounding periods are there for three years and nine months?
2. What is the value of the investment at the end of three years and nine months?
3. How many compounding periods are there for two years and three months?
4. What is the final value of the investment at the end of the six years?

Solution (Compound Interest—Multiple Compounding Periods)

1. The initial term is for three years and nine months (i.e., 3.75 years), and so the total number of compounding periods is given by $n = $ #years $* m$, where #years $= 3.75$ and $m = 4$. Therefore, $n = 3.75 * 4 = 15$.
2. The nominal rate of interest r is $8\% = 0.08$, and so the interest rate i per quarterly compounding period is $0.08/4 = 0.02$.
   Therefore at the end of the term the principal amounts to:

$$A = FV_1 = P(1 + i)^n$$
$$= 10,000(1 + 0.02)^{15}$$
$$= 10,000(1.02)^{15}$$
$$= 10,000(1.3458)$$
$$= £13,458$$

3. The term is for two years and three months (i.e., 2.25 years), and so the total number of compounding periods is given by $n = $ #years $* m$, where #years $= 2.25$ and $m = 12$. Therefore, $n = 2.25 * 12 = 27$.
4. The new nominal interest rate is $6\% = 0.06$ and so the interest rate i per compounding period is $0.06/12 = 0.005$. Therefore at the end of the term the principal amounts to:

$$A = FV_2 = FV_1(1 + i)^n$$
$$= 13,458(1 + 0.005)^{27}$$
$$= 13,458(1.005)^{27}$$
$$= 13,458 * 1.14415$$
$$= £15,398$$

30.3.1 Present Value Under Compound Interest

The time value of money is the concept that the earlier that cash is received the greater its value to the recipient. Similarly, the later that a cash payment is made, the lower its value to the payee, and the lower its cost to the payer.

This is clear if we consider the example of a person who receives $1000 now and a person who receives $1000 five years from now. The person who receives $1000 now is able to invest it and to receive compound interest on the principal, whereas the other person who receives $1000 in five years earns no interest during the period. Further, inflation during the period means that the purchasing power of $1000 is less in five years' time than it is today.

We presented the general formula for what the future value of a principal P invested for n compounding periods at a compound rate r of interest per compounding period as:

$$A = P(1 + r)^n$$

The present value of a given amount A that will be received in future is the principal ($P = \text{PV}$) that will grow to that amount where there are n compounding periods and the rate of interest is r for each compounding period. The present value of an amount A received in n compounding periods at an interest rate r for the compounding period is given by:

$$P = \frac{A}{(1 + r)^n}$$

We can write also write the present value formula as $\text{PV} = P = A(1 + r)^{-n}$.

Example 30.9 (Present Value) Find the principal that will amount to $10,000 in five years at 8% per annum compounded quarterly.

Solution (Present Value)

The term is five years $= 5 * 4 = 20$ compounding period. The nominal rate of interest is $8\% = 0.08$ and so the interest rate i per compounding period is $^{0.08}/_4 = 0.02$. The present value is then given by:

$$\text{PV} = A(1 + i)^{-n} = \text{FV}(1 + i)^{-n}$$
$$= 10,000(1.02)^{-20}$$
$$= \$6729.71$$

The difference between the known future value of $10,000, and the computed present value (i.e., the principal of $6729.71) is termed the compound discount and represents the compound interest that accumulates on the principal over the period of time. It is given by:

$$\text{Compound Discount} = FV - PV$$

For this example the compound discount is $10,000 - 6729.71 = \$3270.29$.

Example 30.10 (Present Value) Elodie is planning to buy a home entertainment system for her apartment. She can pay £1500 now or pay £250 now and £1600 in two years' time. Which option is better if the nominal rate of interest is 9% compounded monthly?

Solution (Present Value)

There are $2 * 12 = 24$ compounding periods and the interest rate i for the compounding period is $^{0.09}/_{12} = 0.0075$. The present value of £1600 in two years time at an interest rate of 9% compounded monthly is:

$$
\begin{aligned}
PV &= FV(1+i)^{-n} \\
&= 1600(1.0075)^{-24} \\
&= 1600/1.1964 \\
&= £1337.33
\end{aligned}
$$

The total cost of the second option is $£250 + 1337.33 = £1587.33$.

Therefore, Elodie should choose the first option since it is cheaper by £87.33 (i.e., $£1587.33 - 1500$).

30.3.2 Equivalent Values

When two sums of money are to be paid/received at different times they are not directly comparable as such, and a point in time (the focal date) must be chosen to make the comparison (Fig. 30.1).

The choice of focal date determines whether the present value or future value formula will be used. That is, when computing equivalent values we first determine the focal date, and then depending on whether the payment date is before or after this reference date we apply the present value or future value formula.

If the due date of the payment is before the focal date then we apply the future value FV formula:

$$FV = P(1+i)^n$$

If the due date of the payment is after the focal date then we apply the present value PV formula:

$$PV = FV(1+i)^{-n}$$

Fig. 30.1 Equivalent weights

Example 30.11 (Equivalent Values) A debt value of €1000 that was due three months ago, €2000 that is due today, and €1500 that is due in 18 months are to be combined into one payment due six months from today at 8% compounded monthly. Determine the amount of the single payment.

Solution (Equivalent Values)

The focal date is six months from today and so we need to determine the equivalent value E_1, E_2, and E_3 of the three payments on this date, and we then replace the three payments with one single payment $E = E_1 + E_2 + E_3$ that is payable six months from today.

The equivalent value E_1 of €1000 which was due three months ago in six months from today is determined from the future value formula where the number of interest periods $n = 6 + 3 = 9$. The interest rate per period is $8\%/12 = 0.66\% = 0.00667$.

$$\begin{aligned} FV &= P(1 + i)^n \\ &= 1000(1 + 0.00667)^9 \\ &= 1000(1.00667)^9 \\ &= 1000 * 1.0616 \\ &= €1061.60 \end{aligned}$$

The equivalent value E_2 of €2000 which is due today in six months is determined from the future value formula, where the number of interest periods $n = 6$. The interest rate per period is $= 0.00667$.

$$FV = P(1 + i)^n$$

$$= 2000(1 + 0.00667)^6$$
$$= 2000(1.00667)^6$$
$$= 2000 * 1.0407$$
$$= €2081.40$$

The equivalent value E_3 of €1500 which is due in 18 months from today is determined from the present value formula, where the number of interest periods n $= 18 - 6 = 12$. The interest rate per period is $= 0.00667$.

$$PV = FV(1 + i)^{-n}$$
$$= 1500(1 + 0.0067)^{-12}$$
$$= 1500(1.0067)^{-12}$$
$$= €1384.49$$

$$E = E_1 + E_2 + E_3$$
$$= 1061.60 + 2081.40 + 1384.49$$
$$= €4527.49$$

Example 30.12 (Equivalent Values—Replacement Payments) Liz was due to make a payment of £2000 today. However, she has negotiated a deal to make two equal payments: the first payment is to be made one year from now and the second payment two years from now. Determine the amount of the equal payments where the interest rate is 9% compounded quarterly and the focal date is today.

Solution (Equivalent Values—Replacement Payments)

Let x be the value of the equal payments. The first payment is made in one year and so there are $n = 1 * 4 = 4$ compounding periods, and the second payment is made in two years and so there are $n = 2 * 4 = 8$ compounding periods. The interest rate i is $9\%/4 = 2.25\% = 0.0225$.
    The present value E_1 of a sum x received in one year is given by:

$$PV = FV(1 + i)^{-n}$$
$$= x(1 + 0.0225)^{-4}$$
$$= x(1.0225)^{-4}$$
$$= 0.9148x$$

The present value E_2 of a sum x received in two years is given by:

$$PV = FV(1 + i)^{-n}$$
$$= x(1 + 0.0225)^{-8}$$
$$= x(1.0225)^{-8}$$

$$= 0.8369x$$

The sum of the present value of E_1 and E_2 is £2000 and so we have:

$$0.9148x + 0.8369x = 2000$$
$$1.7517x = 2000$$
$$x = £1141.75$$

30.4   Basic Mathematics of Annuities

An *annuity* is a sequence of fixed equal payments made over a period of time, and it is usually paid at the end of the payment interval. For example, for a hire purchase contract the buyer makes an initial deposit for an item and then pays an equal amount per month (the payment is generally at the end of the month) up to a fixed end date. Personal loans from banks are paid back in a similar manner but without an initial deposit.

An investment annuity (e.g., a regular monthly savings scheme) may be paid at the start of the payment interval. A pension scheme involves two stages, with the first stage involving an investment of regular payments at the start of the payment interval up to retirement, and the second stage involving the payment of the retirement annuity. The period of payment of a retirement annuity is usually for the remainder of a person's life (life annuity), or it could be for a period of a fixed number of years.

We may determine the final value of an investment annuity by determining the future value of each payment up to the maturity date and then adding them all together. Alternately, as the future values form a geometric series we may derive a formula for the value of the investment by using the formula for the sum of a geometric series.

We may determine the present value of an annuity by determining the present value of each payment made and summing these, or we may also develop a formula to calculate the present value.

The repayment of a bank loan is generally with an amortization annuity, where the customer borrows a sum of money from a bank (e.g., a personal loan for a car or a mortgage for the purchase of a house). The loan is for a defined period of time, and its repayment is with a regular annuity, where each annuity payment consists of interest and capital repayment. The bulk of the early payments go on interest due on the outstanding capital with smaller amounts going on capital repayments. However, the bulk of the later payments go on repaying the capital with smaller amounts going on interest.

An *annuity* is a series of equal cash payments made at regular intervals over a period of time, and they may be used for investment purposes or paying back a loan or mortgage. We first consider the example of an investment annuity.

Example 30.13 (Investment Annuity) Sheila is investing €10,000 a year in a savings scheme that pays 10% interest every year. What will the value of her investment be after five years?

Solution (Invested Annuity)

Sheila invests €10,000 at the start of year 1 and so this earns five years of compound interest of 10% and so its future value in five years is given by $10,000 * 1.1^5 = $ €16,105. The future value of the payments that she makes is presented in Table 30.2.

Therefore, the value of her investment at the end of five years is the sum of the future values of each payment at the end of five years $= 16,105 + 14,641 + 13,310 + 12,100 + 11,000 = $ €67,156.

We note that this is the sum of a geometric series and so in general if an investor makes a payment of A at the start of each year for n years at a rate r of interest then the investment value at the end of n years is:

$$A(1+r)^n + A(1+r)^{n-1} + \ldots A(1+r)$$
$$= A(1+r)\left[1 + A(1+r) + \cdots + A(1+r)^{n-1}\right]$$
$$= A(1+r)\frac{(1+r)^n - 1}{(1+r) - 1}$$
$$= A(1+r)\frac{(1+r)^n - 1}{r}$$

We apply the formula to check our calculation.

$$10000(1+0.1)\frac{(1+0.1)^5 - 1}{0.1}$$
$$= 11000\frac{(1.1)^5 - 1}{0.1}$$
$$= 11000\frac{(1.61051 - 1)}{0.1}$$
$$= €67,156$$

Table 30.2 Calculation of future value of annuity

Year	Amount	Future value ($r = 0.1$)
1	10,000	$10,000 * 1.1^5 = $ €16,105
2	10,000	$10,000 * 1.1^4 = $ €14,641
3	10,000	$10,000 * 1.1^3 = $ €13,310
4	10,000	$10,000 * 1.1^2 = $ €12,100
5	10,000	$10,000 * 1.1^1 = $ €11,000
Total		€67,156

Note 30.1

We assumed that the annual payment was made at the start of the year. However, for ordinary annuities payment is made at the end of the year (or compounding period) and so the formula would be slightly different:

$$FV = A\frac{(1+r)^n - 1}{r}$$

The future value formula is adjusted for multiple (m) compounding periods per year, where the interest rate for the period is given by $i = {}^r/_m$, and the number of payment periods n is given by where $n = tm$ (where t is the number of years). The future value of a series of payments of amount A (made at start of the compounding period) with interest rate i per compounding period, where there are n compounding periods, is given by:

$$FV = A(1+i)\frac{(1+i)^n - 1}{i}$$

The future value of a series of payments of amount A (made at end of the compounding period) with interest rate i per compounding period, where there are n compounding periods, is given by:

$$FV = A\frac{(1+i)^n - 1}{i}$$

An annuity consists of a series of payments over a period of time, and so it is reasonable to consider its present value with respect to a discount rate r (this is applicable to calculating the present value of the annuity for mortgage repayments discussed in the next section).

The net present value of an annuity is the sum of the present value of each of the payments made over the period, and the method of calculation is clear from Table 30.3.

Example 30.14 (Present Value Annuities) Calculate the present value of a series of payments of $1000 with the payments made for five years at a discount rate of 10%.

Table 30.3 Calculation of present value of annuity

Year	Amount	Present value ($r = 0.1$)
1	1000	$909.91
2	1000	$826.44
3	1000	$751.31
4	1000	$683.01
5	1000	$620.92
Total		$3791

Solution (Present Value Annuities)

The regular payment A is 1000, and the rate r is 0.1 and $n = 5$. The present value of the first payment received is $1000/1.1 = 909.91$ at the end of year of year 1; at the end of year 2 it is $1000/(1.1)^2 = 826.45$; and so on. At the end of year 5 its present value is 620.92. The net present value of the annuity is the sum of the present value of all the payments made over the five years, and it is given by the sum of the present values from Table 30.3. That is, the present value of the annuity is $909.91 + 826.44 + 751.31 + 683.01 + 620.92 = \$3791$.

We may derive a formula for the present value of a series of payments A made over a period of n years at a discount rate of r as follows: Clearly, the present value is given by:

$$\frac{A}{(1+r)} + \frac{A}{(1+r)^2} + \cdots + \frac{A}{(1+r)^n}$$

This is a geometric series where the constant ratio is $\frac{1}{1+r}$ and the present value of the annuity is given by its sum:

$$PV = \frac{A}{r}\left[1 - \frac{1}{(1+r)^n}\right]$$

For the example above we apply the formula and get

$$PV = \frac{1000}{0.1}\left[1 - \frac{1}{(1.1)^5}\right]$$
$$= 10000(0.3791)$$
$$= \$3791$$

The annuity formula is adjusted for multiple (m) compounding periods per year, and the interest rate for the period is given by $i = {}^r/m$, and the number of payment periods n is given by where $n = tm$ (where t is the number of years). For example, the present value of an annuity of amount A, with interest rate i per compounding period, where there are n compounding periods, is given by:

$$P = \frac{A}{i}\left[1 - \frac{1}{(1+i)^n}\right]$$

Example 30.15 (Retirement Annuity) Bláithín has commenced employment at a company that offers a pension in the form of an annuity that pays 5% interest per annum compounded monthly. She plans to work for 30 years and wishes to accumulate a pension fund that will pay her €2000 per month for 25 years after she retires. How much does she need to save per month to do this?

Solution (Retirement Annuities)

First, we determine the value that the fund must accumulate to pay her €2000 per month, and this is given by the present value of the 25-year annuity of €2000 per month. The interest rate r is 5% and as there are 12 compounding periods per year there are a total of $25 * 12 = 300$ compounding periods, and the interest rate per compounding period is $0.05/12 = 0.004166$.

$$P = 2000/0.004166[1 - (1.004166)^{-300}]$$
$$= €342,174.$$

That is, her pension fund at retirement must reach €342,174 and so we need to determine the monthly payments necessary for her to achieve this. The future value is given by the formula:

$$FV = A\frac{(1+i)^{n+1} - 1}{i}$$

and so

$$A = FV^*i/[(1+i)^{n+1} - 1]$$

where $m = 12$, $n = 30 * 12 = 360$ and $i = 0.05/12 = 0.004166$ and $FV = 342,174$.

$$A = 342,174 * 0.004166/3.4863$$
$$= €408.87$$

That is, Bláithín needs to save €408.87 per month into her retirement account (sinking fund) for 30 years in order to have an annuity of €2000 per month for 25 years (where there is a constant interest rate of 5% compounded monthly).

30.5 Loans and Mortgages

The purchase of a home or car requires a large sum of money, and so most purchasers need to obtain a loan from the bank to fund the purchase. Once the financial institution has approved the loan, the borrower completes the purchase and pays back the loan to the financial institution over an agreed period of time (the term of the loan). For example, a mortgage is generally paid back over 20–25 years, whereas a car loan is usually paid back in five years (Fig. 30.2).

An interest-bearing debt is *amortized* if both the principal and interest are repaid by a series of equal payments (except for possibly the last payment) made at equal intervals of time. That is, the amortization of loans refers to the repayment of interest-bearing debts by a series of equal payments made at equal intervals of time. The debt is repaid by an amortization annuity, where each payment consists

Fig. 30.2 Loan or mortgage

of both the repayment of the capital borrowed and the interest due on the capital for that time interval.

Mortgages and many consumer loans are repaid by this method, and the standard problem is to calculate what the annual (or monthly) payment should be to amortize the principal P in n years where the rate of interest is r.

The present value of the annuity is equal to the principal borrowed: i.e., the sum of the present values of the payments must be equal to the original principal borrowed. That is:

$$P = \frac{A}{(1+r)} + \frac{A}{(1+r)^2} + \cdots + \frac{A}{(1+r)^n}$$

We may also use the formula that we previously derived for the present value of the annuity to get:

$$P = \frac{A}{r}\left[1 - \frac{1}{(1+r)^n}\right]$$

We may calculate A by manipulating this formula to get:

$$A = \frac{Pr}{\left[1 - \frac{1}{(1+r)^n}\right]}$$

$$A = Pr/\left[1 - (1+r)^{-n}\right]$$

Example 30.16 (Amortization) Joanne has taken out a €200,000 mortgage over 20 years at 8% per annum. Calculate her annual repayment amount to amortize the mortgage.

Solution (Amortization)

We apply the formula to calculate her annual repayment:

$$A = \frac{Pr}{\left[1 - \frac{1}{(1+r)^n}\right]}$$

$$A = \frac{200,000 * 0.08}{\left[1 - \frac{1}{(1+0.08)^{20}}\right]}$$

$$= \frac{16,000}{\left[1 - \frac{1}{4.661}\right]}$$

$$= €20,370$$

We adjust the formula for the more common case where the interest is compounded several times per year (usually monthly), and so $n = $ # years * # compoundings and the interest $i = r/$# compoundings.

$$A = \frac{Pi}{\left[1 - \frac{1}{(1+i)^n}\right]}$$

Example 30.17 (Amortization) A mortgage of £150,000 at 6% compounded monthly is amortized over 20 years. Determine the following:

1. Repayment amount per month
2. Total amount paid to amortize the loan.
3. The cost of financing

Solution (Amortization)

The number of payments $n = $ #years * payments per year $= 20 * 12 = 240$.
The interest rate $i = 6\%/12 = 0.5\% = 0.005$.

1. We calculate the amount of the repayment A by substituting for n and i and obtain:

$$A = \frac{150,000 * 0.005}{\left[1 - \frac{1}{(1+0.005)^{240}}\right]}$$

$$= \frac{750}{\left[1 - \frac{1}{3.3102}\right]}$$

$$= £1074.65$$

2. The total amount paid is the number of payments * amount of each payment $=$ $n * A = 240 * 1074.65 = £257,916$
3. The total cost of financing $=$ total amount paid $-$ original principal $= 257,916 - 150,000 = £107,916$.

Example 30.18 (Amortization) For the previous example determine the following at the end of the first period:

1. The amount of interest repaid.

2. The amount of the principal repaid.
3. The amount of the principal outstanding.

Solution (Amortization)

The amount paid at the end of the first period is £1074.65.

1. The amount paid in interest for the first period is 150,000 * 0.005 = £750.
2. The amount of the principal repaid is £1074 − 750 = £324.65.
3. The amount of the principal outstanding at the end of the first interest period is
   £150,000–324.65 = £149,675.35.

The early payments of the mortgage mainly involve repaying interest on the capital borrowed, whereas the later payments mainly involve repaying the capital borrowed with less interest due. We can create an amortization table which shows the interest paid, the amount paid in capital repayment, and the outstanding principal balance for each payment interval. Often, the last payment is different to the others due to rounding errors introduced and carried through.

Each entry in the amortization table includes interest, the principal repaid, and outstanding principal balance. The interest is calculated by the principal balance * periodic interest rate i; the principal repaid is calculated by the payment amount − interest; and the new outstanding principal balance is given by the principal balance − principal repaid.

30.6   Review Questions

1. Explain the difference between simple and compound interest?
2. Calculate the simple interest payable on an investment of £12,000 placed on deposit for nine months at an interest rate of 8%.
3. An investor places £5000 on a term deposit that earns interest at 8% per annum compounded quarterly for two years and three months. Find the value of the investment on the maturity date.
4. Find the principal that will amount to $12,000 in three years at 6% per annum compounded quarterly.
5. How many years will it take a principal of $5000 to exceed $10,000 at a constant annual growth rate of 6% compound interest?
6. What is the present value of $5000 to be received in five years time at a discount rate of 7%?
7. Explain the concept of equivalent values and when to use the present value/future value in its calculation.
8. A debt value of €2000 due six months ago, €5000 due today and €3000 due in 18 months are to be combined into one payment due three months

from today at 6% compounded monthly. Determine the amount of the single payment.

30.7 Summary

Simple interest is calculated on the principal only whereas compound interest is calculated on both the principal and the accumulated interest of previous periods. Compound interest is generally used for long-term investments and loans, and its calculation is more complicated than that of simple interest.

The time value of money is the concept that the earlier that cash is received the greater its value to the recipient, and vice versa for late payments. The future value of a principal P invested for n-year compounding periods at a compound rate r of interest per compounding period is given by $A = P(1 + r)^n$. The present value of a given amount A that will be received n years in future, at an interest rate r for each compounding period is the principal that will grow to that amount and is given by $P = A(1 + r)^{-n}$.

A long-term promissory note has a term greater than one year and may be bought or sold prior to its maturity date. The calculation of the value of a promissory note is similar to that used in the calculation of the value of short-term promissory notes.

The future value is what the principal will amount to in future at a given rate of interest, whereas the present value of an amount to be received in future is the principal that would grow to that amount at a given rate of interest.

An annuity is a sequence of fixed equal payments made over a period of time, and it is usually paid at the end of the payment interval. An interest-bearing debt is amortized if both the principal and interest are repaid by a series of equal payments (with the exception of possibly the last payment) made at equal intervals of time. The debt is repaid by an amortization annuity, where each payment consists of both the repayment of the capital borrowed and the interest due on the capital for that time interval.

Introduction to Operations Research

31

Key Topics

Linear programming

Constraints

Variables

Optimization

Objective function

Cost volume profit analysis

Game theory

31.1 Introduction

Operations research is a multidisciplinary field that is concerned with the application of mathematical and analytic techniques to assist in decision-making. It includes techniques such as mathematical modelling, statistical analysis, and mathematical optimization as part of its goal to achieve optimal (or near optimal) solutions to complex decision-making problems. The modern field of operations research includes various other disciplines such as computer science, industrial engineering, business practices in manufacturing and service companies, supply chain management, and operations management.

Pascal did early work on operations research in the seventeenth century. He attempted to apply early work on probability theory to solve complex decision-making problems. Babbage's work on the transportation and sorting of mail contributed to the introduction of the uniform *"Penny Post"* in England in the

© The Author(s), under exclusive license to Springer Nature Switzerland AG 2023
G. O'Regan, *Mathematical Foundations of Software Engineering*,
Texts in Computer Science, https://doi.org/10.1007/978-3-031-26212-8_31

nineteenth century. The origins of the operations research field are from the work of military planners during the First World War, and the field took off during the Second World War as it was seen as a scientific approach to decision-making using quantitative techniques. It was applied to strategic and tactical problems in military operations, where the goal was to find the most effective utilization of limited military resources through the use of quantitative techniques. It played an important role in solving practical military problems such as determining the appropriate convoy size in the submarine war in the Atlantic.

Numerous peacetime applications of the field of operations research emerged after the Second World War, where operations research and management science were applied to many industries and occupations. It was applied to procurement, training, logistics, and infrastructure in addition to its use in operations. The progress that has been made in the computing field means that operations research can now solve problems with thousands of variables and constraints.

Operations research (OR) is the study of mathematical models for complex organizational systems, where a *model* is a mathematical description of a system that accounts for its known and inferred properties, and it may be used for the further study of its properties, and a *system* is a functionally related collection of elements such as a network of computer hardware and software. *Optimization* is a branch of operations research that uses mathematical techniques to derive values from system variables that will optimize system performance.

Operations research has been applied to a wide variety of problems including network optimization problems, designing the layouts of the components on a computer chip, supply chain management, critical path analysis during project planning to identify key project activities that effect the project timeline, scheduling project tasks and personnel, and so on. Several of the models used in operations research are described in Table 31.1.

Mathematical programming involves defining a mathematical model for the problem and using the model to find the optimal solution. A mathematical model consists of variables, constraints, the objective function to be maximized or minimized, and the relevant parameters and data. The general form is:

Min or Max $f(x_1, x_2, \ldots x_n)$ (Objective function)

$g(x_1, x_2, \ldots x_n) \leq$ (or $>, \geq, =<$) b_i (Constraints)

$x \in X$

f, g are linear and X is continuous (for linear programming LP)

A *feasible solution* is an assignment of values to the variables such that the constraints are satisfied. An optimal solution is one whose objective function exceeds all other feasible solutions (for maximization optimization). We now discuss linear programming in more detail.

Table 31.1 Models used in operations research

Model	Description
Linear programming	These problems aim to find the best possible outcome (where this is expressed as a linear function) such as to maximize a profit or minimize a cost subject to various linear constraints. The function and constraints are linear functions of the decision variables, and modern software can solve problems containing millions of variables and thousands of constraints
Network flow programming	This is a special case of the general linear programming problem and includes problems such as the transportation problem, the shortest path problem, the maximum flow problem, and the minimum cost problem. There are very efficient algorithms available for these (faster and more efficient than standard linear programming)
Integer programming	This is a special case of the general linear programming problem, where the variables are required to take on discrete values
Nonlinear programming	The function and constraints are nonlinear, and these are much more difficult to solve than linear programming. Many real-world applications require a nonlinear model, and the solution is often approximated with a linear model
Dynamic programming	A dynamic programming (DP) model describes a process in terms of states, decisions, transitions, and a return. The process begins in some initial state, a decision is made leading to a transition to a new state, the process continues through a sequence of states until final state is reached. The problem is to find a sequence that maximizes the total return
Stochastic processes	A stochastic process models practical situations where the attributes of a system randomly change over time (e.g., number of customers at an ATM machine, the share price), and the state is a snapshot of the system at a point in time that describes its attributes. Events occur that change the state of the system
Markov chains	A stochastic process that can be observed at regular intervals (such as every day or every week) can be described by a matrix, which gives the probabilities of moving to each state from every other state in one-time interval. The process is called a Markov Chain when this matrix is unchanging over time
Markov processes	A Markov process is a continuous time stochastic process in which the duration of all state-changing activities is exponentially distributed
Game theory	Game theory is the study of mathematical models of strategic interaction among rational decision-makers. It is concerned with logical decision-making by humans, animals, and computers
Simulation	Simulation is a general technique for estimating statistical measures of complex systems
Time series and forecasting	A time series is a sequence of observations of a periodic random variable and is generally used as input to an OR decision model

(continued)

Table 31.1 (continued)

Model	Description
Inventory theory	Aims to optimize inventory management: e.g., determining when and how much inventory should be ordered
Reliability theory	Aims to model the reliability of a system from probability theory

31.2 Linear Programming

Linear programming (LP) is a mathematical model for determining the best possible outcome (e.g., maximizing profit or minimizing cost) of a particular problem. The problem is subject to various constraints such as resources or costs, and the constraints are expressed as a set of linear equations and linear inequalities. The best possible outcome is expressed as a linear equation. For example, the goal may be to determine the number of products that should be made to maximize profit subject to the constraint of limited available resources.

The constraints for the problem are linear, and they specify regions that are bounded by straight lines. The solution will lie somewhere within the regions specified, and a feasible region is a region where all of the linear inequalities are satisfied. Once the feasible region is found the challenge is then to find where the best possible outcome may be maximized in the feasible region, and this will generally be in a corner of the region. The steps involved in developing a linear programming model include:

- Formulation of the problem
- Solution of the problem
- Interpretation of the solution

Linear programming models seek to select the most appropriate solution from the alternatives that are available subject to the specified constraints. Often, graphical techniques are employed to sketch the problem and the regions corresponding to the constraints.

The graphical techniques identify the feasible region where the solution lies, and then the maximization or minimization function is employed within the region to search for the optimal value. The optimal solution will lie at one or more of the corner points of the feasible region.

31.2.1 Linear Programming Example

We consider an example in an industrial setting where a company is trying to decide how many of each product it should make to maximize profits subject to the constraint of limited resources.

Table 31.2 Square deal furniture

	Table	Chair	Hours available
Carpentry	3 h	4 h	2400
Painting	2 h	1 h	1000
Profit contribution	7 Euros	5 Euros	

Square Deal Furniture produces two products, namely chairs and tables, and it needs to decide on how many of each to make each month in order to maximize profits. The amount of time to make tables and chairs and the maximum hours available to make each product, as well as the profit contribution of each product, is summarized in Table 31.2. There are additional constraints that need to be specified:

- At least 100 tables must be made
- The maximum number of chairs to be made is 450

We use variables to represent tables and chairs and formulate an objective function to maximize profits subject to the constraints.

$$T = \text{Number of tables to make}$$
$$C = \text{Number of chairs to make}$$

The objective function (to maximize profits) is then specified as

$$\text{Maximize the value of } 5C + 7T$$

The constraints on the hours available for carpentry and painting may be specified as:

$$3T + 4C \leq 2400 \quad \text{(carpentry time available)}$$
$$2T + C \leq 1000 \quad \text{(painting time available)}$$

The constraints that at least 100 tables must be made and the maximum number of chairs to be made is 450 may be specified as:

$$T \geq 100 \quad \text{(number of tables)}$$
$$C \leq 450 \quad \text{(number of chairs}$$

Finally, it is not possible to produce a negative number of chairs or tables and this is specified as:

$$T \geq 0 \quad \text{(non-negative)}$$
$$C \geq 0 \quad \text{(non-negative)}$$

The model is summarized as

$$
\begin{array}{ll}
\text{Max } 5C + 7T & \text{(Maximation problem)} \\
3T + 4C \leq 2400 & \text{(carpentry time available)} \\
2T + C \leq 1000 & \text{(painting time available)} \\
T \geq 100 & \text{(number of tables)} \\
C \leq 450 & \text{(number of chairs)} \\
T \geq 0 & \text{(non-negative)} \\
C \geq 0 & \text{(non-negative)}
\end{array}
$$

We graph the LP model and then use the graph to find a feasible region for where the solution lies, and we then identify the optimal solution. The feasible region is an area where all of the constraints for the problem are satisfied, and the optimal solution lies at one or more of the corner points of the feasible region.

First, for the constraints on the hours available for painting and carpentry $3T + 4C \leq 2400$ and $2T + C \leq 1000$, respectively, we draw the two lines $3T + 4C = 2400$ and $2T + C = 1000$. We choose two points on each line and then join both points to form the line, and we choose the intercepts of both lines as the points.

For the line $3T + 4C = 2400$ when T is 0 C is 600 and when C is 0 T is 800. Therefore, the points (0, 600) and (800, 0) are on the line $3T + 4C = 2400$. For the line $2T + C = 1000$ when $T = 0$ then $C = 1000$ and when $C = 0$ then $T = 500$. Therefore, the points (0, 1000) and (500, 0) are on the line $2T + C = 1000$.

Figure 31.1 is the first step in developing a graphical solution and we note that for the first two constraints $3T + 4C \leq 2400$ and $2T + C \leq 1000$ that the solution lies somewhere in the area bounded by the lines $3T + 4C = 2400$, $2T + C = 1000$, the T axis and the C axis.

Next, we add the remaining constraints ($T \geq 100$, $C \leq 450$, $T \geq 0$, $C \geq 0$) to the graph, and this has the effect of reducing the size of the feasible region in Fig. 31.1 (which placed no restrictions on T and C). The feasible region can be clearly seen in Fig. 31.2, and the final step is to find the optimal solution in the feasible region that maximizes the profit function $5C + 7T$.

Fig. 31.1 Linear programming—developing a graphical solution

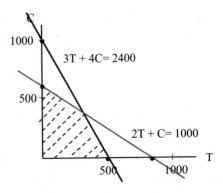

Fig. 31.2 Feasible region of solution

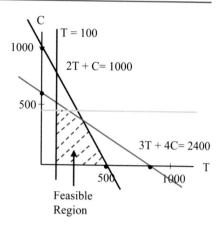

Figure 31.3 shows how we find the optimal solution by drawing the line $7T + 5C = k$ within the feasible region, and this forms a family of parallel lines where the slope of the line is $-5/7$ and k represents the profit. Each point in the feasible region is on one of the lines in the family, and to determine the equation of that line we just input the point into the equation $7T + 5C = k$. For example, the point (200, 0) is in the feasible region and it satisfies the equation $7T + 5C = 7 * 200 + 5 * 0 = 1400$.

We seek to maximize k and it is clear that the value of k that is maximal is at one of the corner points of the feasible region. This is the point of intersection of the lines $2T + C = 1000$ and $3T + 4C = 2400$, and we solve for T and C to get $T = 320$ and $C = 360$. This means that the equation of the line containing the optimal point is $7T + 5C = 2240 + 1800 = 4040$. That is, its equation is $7T + 5C = 4040$ and so the maximum profit is €4040.

Fig. 31.3 Optimal solution

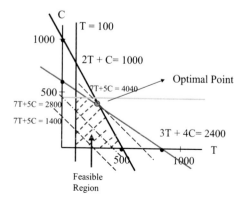

31.2.2 General Formulation of LP Problem

The more general formulation of the linear programming problem can be stated as follows. Find variables $x_1, x_2, ..., x_n$ to optimize (i.e., maximize or minimize) the linear function:

$$Z = c_1x_1 + c_2x_2 + \cdots, c_nx_n$$

where the problem is subject to the linear constraints:

$$a_{11}x_1 + a_{12}x_{12} + \cdots ., a_{1j}x_j + \cdots . + a_{1n}x_n \quad (\leq=\geq)b_1$$
$$a_{21}x_1 + a_{22}x_{12} + \cdots ., a_{2j}x_j + \cdots . + a_{2n}x_n \quad (\leq=\geq)b_2$$
$$\vdots \quad \vdots \qquad \vdots \qquad \vdots \qquad \vdots$$
$$a_{m1}x_1 + a_{m2}x_{12} + \cdots ., a_{mj}x_j + \cdots . + a_{mn}x_n \quad (\leq=\geq)b_m$$

and to non-negative constraints on the variables such as:

$$x_1, x_2, \ldots, x_n \geq 0$$

where a_{ij}, b_j, and c_i are constants and x_i are variables.

The variables $x_1, x_2, ..., x_n$ whose values are to be determined are called decision variables.

The coefficients $c_i, c_2, ..., c_n$ are called cost (profit) coefficients.

The constraints $b_1, b_2, ..., b_m$ are called the requirements.

A set of real values $(x_1, x_2, ..., x_n)$ which satisfies the constraints (including the non-negative constraints) is said to be a *feasible* solution.

A set of real values $(x_1, x_2, ..., x_n)$ which satisfies the constraints (including the non-negative constraints) and optimizes the objective function is said to be an *optimal* solution.

There may be no solution, a unique solution, or multiple solutions.

The constraints may also be formulated in terms of matrices as follows:

$$\begin{pmatrix} a_{11} & a_{12} & a_{13} & \cdots & a_{1n} \\ a_{21} & a_{22} & a_{23} & \cdots & a_{2n} \\ a_{31} & a_{32} & a_{33} & \cdots & a_{3n} \\ \cdots & \cdots & \cdots & \cdots & \cdots \\ \cdots & \cdots & \cdots & \cdots & \cdots \\ a_{m1} & a_{m2} & a_{m3} & \cdots & a_{mn} \end{pmatrix} \begin{pmatrix} x_1 \\ x_2 \\ x_3 \\ \cdots \\ \cdots \\ \cdots \\ x_n \end{pmatrix} (\leq=\geq) \begin{pmatrix} b_1 \\ b_2 \\ b_3 \\ \cdots \\ \cdots \\ \cdots \\ b_m \end{pmatrix}$$

This may also be written as $AX (\leq = \geq)$ B *and* the optimization function may be written as $Z = CX$ where $C = (c_1, c_2, ..., c_n)$ and $X = (x_1, x_2, ..., x_n)^T$ and $X \geq 0$.

Linear programming problems may be solved by graphical techniques (when there are a small number of variables) or analytic techniques using matrices. There are techniques that may be employed to find the solution of the LP problem that are similar to finding the solution to a set of simultaneous equations using Gaussian elimination (see Chap. 29).

31.3 Cost Volume Profit Analysis

A key concern in business is profitability, and management need to decide on the volume of products to produce, including the costs and total revenue. Cost volume profit analysis (CVPA) is a useful tool in the analysis of the relationship between the costs, volume, revenue, and profitability of the products produced. The relationship between revenue and costs at different levels of output can be displayed graphically, with revenue behaviour and cost behaviour shown graphically.

The *breakeven point* (BP) is where the total revenue is equal to the total costs, and breakeven analysis is concerned with identifying the volume of products that need to be produced to break even.

Example (CVPA)

Pilar is planning to set up a business that makes pottery cups, and she has been offered a workshop to rent for €800 per month. She estimates that she needs to spend €10 on the materials to make each pottery cup and that she can sell each cup for €25. She estimates that if she is very productive that she can make 500 pottery cups in a month.

Prepare a table that shows the profit or loss that Pilar makes based on the sales of 0, 100, 200, 300, 400, and 500 pottery cups.

Solution (CVPA)

Each entry in the table consists of the revenue for the volume sold, the material costs per volume of the pottery cups, the fixed cost of renting the workshop per month, the total cost per month, and the net income per month (Table 31.3).

The total sales (revenue) are determined from the volume of sales multiplied by the unit sales price of a pottery cup (€25). There are two types of cost that may be incurred namely *fixed costs* and *variable costs*.

Fixed costs are incurred irrespective of the volume of items produced, and so the cost of renting of the workshop is a fixed cost. Variable costs are constant per unit of

Table 31.3 Projected profit or loss per volume

#Cups	0	100	200	300	400	500
Revenue (sales)	0	2500	5000	7500	10,000	12,500
Materials (var cost)	0	1000	2000	3000	4000	5000
Workshop (fix cost)	800	800	800	800	800	800
Total cost	800	1800	2800	3800	4800	5800
Net income	− €800	€700	€2200	€3700	€5200	€6700

Table 31.4 Revenue and costs

Item	Amount
Total revenue (TR)	SP * X
Total variable cost (TVC)	VC * X
Fixed cost (FC)	FC
Total cost (TC)	FC + TVC = FC + (VC*X)
Net income (profit)	TR – TC – (SP * X) – FC – (VC * X)

Fig. 31.4 Breakeven point

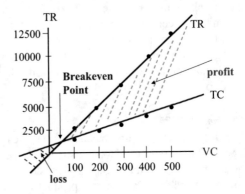

output and include the direct material and labour costs, and so the total variable cost increases as the volume increases. That is, the total variable cost is directly related to the volume of items produced, and Table 31.4 summarizes the revenue and costs.

We may represent the relationships between volume, cost, and revenue graphically and use it to see the relationship between revenue and costs at various levels of output (Fig. 31.4). We may then use the graph to determine the breakeven point for when total revenue is equal to total cost.

We may also determine the breakeven point algebraically by letting X represent the volume of cups produced for breakeven. Then breakeven is when the total revenue is equal to the total cost. That is,

$$SP * X = FC + VC * X$$
$$\Rightarrow 25X = 800 + 10X$$
$$\Rightarrow 15X = 800$$
$$\Rightarrow X = {}^{800}/_{15} = 53.3 \text{ units}$$

The breakeven amount in revenue is 25 * 53.3 = €1333.32.

Next, we present an alternate way of calculating the breakeven point in terms of *contribution margin* and sales. Contribution margin is the monetary value that each extra unit of sales makes towards profitability, and it is given by the selling price per unit minus the variable cost per unit.

Each additional pottery cup sold increases the revenue by €25 whereas the increase in costs is just €10 (the materials required). This, the contribution margin per unit is the selling price minus the variable cost per unit (i.e., SP – VC = €15), and so the total contribution margin is the total volume of units sold multiplied by the contribution margin per unit (i.e., $X * (SP - VC) = 15X$).

The breakeven volume is reached when the total contribution margin covers the fixed cost (i.e., the cost of renting the workshop which is €800 is covered by the total contribution). That is, the breakeven volume is reached when:

$$X * (SP - VC) = FC$$
$$X = \frac{FC}{SP - VC}$$
$$X = {}^{800}\!/_{(25 - 10)} = {}^{800}\!/_{15} = 53.3$$

Example Suppose that the rent of the workshop is increased to €1200 per month and that it also costs €12 (more than expected) to make each cup, and that she can sell each cup for just €20. What is her new breakeven volume and revenue?

Solution
The breakeven volume is reached when:

$$X * (SP - VC) = FC$$
$$X * (20 - 12) = 1200$$
$$8X = 1200$$
$$X = 150$$

Further, the breakeven revenue is:

$$X * SP$$
$$= 150 * 20$$
$$= €3000.$$

31.4 Game Theory

Game theory is the study of mathematical models of strategic interaction among rational decision-makers, and it was originally applied to zero sum games where the gains or losses of each participant are exactly balanced by those of the other participants.

Modern game theory emerged as a field following John von Neumann's 1928 paper on the theory of games of strategy [1]. The Rand corporation investigated possible applications of game theory to global nuclear strategy in the 1950s. Game

Table 31.5 Network
viewing figures

Network 1	Network 2		
	Western	Soap opera	Comedy
Western	35	15	60
Soap Opera	45	58	50
Comedy	38	14	70

theory has been applied to many areas including economics, biology, and the social sciences. It is an important tool in situations where a participant's best outcome depends on what other participants do, and their best outcome depends on what he/she does. We illustrate the idea of game theory through the following example.

Example (Game Theory)
We consider an example of two television networks that are competing for an audience of 100 million on the 8 to 9 pm night-time television slot. The networks announce their schedule ahead of time, but do not know of the other network's decision until the program begins. A certain number of viewers will watch Network 1, with the remainder watching Network 2. Market research has been carried out to show the expected number of viewers for each network based on what will be shown by the networks (Table 31.5).

Problem to Solve (Viewing Figures)
Determine the best strategy that both networks should employ to maximize their viewing figures.

Table 31.5 shows the number of viewers of Network 1 for each type of film that also depends on the type of film that is being shown by Network 2. For example, if Network 1 is showing a western while Network 2 is showing a comedy then Network 1 will have 60 million viewers, and Network 2 will have $100 - 60 = 40$ million viewers. However, if Network 2 was showing a soap opera then the viewing figures for Network 1 are 15 million, and $100 - 15 = 85$ million will be tuned into Network 2.

Solution (Game Theory)

Network 1 is a *row player* whereas Network 2 is a *column player*, and the table above is termed a *payoff matrix*. This is a *constant-sum game* (as the outcome for both players always adds up to a constant 100 million).

The approach to finding the appropriate strategy for Network 1 is to examine each option. If Network 1 decides to show a western then it can get as many as 60 million viewers if Network 2 shows a comedy, or as few as 15 million viewers if Network 2 shows a soap opera. That is, this choice cannot guarantee more than 15 million viewers. Similarly, if Network 1 shows a soap opera it may get as many as 58 million viewers if Network 2 shows a soap opera as well, or as few as 45 million viewers should Network 2 show a western. That is, this choice cannot guarantee more than 45 million viewers. Finally, if Network 1 shows a comedy it would get 70 million viewers if Network 2 is showing a comedy as well, or as few as 14 million viewers if Network 2 is showing a soap opera. That is, this option cannot guarantee more than 14 million viewers. Clearly, the best option for Network 1 would be to show a soap opera, as at least 45 million viewers would tune into Network 1 irrespective of what Network 2 does.

In other words, the strategy for Network 1 (being a row player) is to determine the row minimum of each row and then to choose the row with the largest row minimum.

Similarly, the best strategy for Network 2 (being a column player) is to determine the column maximum of each column and then to choose the column with the smallest column maximum. For Network 2 the best option is to show a western and so 45 million viewers will tune into Network 1 to watch the soap opera, and 55 million will tune into Network 2 to watch a western.

It is clear from the table that the two outcomes satisfy the following inequality:

$$\text{Max}_{(\text{rows})} \left(\text{row minimum}\right) \leq \text{Min}_{(\text{cols})} (\text{col maximum}))$$

This choice is simultaneously best for Network 1 and Network 2 [as max(row minimum) = min(col maximum)], and this is called a *saddle point*, and the common value of both sides of the equation is called the *value* of the game. An equilibrium point of the game is where there is a choice of strategies for both players where neither player can improve their outcome by changing their strategy, and a saddle point of a game is an example of an equilibrium point.

Example (Prisoner Dilemma)

The police arrest two people who they know have committed an armed robbery together. However, they lack sufficient evidence for a conviction for armed robbery, but they have sufficient evidence for a conviction of two years for the theft of the getaway car. The police make the following offer to each prisoner:

If you confess to your part in the robbery and implicate your partner and he does not confess, then you will go free and he will get ten years. If you both confess you will both get five years. If neither of you confess you will get each get two years for the theft of the car

Model the prisoners' situation as a game and determine the rational (best possible) outcome for each prisoner.

Solution (Prisoner's Dilemma)

There are four possible outcomes for each prisoner:

- Go Free (He confesses, Other does not)
- 2-year sentence (Neither confess)
- 5-year sentence (Both confess)
- 10 years (He does not confess. Other does)

Table 31.6 Outcomes in prisoners' dilemma

Prisoner 1	Prisoner 2	
	Confess	Refuse confess
Confess	5, 5	0, 10
	10, 0	2, 2

Each prisoner has a choice of confessing or not, and Table 31.6 summarizes the various outcomes depending on the choices that the prisoners make. The first entry in each cell of the table is the outcome for prisoner 1, and the second entry is the outcome for prisoner 2. For example, the cell with the entries 10, 0 states that prisoner 1 is sentenced for 10 years and prisoner 2 goes free.

It is clear from Table 31.6 that if both prisoners confess they both will receive a 5-year sentence; if neither confesses then they will both receive a 2-year sentence; if prisoner 1 confesses and prisoner 2 does not then prisoner 1 goes free whereas prisoner 2 gets a 10-year sentence; and finally, if prisoner 2 confesses and prisoner 1 does not then prisoner 2 goes free and prisoner a 1 gets a 10-year sentence.

Each prisoner evaluates his two possible actions by looking at the outcomes in both columns, as this will show which action is better for each possible action of their partner. If prisoner 2 confesses then prisoner 1 gets a 5-year sentence if he confesses or a 10-year sentence if he does not confess. If prisoner 2 does not confess then prisoner 1 goes free if he confesses or 2 years if he does not confess. Therefore, prisoner 1 is better off confessing irrespective of the choice of prisoner 2. Similarly, prisoner 2 comes to exactly the same conclusion as prisoner 1, and so the best outcome for both prisoners is to confess to the crime, and both will go to prison for 5 years.

The paradox in the prisoners' dilemma is that two individuals acting in their own self-interest do not produce the optimal outcome. Both parties choose to protect themselves at the expense of the other, and as a result both find themselves in a worse state than if they had cooperated with each other in the decision-making process and received two years. For more detailed information on operations research see [2].

31.5   Review Questions

1. What is operations research?
2. Describe the models used in operations research.
3. What is linear programming and describe the steps in developing a linear programming model?
4. What is cost volume profit analysis?
5. Suppose the fixed costs are rent of £1,500 per month and that the cost of making each item is £20 and it may then be sold for £25. How many items must be sold to breakeven and what is the breakeven revenue?

6. What is game theory?
7. What is a zero sum game?

31.6 Summary

Operations research is a multidisciplinary field concerned with the application of mathematical and analytic techniques to assist in decision-making. It employs mathematical modelling, statistical analysis, and mathematical optimization to achieve optimal (or near optimal) solutions to complex decision-making problems. The modern field of operations research includes other disciplines such as computer science, industrial engineering, business practices in manufacturing and service companies, supply chain management, and operations management.

Linear programming is a mathematical model for determining the best possible outcome such as maximizing profit or minimizing cost of a particular problem. The problem is subject to various constraints such as resources or costs, and the constraints are expressed as a set of linear equations and linear inequalities. The best possible outcome is expressed as a linear equation. For example, the goal may be to determine the number of products that should be made to maximize profit subject to the constraint of limited available resources.

Cost volume profit analysis (CVPA) is used in the analysis of the relationship between the costs, volume, revenue, and profitability of the products produced. The relationship between revenue and costs at different levels of output can be displayed graphically, with revenue behaviour and cost behaviour shown graphically.

Game theory is the study of mathematical models of strategic interaction among rational decision-makers. Von Neumann was the founder of modern game theory with his 1928 paper on the theory of games of strategy. The Rand Corporation applied game theory to global nuclear strategy in the 1950s, and the original applications of game theory were to zero sum games.

References

1. von Neumann J (1928) On the theory of games of strategy. Math Ann (in German) 100(1):295–320
2. Taher H (2016) Operations research. An introduction, 10th edn. Pearson, London

Mathematical Software for Software Engineers

32

32.1 Introduction

The goal of this appendix is to introduce essential software to support software engineering mathematics. We discuss a selection of software including Excel, Python, Maple, Mathematica, MATLAB, Minitab, and R (Table 32.1).

32.2 Microsoft Excel

Microsoft Excel is a spreadsheet program created by the Microsoft Corporation, and it consists of a rectangular grid of cells in rows and columns that may be used for data manipulation and arithmetic operations. It includes functionality for statistical, engineering, and financial applications, and it has graphical functionality to display lines, histograms, and charts (Fig. 32.1).

This software program was initially called MultiPlan when it was released in 1982, and it was Microsoft's first Office application. It was developed as a competitor to Apple's VisiCalc, and it was initially released on computers running the CP/M operating system.[1] It was renamed to *Excel* when it was released on the Macintosh in 1985, and the first version of Excel for the IBM PC was released in 1987.

It provides support for user-defined macros, and it also allows the user to employ Visual Basic for Applications (VBA) to perform numeric computation and report the results back to the Excel spreadsheet. Lotus 1–2–3 was the leading spreadsheet tool of the 1980s, but Excel overtook it from the early 1990s.

---

[1] The CP/M operating system was developed by Gary Kildall at Digital Research, and Kildall did a lot of early work on operating systems for microprocessors. The award of the operating system for the original IBM PC to Microsoft was highly controversial, as the operating system that Microsoft provided to IBM was essentially Kildall's CP/M.

© The Author(s), under exclusive license to Springer Nature Switzerland AG 2023
G. O'Regan, *Mathematical Foundations of Software Engineering*,
Texts in Computer Science, https://doi.org/10.1007/978-3-031-26212-8_32

Table 32.1 Software for business mathematics

Software	Description
Excel	This is a spreadsheet program created by Microsoft that consists of a grid of cells in rows and columns
Python	Python is an interpreted high-level programming language that has been applied to many areas including web development, game development, machine learning and artificial intelligence, and data science and visualization
Maple	Maple is a computer algebra system that can manipulate mathematical expressions and find symbolic solutions to certain kinds of problems in calculus, linear algebra, etc.
Minitab	Minitab is a statistical software package that provides a powerful and comprehensive set of statistics to investigate the data
R	R is an open-source statistical computing environment that is used for developing statistical software and for data analysis
Mathematica	Mathematica is a computer algebra program that is used in the scientific, engineering, and computer fields
MATLAB	MATLAB is a numeric computing environment that supports matrix manipulation, plotting of data and functions, and the implementation of algorithms

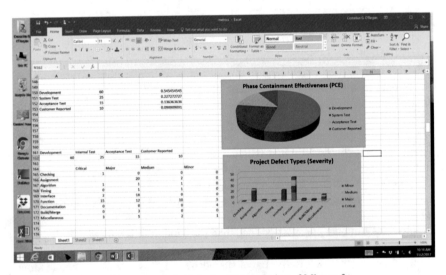

Fig. 32.1 Excel spreadsheet screenshot. Used with permission of Microsoft

Excel is used to organize data and to perform financial analysis. It is used by both small and large companies and across all business functions. The main uses of Excel include:

Data entry
Data management
Accounting
Financial analysis

Charts and graphs
Financial modelling.

Excel is used extensively for financial analysis, and many businesses use Excel for budgeting, forecasting, and accounting. Spreadsheet software may be used to forecast future performance, as well as calculating revenue and tax due and completing the payroll. Excel may be used to generate financial reports and charts.

An Excel workbook consists of a collection of worksheets, where each worksheet is a spreadsheet page (i.e., a collection of cells organized in rows and columns). A workbook may contain as many sheets as required, and the columns in a sheet are generally labelled with letters, whereas the rows are generally labelled with numbers. Each cell contains one piece of data or information.

A cell may contain a formula that refers to values in other cells (e.g., the effect of the formula = B3 + C3 in cell B1 is to add cells B3 and C3 together and to place the result in the cell B1). A formula may include a function, a reference to other cells, constants, as well as arithmetic operators. Excel uses standard mathematical operators such as + , −, *, /, and it employs ^ for the exponential function. An Excel formula always commences with an equals sign (=), and some of the functions employed include:

AVERAGE, COUNT, SUM, MAX, MIN, and IF.

Excel is very useful in recording, analysing, and storing numeric data, and various calculations may be performed on the numeric data, or graph tools may be employed for visualization of the data. That is, it allows easy manipulation of the data and graphing of the data for visualization. It may be used to create a wide range of graphs and charts from the data in the spreadsheet, and there is a *chart wizard* to assist with building the desired chart. Some of the charts that may be displayed include:

Bar charts
Histograms
Pie charts
Scatter plots
Lines.

Excel may be used for data analysis, and the ability to analyse data is essential in order to make better decisions. This generally involves the use of pivot tables, and pivot tables are a technique in data processing that is used to arrange or rearrange statistics in order to identify useful information. They may be employed to aggregate the individual items of a more extensive table (e.g., a database or another spreadsheet) within one or more discrete categories.

32.3 Python

Python is an interpreted high-level programming language that was designed by Guido van Rossum in the Netherlands in the early 1990s. The design of Python was influenced by the ABC programming language, which was designed as a teaching language, and developed at CWI in Amsterdam. Python has become a very popular programming language, and today the Python Software Foundation (PSF) is responsible for the language and its development. It is an object-oriented language that is based on the C programming language, and the language is versatile with applications in many areas including:

Rapid web development
Scientific and numeric computing
Machine learning applications
Image processing applications
Game development
Artificial intelligence
Gathering data from websites
Data science and data analytics
Data visualization
Business applications such as ERP
Education on programming.

The language has become very popular especially for machine learning and artificial intelligence, as it is reasonably easy to use especially for those who are new to programming. Its syntax is relatively simple, and the language is readable and easy to understand. Python applications can run on any operating system for which a Python interpreter exists.

Python includes many libraries such as libraries for web development, libraries for the development of interactive games, and libraries for machine learning and artificial intelligence. Python includes libraries that enable the development of applications that can multitask and output video and audio media.

Python supports data science and data visualization, and its libraries allow the data to be studied and information to be extracted. The data may be visualized such as in plotting graphs.

Python is able to gather a large amount of data from websites, which enables operations such as price comparison and job listings to be performed. For more information on Python see https://www.python.org/.

32.4   Maple

Maple is math software that includes a powerful math engine and a user interface that makes it easy to analyse, explore, visualize, and solve mathematical problems. It allows problems to be solved easily and quickly in most areas of mathematics. It is a commercial general-purpose computer algebra system that can manipulate mathematical expressions and find symbolic solutions to certain kinds of problems including those that arise in ordinary and partial differential equations. It supports symbolic mathematics, numerical analysis, data processing, and visualization. The Canadian software company, Maplesoft™, developed Maple, and its initial release was in the early 1980s (Fig. 32.2).

Maple supports several areas of mathematics including calculus, linear algebra, differential equations, equation solving, and symbolic manipulation of expressions. Maple has powerful visualization capabilities including support for two-dimensional or three-dimensional plotting as well as animation. Further, Maple includes a high-level programming language that enables users to create their own applications.

Maple makes use of matrix manipulation tools along with sparse arrays, and it has a wide range of special mathematical libraries. Maple supports 2D image processing and supports several probability distributions. It has functionality for code generation in languages such as C, Fortran, Python, and Java.

Maple includes the standard arithmetic operators such as $+$, $-$, $*$, $/$, and $^\wedge$ for exponential; it includes the relational operators $<$, $>$, $<=$, $>=$, $<>$, and $=$; the logical operators AND, OR, XOR, implies, and NOT; and the set operators intersect,

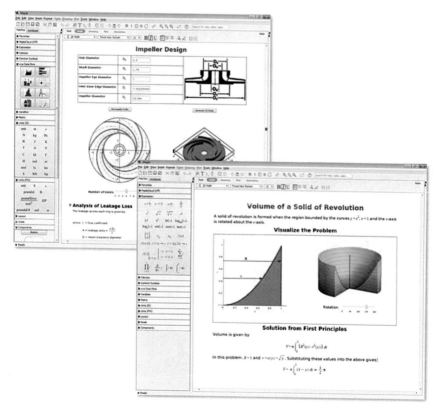

Fig. 32.2 Maple user interface. Creative commons

union, minus, subset, and member. A value may be assigned to a variable with the assign command (:=). It includes special constants such as Pi, infinity, and the complex number I. For more information on Maple see https://www.maplesoft.com/.

32.5   Minitab Statistical Software

Minitab is a statistical software package that was originally developed at the University of Pennsylvania in the early 1970s. Minitab, LLC, was formed in the early 1980s, and the company is based in Pennsylvania. It is responsible for the Minitab statistical software and its associated products, and it distributes the suite of commercial products around the world (Fig. 32.3).

Minitab statistical software is used by thousands of organizations around the world. The software helps companies and institutions to identify trends, solve problems, and discover valuable insights in data by delivering a comprehensive suite of data analysis and process improvement tools. The software is easy to use and makes it easy for business and organizations to gain insights from the data and to discover trends and predict patterns. It assists in identifying hidden relationships between variables as well as providing dazzling visualizations. Minitab has a team of data analytic experts and services to ensure that users get the most out of their analysis, enabling them to make better, faster, and more accurate decisions.

Minitab includes a complete set of statistical tools including descriptive statistics, hypothesis testing, confidence intervals, and normality tests. It provides a powerful and comprehensive set of statistics to investigate the data. It includes functionality to support regression thereby identifying relationships between variables, as well as functionality·to support the analysis of variance (ANOVA).

Minitab supports several statistical tests such as t tests, one and two proportions tests, normality test, chi-square, and equivalence tests. Minitab's advanced analytics provides modern data analysis and allows the data to be explored further. Minitab's predictive analytics techniques allow predictions and forecasts to be made, and Minitab's powerful visualizations allow the user to decide which graph that best displays the data and supports the analysis.

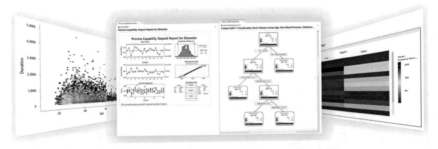

Fig. 32.3   Minitab screenshot. Created by and used with permission of Minitab LLC

Minitab includes functionality for control charts that allows processes to be monitored over time, thereby ensuring that they are performing between the lower and upper control limits for the process. That is, Minitab may be used for statistical process control thereby ensuring process stability and for data-driven process improvement to transform the business. Minitab Engage is a tool for managing innovation and may be used for managing six-sigma and lean manufacturing deployments.

For more detailed information on Minitab LLC (the makers of Minitab Statistical Software) see https://www.minitab.com/.

32.6   R Statistical Software Environment

R is a free open-source statistical computing environment that is used by statisticians and data scientists for developing statistical software and for data analysis. R includes various libraries that implement various statistical and graphical techniques such as statistical tests, linear and nonlinear modelling, and time series analysis. It allows the user to clean, analyse, plot, and communicate all of their data all in one place (Fig. 32.4).

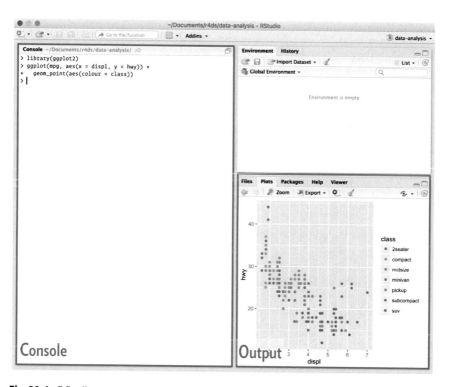

Fig. 32.4   RStudio

R is an interpreted language, and the user generally accesses it through a command line interpreter, and it has thousands of packages to assist the user. R is a popular statistical software tool, and it is widely used in academia and industry. It can produce high-quality graphs, and the advantages of R include:

Free open-source software
Large community
Integrates with other languages (e.g., C and C++).

R was created in the 1990s by Ross Ihaka and Robert Gentleman at the University of Auckland in New Zealand, and it was based on the S statistical programming language that was developed by John Chambers and others at Bell Labs in the 1970s. The R Development Core Team is now responsible for its development, and R programming plays a key role in statistics, machine learning, and data analysis.

RStudio is an integrated development environment (IDE) for R, and its functional user interface provides an easier way of using R. Programs may be written using the RStudio IDE. R may be downloaded from the Comprehensive R Archive Network (CRAN) https://cloud.r-project.org, and RStudio may be downloaded from http://www.rstudio.com/download. After installing RStudio there will be two key regions in the interface, and R code is typed in the console panel.

R packages may then be installed where an R package consists of functions, data, and documentation that extend the capabilities of R. The use of packages is the key to the successful use of R in data science, as it has a large number of packages available, and it is easy to install and use them. A core set of packages is included with the basic installation, and other packages may be installed as required. For example, the package tidyverse may be installed with a single line of code that is typed in the console.

```
install.packages("tidyverse")
library("tidyverse")
```

The package must then be loaded with the library command before the functions, objects, and help files may be used. The statistical features of R include:

Basic statistics including measures of central tendency
Static graphics
Probability distributions (e.g., binomial and normal)
Data analytics (tools for data analysis).

For more information on R and RStudio see https://www.r-project.org/.

32.7   Mathematica

Mathematica is a powerful tool for problem solving, and it includes symbolic calculations with nice graphical output. It is a way for doing mathematics with a computer, and this powerful computer algebra program is used in the scientific, engineering, and computer fields. Symbolic mathematics involves the use of computers to manipulate equations and expressions in symbolic form, as distinct from manipulating the numerical quantities represented by the symbols (Fig. 32.5).

Mathematica was developed by Stephan Wolfram of Wolfram Research in the late 1980s, and it supports many areas of mathematics including basic arithmetic, algebra, geometry and trigonometry, calculus, complex analysis, vector analysis, matrices and linear algebra, and probability and statistics.

It has a large number of predefined functions for mathematics and other disciplines and includes functionality for the visualization of data and functions. This includes good graphical capabilities that are useful in plotting functions and data in two or three dimensions. For example, the Mathematica command *RevolutionPlot3D* constructs the surface formed by revolving an expression around an axis, and Fig. 32.6 is generated from the command:

RevolutionPlot3D$[x^4 - x^2, \{x, -1, 1\}]$

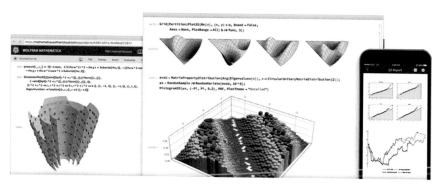

Fig. 32.5 Mathematica in operation. Provided courtesy of Wolfram Research, Inc., the makers of Mathematica, www.wolfram.com

Fig. 32.6 Surface generated with RevolutionPlot3D. Provided courtesy of Wolfram Research, Inc., the makers of Mathematica, www.wolfram.com

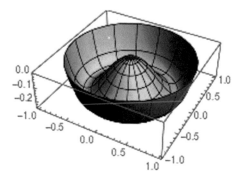

Mathematica may be used to solve very simple arithmetic problems, as well as solving complex problems in differential equations. It has approximately 5000 built-in functions covering the vast majority of areas in technical computing and that work together in the integrated system. The Wolfram programming was developed by Wolfram Research, and this multiparadigm language emphasizes functional programming and symbolic computation.

The language can perform integration, differentiation, matrix manipulation and solve differential equations using a set of rules. Mathematica has been applied to many areas of computing including:

Machine learning
Neural networks
Image processing
Data science
Geometry
Visualization.

Mathematica has also been applied to many other areas, and it produces documents as well as code. Its visualizations of results are aesthetically pleasing and powerful, and Mathematica also produces publication quality documents, and it has thousands of examples in its documentation centre. It has built-in powerful algorithms across many areas that aim to be of industrial strength. Finally, Mathematica is integrated with the cloud, and this allows sharing as well as cloud computing. There is more detailed information on Mathematica at Wolfram Research, Inc. (see https://www.wolfram.com/mathematica/).

32.8 MATLAB

MATLAB is a commercial high-level programming language that is used to perform mathematical computing, and this numeric computing environment was developed by MathWork. It is used by engineers and scientists to organize, explore, and analyse the data, and the MATLAB language may be employed to develop programs based on algorithms from a variety of domains. MATLAB allows the user to create customized visualizations and to automatically generate the MATLAB code to reproduce them with new data.

MATLAB manages array and matrix problems, and it may be used to solve complex algebraic equations as well as analysing data and plotting graphs. It allows the user to create customized visualizations, as well as using the built-in charts. MATLAB has many applications including:

Machine learning
Deep learning
Robotics
Computer vision

Image processing
Control systems.

At the heart of MATLAB is a high-level programming language that allows engineers and scientists to express matrix and array mathematics directly. MATLAB has a large library of toolboxes from everything from wireless communication, to control systems, to signal and image processing, to robotics, and to AI. It is easy to use and learn, and it allows ideas to be explored with the results and visualizations seen quickly.

It includes pre-built apps and allows the user to create their own apps. It includes App Designer, which allows a non-specialist to create professional apps by laying out the visual components of the GUI as well as programming the app behaviour. MATLAB may be extended with thousands of packages and tools, and its capabilities include:

Data analysis
Graphics
Algorithm development
App building.

There is more detailed information on MATLAB at MathWorks (see https://www.mathworks.com/products/matlab.html).

32.9 Summary

The goal of this chapter was to discuss a selection of software available to support software engineering mathematics, including Excel, Python, Maple, Mathematica, MATLAB, Minitab, and R.

Excel is a spreadsheet program that consists of a grid of cells in rows and columns. Python is an interpreted high-level programming language with applications in web development, artificial intelligence, and data science. Maple is a computer algebra system that can find symbolic solutions to certain kinds of problems. Minitab is a statistical software package with a comprehensive set of statistics to investigate the data. R is an open-source statistical computing environment that is used for developing statistical software and for data analysis. Mathematica is a computer algebra program that is used in the scientific, engineering, and computer fields.

MATLAB is a computing environment that supports matrix manipulation, plotting of data and functions, and the implementation of algorithms.

Index

Printed in the United States
by Baker & Taylor Publisher Services